Human Spatial Cognition and Experience

This book offers students an introduction to human spatial cognition and experience and is designed for graduate and advanced undergraduate students who are interested in the study of maps in the head and the psychology of space.

We live in space and space surrounds us. We interact with space all the time, consciously or unconsciously, and make decisions and actions based on our perceptions of that space. Have you ever wondered how some people navigate perfectly using maps in their heads while other people get lost even with a physical map? What do you mean when you say you have a poor "sense of direction"? How do we know where we are? How do we use and represent information about space?

This book clarifies that our knowledge and feelings emerge as a consequence of our interactions with the surrounding space, and show that the knowledge and feelings direct, guide, or limit our spatial behavior and experience. Space matters, or more specifically space we perceive matters. Research into spatial cognition and experience, asking fundamental questions about how and why space and spatiality matter to humans, has thus attracted attention. It is no coincidence that the 2014 Nobel Prize in Physiology or Medicine was awarded for research into a positioning system in the brain or "inner GPS" and that spatial information and technology are recognized as an important social infrastructure in recent years.

This is the first book aimed at graduate and advanced undergraduate students pursuing this fascinating area of research. The content introduces the reader to the field of spatial cognition and experience with a series of chapters covering theoretical, empirical, and practical issues, including cognitive maps, spatial orientation, spatial ability and thinking, geospatial information, navigation assistance, and environmental aesthetics.

Toru Ishikawa is a Professor in the Department of Information Networking for Innovation and Design (INIAD) at Toyo University, Tokyo, Japan. He has a PhD in geography from the University of California, Santa Barbara, and is an associate member of the Science Council of Japan. He specializes in cognitive-behavioral geography, geographic information science, and urban residential environments and planning. His research interests include cognitive maps and mapping, wayfinding and navigation, spatial thinking in geoscience, and geospatial awareness and technology.

Human Spatial Cognition and Experience

Mind in the World, World in the Mind

Toru Ishikawa

Routledge
Taylor & Francis Group
LONDON AND NEW YORK

First published 2021
by Routledge
2 Park Square, Milton Park, Abingdon, Oxon OX14 4RN

and by Routledge
52 Vanderbilt Avenue, New York, NY 10017

Routledge is an imprint of the Taylor & Francis Group, an informa business

British Library Cataloguing-in-Publication Data
A catalogue record for this book is available from the British Library

Library of Congress Cataloging-in-Publication Data
Names: Ishikawa, Toru, author.
Title: Human spatial cognition and experience : mind in the world, world in the mind/Toru Ishikawa.
Description: First Edition. | New York : Routledge, 2020. | Includes bibliographical references and index.
Identifiers: LCCN 2020009307 (print) | LCCN 2020009308 (ebook) | ISBN 9780815369851 (hardback) | ISBN 9780815369868 (paperback) | ISBN 9781351251297 (ebook)
Subjects: LCSH: Space perception. | Spatial behavior. | Spatial ability.
Classification: LCC BF469 .I74 2020 (print) | LCC BF469 (ebook) | DDC 153.7/52—dc23
LC record available at https://lccn.loc.gov/2020009307
LC ebook record available at https://lccn.loc.gov/2020009308

ISBN: 978-0-8153-6985-1 (hbk)
ISBN: 978-0-8153-6986-8 (pbk)
ISBN: 978-1-351-25129-7 (ebk)

Typeset in Bembo and Futura
by Apex CoVantage, LLC

To my family and mentors

Contents

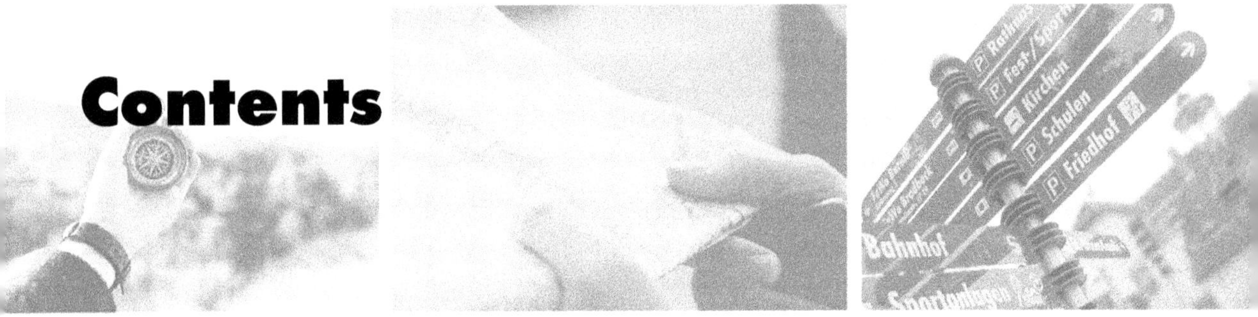

Preface

It is fun to think about space that surrounds us. It offers a variety of interesting questions that can be addressed from a social, cultural, historical, natural, or physical perspective. Different cultures, political systems, industries, agricultural activities, natural landscapes, built environments, and climates make different places. What about considering space from a *psychological* perspective? The space around you is not a uniform, empty space; it is your neighborhood, a shop you frequent, a café you enjoy spending time in, and an environment you are emotionally attached to. Importantly, through experience in it, a space becomes a *place* for you. It is therefore no less fun, intellectually, to think about a psychological space, or space that we perceive and experience.

I came across the ideas of *maps in the head*, *image of the city*, and *experience of space* when I was an undergraduate student, and have since been intrigued by them. These terms cogently point to the existence and significance of mental representations of environments. That is, our behavior is based on subjective space, as well as physical space, and environments represented in our minds direct our spatial behavior. Space matters; or, more specifically, space we perceive matters. This book is intended to be a guide for students who are interested in, and beginning to explore, the fascinating research field of spatial cognition and experience; a book I wish I had had in my college years.

In studying spatial cognition and experience, I was lucky enough to be apprenticed to two mentors in the UCSB Geography Department, Reg Golledge and Dan Montello. Reg Golledge guided me through the field of behavioral geography. When I first visited his office as a graduate student and told him, with a sense of awe and a little hesitation, that I was interested in the question of spatial knowledge acquisition and would like to examine it in my dissertation, he explained in every detail, with the grace and dignity that were characteristic of him, what cognitive maps are, why it is important to study them, how distance and direction are conceived, and how distortions may be introduced in cognitive maps. In particular, he stressed the necessity of considering the development of spatial knowledge from the perspective of knowledge integration, which I eventually pursued in my dissertation. It was an experience that I still remember vividly. I also cannot forget to mention that Reg showed me how a geographer should conduct scientific research into space, or what it means to be a geographer.

Dan Montello paved the way for my career as a cognitive-behavioral geographer (or a psychogeographer). It was more than coincidence that his article on an alternative framework of spatial microgenesis was published while I was working toward a PhD. The article was a big inspiration for the experimental design of my doctoral research, which is now referred to as a route integration paradigm. I enjoyed, and still appreciate, his stimulating ideas and constructive criticism and generous introduction to relevant studies in the literature in our biweekly discussions in his office. Above all, his high standards for science are a guiding light for me. I note that his lecture on spatial cognition and experience, with the title of "Environmental Perception and Cognition," formed the basis of this book. I took the lecture as a graduate student and had the privilege of participating in it as a teaching assistant: what I learned from it, and conceivably misrepresented in my mind, underlies many topics discussed in the book. I acknowledge his encouragement to write this book and the various suggestions that I received from him, including ideas for the title.

This book consists of ten chapters discussing theoretical, empirical, and practical issues of spatial cognition and experience. Students may read it from the first chapter, or look at individual chapters that they are interested in and topics that they want to learn more about. Instructors may use the book in a quarter course or a semester course, for example as a main textbook in an advanced undergraduate course or as a reader in a graduate seminar. I hope that the book particularly helps young scholars to design a new class in spatial cognition and experience in their various related disciplines. Part of the text in Chapters 6 and 10 is based on my articles published in the *Handbook of Behavioral and Cognitive Geography* the *Handbook on Planning and Complexity*, and *The Professional Geographer*. I appreciate the permission from Edward Elgar Publishing and Taylor & Francis to reproduce the materials.

I am grateful to many people for their interest and support. Some of my respected colleagues were kind enough to read early manuscripts of the book and provided invaluable comments. I want to thank Nora Newcombe, Christian Freksa, Sara Fabrikant, and Dan Montello. I also want to mention two more mentors of mine: Atsu Okabe, who first introduced me to the research area of spatial cognition, and Ken Sakamura, who encouraged me to continue research into spatial cognition in a broader context of information science.

Last, but not least, I acknowledge the support from people at Routledge, especially Katie Peace and ShengBin Tan. Their enthusiasm, encouragement, and experienced guidance enabled me to begin, continue, and finish the ambitious and enjoyable project of writing this book. Thank you very much.

Figures

Tables

Chapter 1

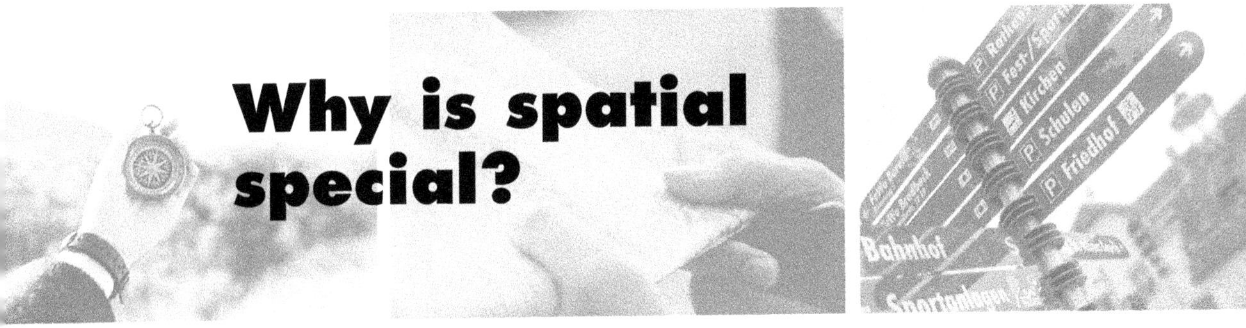

Why is spatial special?

To know where you are in space

You are now starting to read a book about space. If it is morning time, think about what you will be doing today. You go to school after stopping by the post office, have lunch with your roommate at Highland Cafe, and go downtown to see the latest movie. If it is evening time, consider what you did today. You took a midterm exam in calculus in Columbus Hall, talked about your thesis with your advisor in her office, and did a literature search in the Main Library. You note in these examples the fact that your behavior takes place (literally) at a specific place, at a specific time. That is, you are located in space and time (Figure 1.1). Where are you reading this book now?

Space and time, or *where* and *when*, are found in almost all our behaviors as fundamental and integral components. What do we know, then, about space and time? We can conceptualize time as a temporal flow within which we are located, going from the past into the future. Time intersects us in a plane called the *present*. If we look back, we see the arrow of time extending back into the *past*; if we look ahead, we see the arrow moving into the *future*. The future becomes the present; the present becomes the past. What about space? Everyday activities that we engage in, as varied as driving from home to work, taking a detour to avoid traffic congestion, choosing where to be seated in a classroom, or arranging furniture in a room, involve cognitive processing concerning *where*, whether we are conscious of it or not. We live in space and space surrounds us; a fact that seems so plain that we usually do not pay close attention to it, but it has important consequences for our everyday cognition and behavior. St. Augustine stated in his discussion of the nature of time:

> Yet is it not true that in conversation we refer to nothing more familiarly or knowingly than time? And surely we understand it when we speak of it; we understand it also when we hear another speak of it. What, then, is time? If no one asks me, I know what it is. If I wish to explain it to him who asks me, I do not know.
>
> (*The Confessions*, Book XI, Chapter XIV, trans. A. C. Outler)

Can we ask, then, What is space?

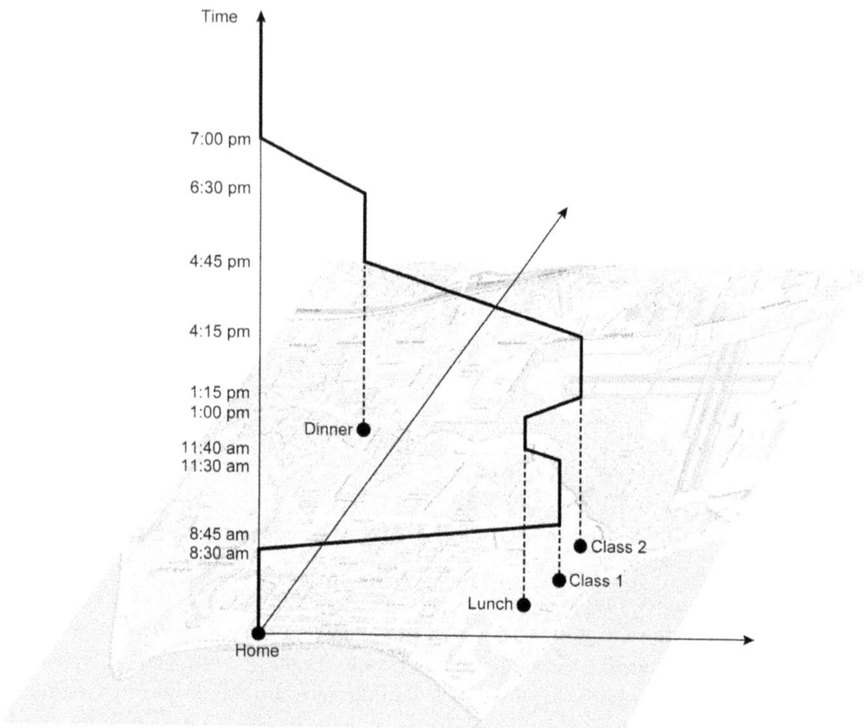

Figure 1.1 **Trace of your activities on a day. This diagram, called a space–time prism, shows where you were that day. Human behavior thus takes place in space and time, and we will look at the spatial–temporal nature (particularly spatial components) of human cognition and experience in this book. © OpenStreetMap contributors, https://www.openstreetmap.org/**

Now consider what it means that we live in space and space surrounds us. We move around in the environment and perceive the surrounding space, or we interact with the environment. Through interaction, we know, feel, and think about the environment. Put differently, the environment *affords* us the opportunities to learn about and attend to the information relevant to our behavior. Importantly, the knowledge and feelings about the environment guide, mold, or stimulate our behavior in an adaptive way. Our knowledge and feelings about surrounding environments emerge through interactions with them, and our behavior builds on the knowledge and feelings that we have (or represent) in our minds. For example, we go and drop a letter in a certain mailbox; we point to a specific direction when asked for navigational directions; we take a seat in the front or back row in a classroom; or we choose among alternative routes from home to work. All these behaviors reflect our thoughts and feelings about the places. We behave that way because the mailbox we choose is closest to home; the destination is located in that direction; the front or back seat fits the purpose of studying or sleeping; or the route is more scenic or less congested. More correctly, *we think* that the mailbox is closest to home; *we think* that the destination is in that direction; *we prefer* that seat; or we *like* that route. These observed behaviors are thus based on what we think, what we know, and what we feel about the places. Golledge and Stimson (1997, pp. 4–7) used the term *spatial behavior* to emphasize the process of thinking (information processing) underlying observed behavior and distinguished it from mere *behavior in space* (see also Rushton, 1969).

Thinking about space is therefore indispensable for us, because we live in space—*space matters*. For research into spatial cognition, the phrase "live *in* space" has important implications. The space we live in, or the environment, is larger than and surrounds the human body (Ittelson, 1973). We therefore need to explore the environment to acquire knowledge about it, which is in stark contrast to the case of object perception, where we view an object as an external observer (Figure 1.2). The distinction between exploration and observation, or being an explorer and an observer, has significant consequences for the learning of the environment, which we will see in detail in the following chapters.

Space matters, but in what ways? Most notably, our spatial behavior is influenced by the notion of distance. Places are located in space, being separated from each other by a distance, which we need to overcome by physically traveling, such as commuting to work, going to a grocery store, or dropping a letter in a mailbox. In such behavior of traveling over a distance, do people choose where to go haphazardly or do they show a systematic tendency in their choice of places and routes? Research has shown that the latter is the case, and a well-known statement describes this phenomenon as: "Everything is related to everything else, but near things are more related than distant things" (Tobler, 1970, p. 236). The degree of interaction between places decreases as the places are separated by a longer distance. Simply put, people go to nearer places rather than distant places, other things being equal.

To fully appreciate the influence of distance on spatial behavior, we should note that what matters is not physical distance but rather the distance that we perceive (i.e., how far *you think* it is, not necessarily how far it physically is). Hence, it is more correct to say, "Space we perceive matters," as seen in the example of going to a mailbox that *we think* is closest. That is, our spatial behavior is based on perceived environments, as well as physical

Figure 1.2 **Perception of an object (*left*) and perception of the environment (*right*). You are an observer in the former case and an explorer in the latter case, an important distinction to be made in the discussion of cognition and experience of space. Photographs from https://pixabay.com/**

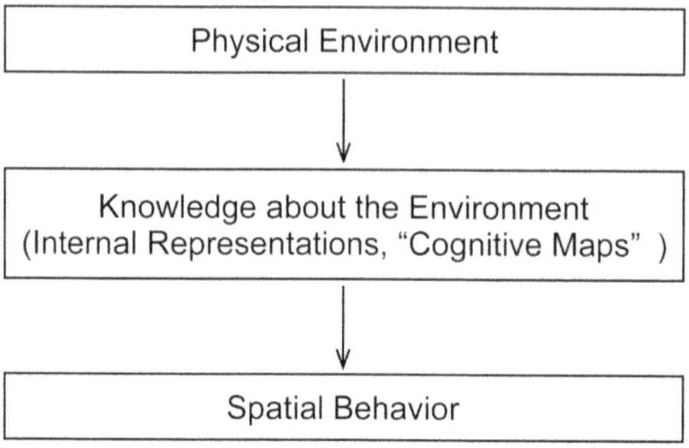

Figure 1.3 **Relationship between a physical space, a subjective space, and human behavior. The subjective space, or mental map of the environment, plays an important role in directing our behavior (cognition, thinking, and feelings) in the environment: Space we perceive matters. Reprinted with permission of Springer Nature from Ishikawa, 2016a, p. 117.**

environments (Figure 1.3). Physical environments are out there in the world, and perceived environments are here in the mind, showing what we think and how we feel about the environments and guiding our behavior in the environments.

Space for humans

Humans have always been interested in space, and in knowing about space. Humans have a long history of exploring, modifying, and adapting to the environment. They evolved in savannas in Africa and expanded their habitats outwards. They began to use tools, hunt for food, and communicate verbally with each other. They expanded their knowledge about the earth through explorations and scientific observations; for example, they discovered new lands through rough and long voyages, made empirical observations to reveal the earth's rotation around its axis and revolution around the sun, and explored outer space by spacecraft. As well as learning about the environment, humans represent the knowledge externally to communicate it to other people, notably in the form of maps. We will see examples of ancient maps, such as Lascaux cave paintings and the Babylonian Map of the World, below.

Along with the basic human desire, knowledge about space has been sought through academic approaches in the discipline of geography. The geographer's interest spans all areas of the earth's environments and human activities that take place on the earth (you may think of the Vermeer painting *The Geographer*, which we will see in Chapter 2). Within the discipline of geography is a branch called physical geography, which mainly studies elements of the natural environment such as landforms, soils, rocks, rivers, oceans, ecosystems, and climates. Physical geographers thus seek to reveal the structures and processes of the earth's environment. Another branch within geography is human geography. This studies people and their social, cultural, and economic activities, with a particular focus on the interaction and interrelationships between people and the environment. Topics of interest to human geographers include population, migration, socio-political systems, economic development, settlement patterns, and urbanization. With academic interest in these issues, human geographers have developed a large inventory of knowledge about spatial distributions and patterns of human activities, systematically detailed and synthesized for individual places.

Together with such a synthetic approach to particular places, human geographers also noted the significance of systematic investigation of human activities from the perspective of individuals, instead of groups of people, based on countries, cultures, or communities. Particularly in the 1960s, human geographers shifted the focus from aggregate to disaggregate data analysis, and began to look at the individual's knowledge and decision making in pursuit of a process-oriented understanding of spatial phenomena. Geographers now look at individual persons as sensate beings who actively gather information about their surroundings and make decisions on the basis of their processing and interpretations of the acquired information. This shift in interest is characterized by the recognition of the importance of explaining processes behind observed spatial phenomena (the questions of *how* and *why*) in addition to describing the forms of distributions (*what* and *where*). In other words, geographers go beyond "knowing the names of the capitals, rivers and mountains in each country of the world" (Gersmehl, 2008, p. 2). We will discuss this in Chapter 3.

As geographers have always been interested in space and place, psychologists have long been interested in the human mind, with specific topics of interest ranging from consciousness, sensation, perception, attention, learning, and personality, to language, memory, mental imagery, and reasoning. Similar to the shift of focus in geography from aggregate forms to individual processes, psychology witnessed a renewed emphasis on mental processes in the late 1950s. Suppose that we observed a rat moving between two places (a start and a goal) along specific paths (Figure 1.4). Did the rat learn the sequence of turns (i.e., take path 1, turn left, go along path 2, turn right, move on to path 3, turn

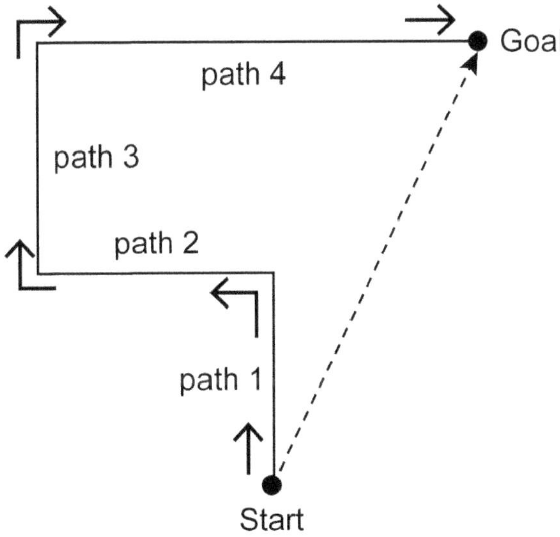

Figure 1.4 **A rat going from a start to a goal. What does the rat learn about the space, a *sequence* or a *place*? It offers an interesting question about the structure and process of spatial knowledge.**

right, and go along path 4 to the goal) or did it know the spatial relations between the start and goal locations? These two views differ in the degree of "inference" involved in the observer's interpretation of the rat's behavior: the former *sequence* view is based strictly on the observed behavior and involves no inference, while the latter *relation* view assumes internal processing and knowledge representations (reasoning based on a conceptual entity of something like a "map in the head"). Psychologists now look into the mind for explanations of observed behavior in terms of underlying processes. We will see this in more detail in Chapter 2 by comparing the approaches of behaviorism and cognitivism.

Another movement among psychologists around that time concerned the recognition of the importance of studying human learning and memory in a real-world context, or in a setting that reflects how humans engage in the activity in question in real life. For example, a researcher can study people's preference and choice of routes by presenting pictures of artificially constructed environmental scenes of a set of equal length paths differing in shape, the number of turns, the types of land uses, and the amount of greenery. Data on the rank order of preferences that participants give to the pictures allow the researcher to identify variables that influence route preference. This experimental setting has the strength of manipulating the variables of interest (shape, turns, land use, and greenery) and controlling for other variables. In a real-world setting, it is difficult to select paths that are varied systematically on these dimensions and equivalent on other dimensions. On the other hand, in the controlled laboratory setting of viewing picture-like representations seen from above, people do not walk the paths in the real world or feel the "ambience" of the environment. Does that affect preference judgments in any significant way? In other words, can we always generalize the findings in the laboratory to real-world settings? Can we examine people's spatial cognition and behavior in the real world rigorously and meaningfully? Do we have to make a trade-off between experimental control and generalizability? These issues will be discussed in relation to the question of ecological validity in Chapter 3.

People and the environment

The previous section saw geography and psychology as two major disciplines in the research into spatial cognition. The discipline that has traditionally been interested in the environment and the discipline that has traditionally studied the human mind interact in the field of spatial cognition to look at people in the environment and environments in the human mind, acknowledging the importance of ecological mutuality between humans and the environment. To clarify the concept of human–environment interaction for the study of spatial cognition, think about one of the environments we live in: the city. A city attracts people and the gathering of people makes a city, as exemplified by the emergence of poleis in ancient Greece, urban problems caused by the concentration of people in the city during the Industrial Revolution, and city shrinkage associated with aging population and decreased birthrates. As seen in these examples, the city is not simply an empty space: the mutual dependence of the physical environment and the people living in the environment makes up the city.

Importantly, people do not live in isolation from other people or independently of the environment; nor do they live in an abstract space conceived, for example, as an isotropic space with an evenly distributed population. Conceptualization of space of this sort advanced a theoretical discussion of city form in a spatial modeling approach (e.g., von Thünen's (1826/1966) model of agricultural land use, Christaller's (1933/1966) central place theory, and Alonso's (1960) bid rent theory). An equally important approach to studying the city focuses on the interaction between people and the environment and looks at the process of learning and decision making by people in the environment. People experience and perceive the environment, attach values to it, and sometimes alter it.

St. Augustine's question is revisited here with respect to space—What do we know about space? What do we know about our perceptions of the environment? How is the environment learned and represented in our minds? How is a mentally represented environment related to spatial behavior? How do perceived environments differ from physical environments? How is the environment represented in your mind different from that in another person's mind? To think about these questions, it is worth looking at one type of spatial representation that is familiar to us: the map.

We saw above that a spatial modeling approach simplifies the real world with a view to describing the spatial distributions of phenomena of interest. Simplification is also seen at the heart of the map. We all know that a map represents part of the earth's surface. Technically speaking, a map is a two-dimensional representation of the earth projected on a flat surface. But what do you mean by a "representation"? Is a map another type of "reality," similar to the environment that surrounds us? Yes it is; a map exists out there in your hand as a physical representation. But the map in your hand is a result of projection: part of the three-dimensional, or spherical, globe is projected onto a piece of paper (or nowadays a tablet screen), which inevitably causes distortions of area, shape, direction, or distance. In this sense, a map is not the "truth," as many people unconsciously think (Monmonier, 2018). Look at the maps in Figure 1.5, projected in two different ways. The map on the left is probably familiar to you, a map often put on the wall in a classroom. On this map, compare the areas for Greenland and Africa: you see that they are of a similar size. Now look at the map on the right. How big are the two land areas on this map? Africa is much larger than Greenland now. Which is correct? The map on the right depicts the areas of continents and islands more or less faithfully, while the map on the left distorts the size of geographical objects, the degree of distortion being greater toward the poles. If you look at other properties, you will find that the left map preserves the angles with the meridians, whereas the right map distorts shape. Maps are thus representations of the world (not reality in this sense), faithful in some respects and not in other respects.

Maps also entail scale reduction: a map with a scale of 1:1 is not conceivable, or at least not handy (see Lewis Carroll's, 1893, *Sylvie and Bruno Concluded*: "We actually made a map of the country, on the scale of a mile to the mile!" "Have you used it much?" I enquired. "It has never been spread out, yet," said Mein Herr: "the farmers objected: they said it would cover the whole country, and shut out the sunlight!", chapter XI, p. 169). A reduction in scale necessitates the selection of information: not everything can be shown on a map with every detail. Cartographers address it through the process of *generalization*, for example, by selecting a subset of critical points along a linear feature so that its geometry can be retained and represented with a minimum number of coordinates (Slocum et al., 2009, p. 101). In map making, scale, detail, and resolution are thus closely related. A large-scale map (e.g., a 1:2,500 map) covers a smaller extent of the earth's surface with more detail and a finer resolution than a small-scale map (e.g., a 1:25,000 map; Figure 1.6).

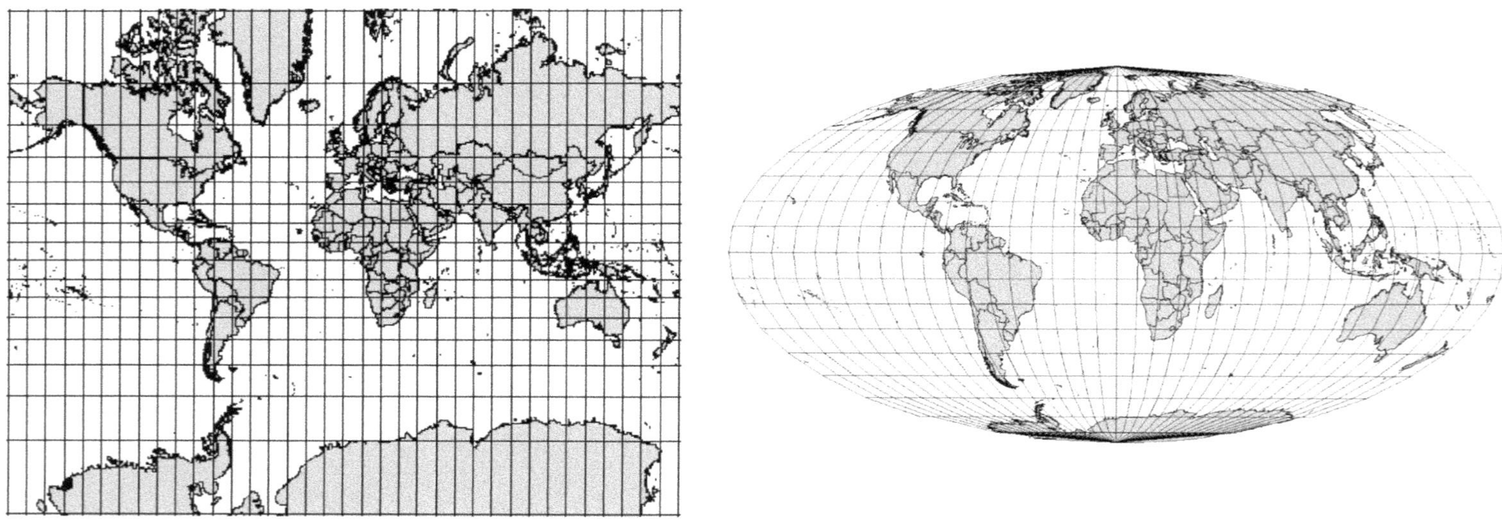

Figure 1.5 Maps of the world. The Mercator projection (*left*) and the Mollweide projection (*right*). Do they depict *the truth* of the world or are they two different *representations* of the earth? The term "representation" has specific meanings in the research into spatial cognition. Illustrations by QGIS, CC BY-SA, from https://docs.qgis.org/2.18/en/docs/gentle_gis_introduction/coordinate_reference_systems.html

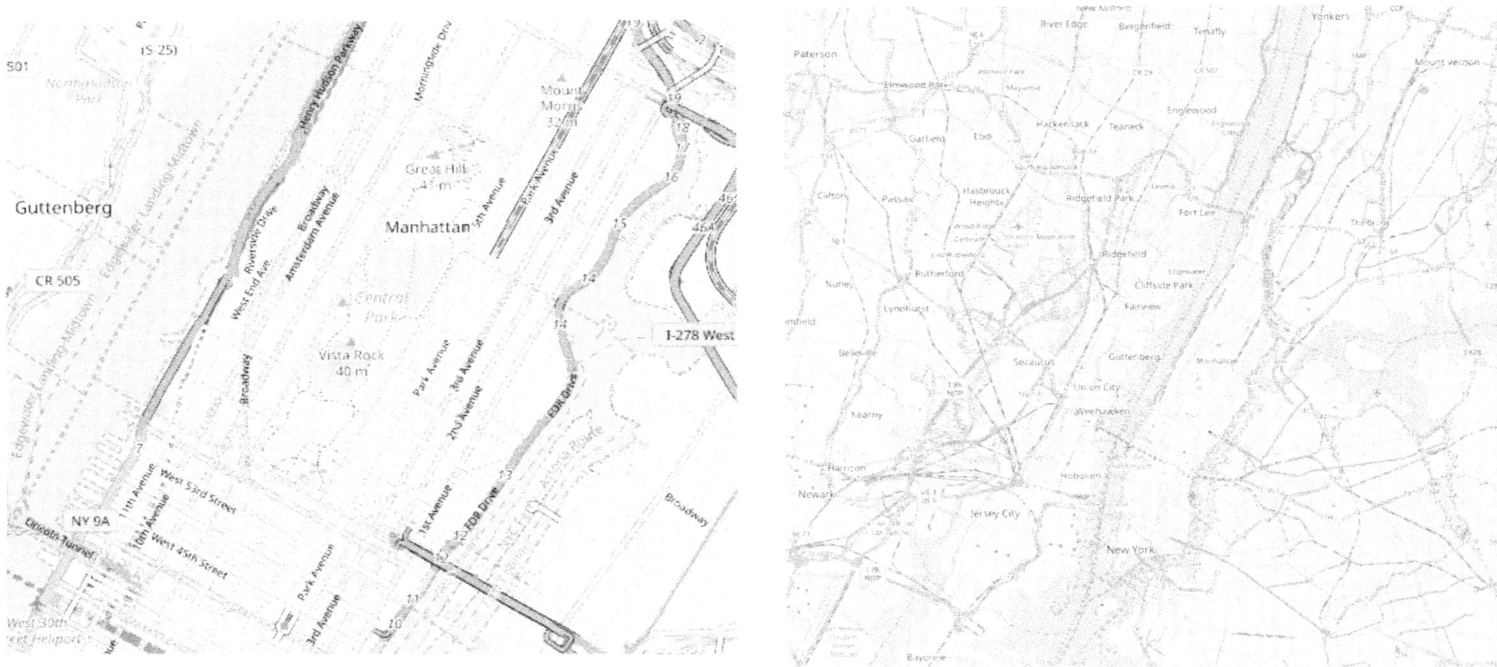

Figure 1.6 **A large-scale map and a small-scale map. The former map depicts a smaller extent of the earth's surface with more detail and a finer resolution; the latter depicts a larger extent with less detail and a coarser resolution. © OpenStreetMap contributors.**

Now we see that a map is a representation that is simplified, generalized, and distorted. What about a knowledge representation in our minds? Is it a faithful *copy* of the physical environment out there, or is it a simplified *representation* of reality? If it is simplified, how? Consider the question with one example: Which city do you think is further north, Rome or New York City? People tend to be surprised to learn that Rome is actually further north than New York (Rome is at 41° 54′ N and New York is at 40° 42′ N; see Figure 1.7). We find it counterintuitive, as we envision, for example, a milder climate in Rome in winter than in New York. Thus the locations and spatial relations of the two cities are distorted in our minds, aligned with our misconceptions, and do not match reality (Tversky, 1981).

We described distortions in our knowledge above in terms of judgments of spatial relations. Is there another way to "look into" a knowledge representation in your mind? What about drawing a sketch map of your neighborhood? What do you put on your map? Your home, a school, a grocery store, and others? Do you put any paths? Does your map show every element that physically exists in your neighborhood or does it select a subset of them? What is the criterion of your choosing some of them and not others? Are the places and paths on your map located accurately in relation to each other (Figure 1.8)?

Now what about describing your neighborhood verbally? Do you mention any "landmarks"? Are there commonalities and differences among people in the choice of landmarks? How do you describe spatial relations in words? Do you use such phrases as "in front of," "to the left," or "on your right"? Do you mention cardinal directions, north, south, east, and west? Do you identify the location of a place using latitude and longitude in a daily conversation (for example, do you address a postcard to a specific latitude and longitude)? How do your verbal descriptions differ from other people's descriptions? What do you think of these different ways of describing space (places, locations, and spatial relations) in words? We will see language and space in Chapter 7.

Functions of spatial cognition

Our daily cognition and behavior thus are substantially concerned with *spatiality*, and involve reasoning about and thinking with space and spatial representations. Montello and Raubal (2013) discussed six major functions of spatial cognition, by explicitly concentrating on cognitive activities that centrally involve mental representations of space.

1. They first looked at the act of establishing one's orientation in coordination with the distal environment, which is generally called wayfinding. This includes tasks such as locating a position in relation to external features of the environment, comprehending the locations and spatial relations of landmarks, judging distances and directions, remembering the sequences of turns along a route, making detours and finding shortcuts, and planning multiple activities distributed in space.
2. A second task of spatial cognition concerns the acquisition of spatial knowledge about large-scale environments and the use of the knowledge in spatial behavior in a creative and adaptive manner. Specific types of spatial knowledge acquired and used include connections, sequences, distances, directions, shapes, and configurations; and, importantly, the acquisition of such knowledge is primarily based on the direct experience of locomotion in the surrounding environments. Knowledge about

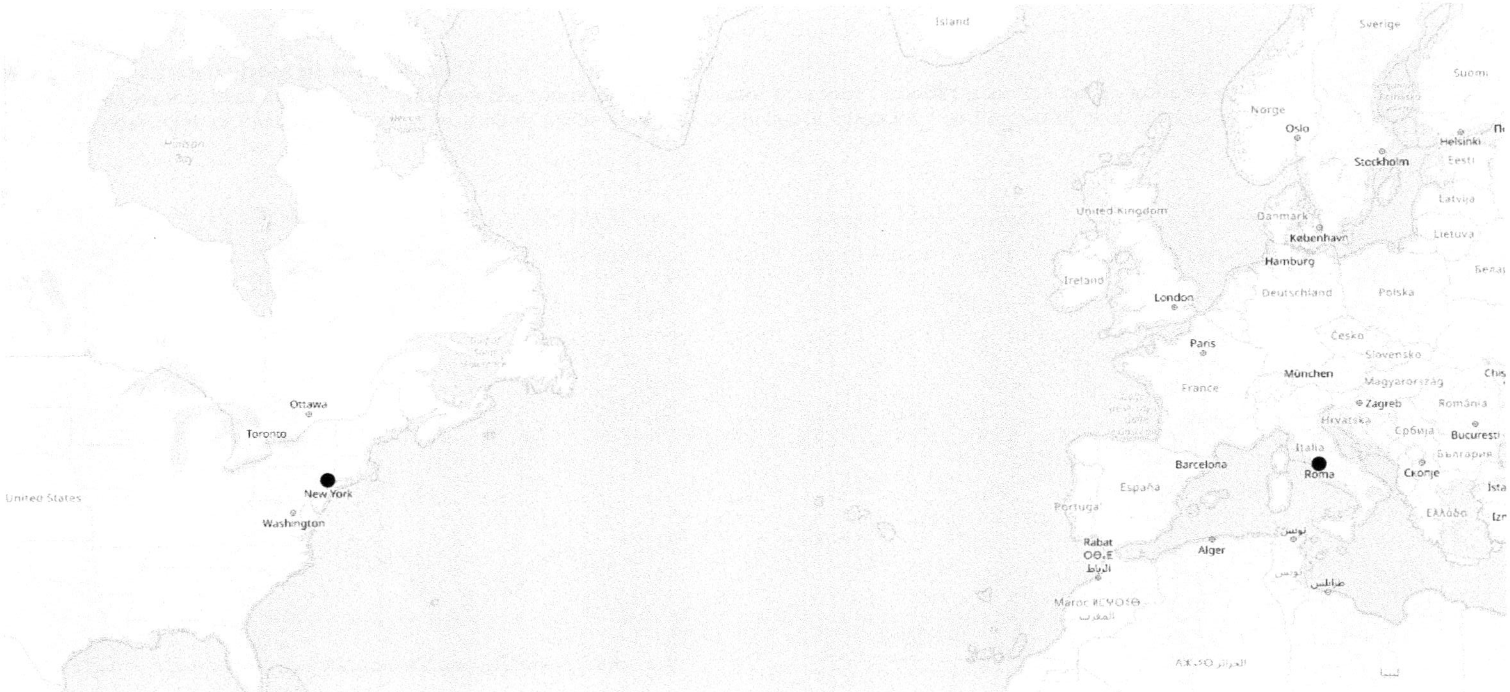

Figure 1.7 **Which city is located further north, Rome or New York City? Rome is at 41° 54′ N and New York is at 40° 42′ N. Was your answer correct? Why or why not? © OpenStreetMap contributors.**

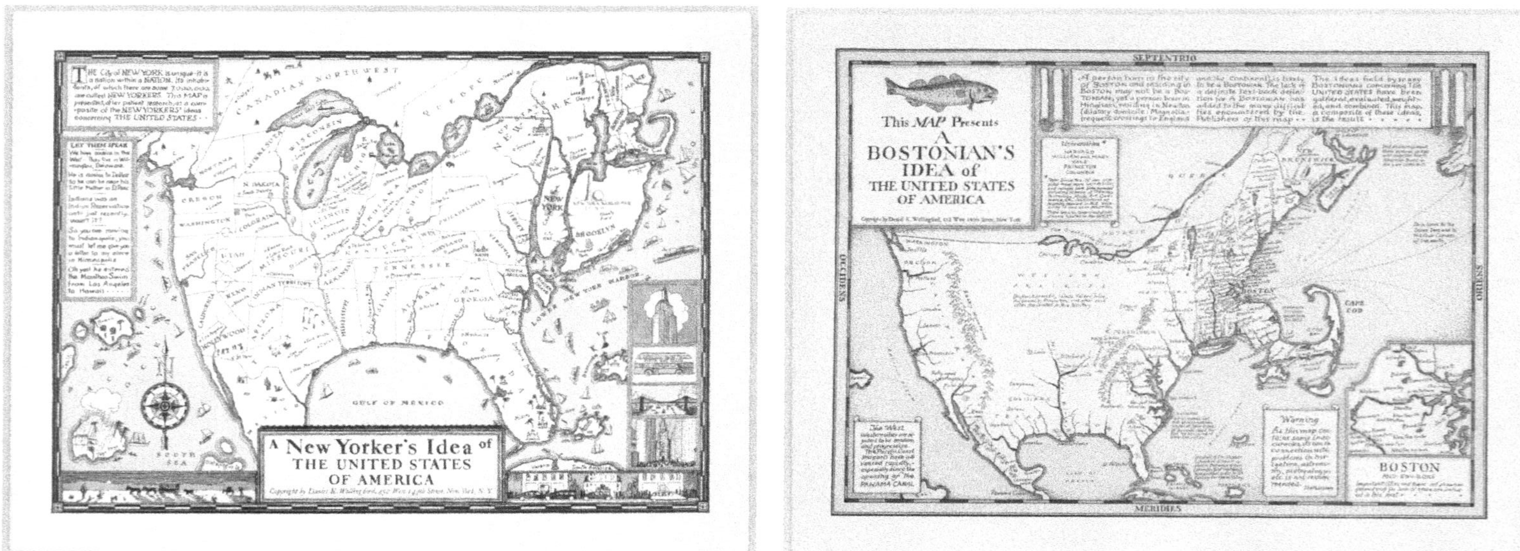

Figure 1.8 **Knowledge of our environments, or *mental maps*. Are they accurate or distorted? Are different people's mental representations similar or different? What is represented in your knowledge about the environment, and how? Images from the David Rumsey Map Collection, https://www.davidrumsey.com/. Used by permission.**

environments is also acquired from their spatial representations, especially for environments that extend beyond our direct sensorimotor experience.

3. Thus Montello and Raubal identified the use of iconic spatial representations as a third task of spatial cognition, and discussed the production and interpretation of maps, graphs, drawings, diagrams, photographs, movies, or physical models as major examples.
4. A fourth task of spatial cognition is describing spatial environments through the use of spatial language, a non-iconic form of spatial symbolic representations. For example, describing environmental scenes, communicating spatial properties and relations of environmental objects, or giving verbal navigational directions.
5. Montello and Raubal also discussed the task of imagining places and reasoning about them with mental models as another important component of spatial cognition. We construct mental models of spatial entities and relations from direct experiences, iconic representations, and linguistic expressions, to process spatial information mentally and to reason about nonspatial information spatially in a variety of tasks.
6. Relatedly, Montello and Raubal identified the task of locational allocation as a sixth task of spatial cognition, in which people reason about their spatial activities from the perspective of facility location, for example, minimizing distance to travel, planning a sequence of activities along an efficient route, reasoning about the cost of travel in terms of money or effort, and locating urban facilities in an economically optimal spatial configuration.

These tasks all entail a great deal of *spatial thinking*; some are conducted consciously, some unconsciously, and some pose difficulty for particular people in specific situations. An understanding of spatial cognition, therefore, contributes to the understanding of various human activities and thinking and helps to solve problems associated with the behavior in, and use of, space by humans. Application areas of spatial cognition that Montello and Raubal discussed include location-based services, geographic and other information systems, information display, architecture and planning, personnel selection, and spatial education.

Human-environment interaction

In this book, we will look at the above issues from the perspective of interaction between people and the environment. People perceive the environment, and the knowledge acquired through interactions with the environment is represented in their minds. A look at mentally represented environments tells us what people think and how they feel about their environments. Importantly, the major premise for the study of spatial cognition is that people's behavior is based on perceived environments, as well as physical environments. We react to the environment as we perceive it, and accordingly, as we think and feel about it—our cognition and behavior are closely related. We will discuss this at length in Chapter 3 in relation to cognitive-behavioral approaches in geography to spatial cognition. To live adaptively, or to survive in the environment, you need to know where you are. You must become oriented in the environment and able to navigate flexibly within it. In so doing, your knowledge about the environment, or "the map in your head," guides and directs your behavior. Inaccurate and distorted inner maps lead to disorientation and inefficient wayfinding behavior in the environment. Schematized knowledge representations lead to biased spatial judgments based on the limited and partial information available. We will consider this in Chapter 4 by looking at people's judgments about the spatial relation, for example, between San Diego and Reno (Stevens and Coupe, 1978). We will also look at the characteristics of mental representations. In Chapter 5, we will look specifically at the structures and processes of maps in the head in the context of spatial microgenesis, and, in Chapter 6, in reference to underlying spatial aptitudes.

Spatial cognition and behavior are also conceived in relation to external representations including maps and verbal descriptions, as seen briefly in the previous section. Maps and spatial language assist our wayfinding and spatial orientation, and also influence our cognition and behavior. For example, familiarity with maps in a north-up orientation correlates with the use of linguistic expressions of "up north" and "down south," and even the tendency to perceive northern routes on a map as more effortful than southern routes (Brunyé et al., 2010). We will look at how linguistic coding of spatial relations relates to nonlinguistic spatial reasoning in Chapter 7.

Just as cognition and behavior are interrelated, internally represented environments and physical environments are dependent on each other. People learn about the environment, think about the environment, and feel about the environment as they perceive and explore (or interact with) the environment. The process of interaction is bidirectional and iterative, as Piaget conceived of it in terms of *assimilation* and *accommodation*. People understand and interpret information from the environment by assimilating it to the structure of existing knowledge; and, at the same time, the existing knowledge is modified, restructured, and enriched by accommodating to the new information (Piaget, 1983). We will see how it works in the context of knowledge acquisition in a new environment in Chapter 5 and the development of spatial knowledge in children in Chapter 8. The reciprocal process of assimilation and accommodation, or adaptation to the environment, shows that a human being is not a passive receiver of information from the environment, and hence external environments are not simply copied onto the human mind. People interpret and select a subset of information that fits their purposes of survival. We will look at how spatial cognition and environmental perception can be considered from an evolutionary perspective, in relation to spatial ability in Chapter 6 and environmental aesthetics in Chapter 9.

Next, think about wayfinding and spatial orientation in the environment. Do you like visiting a new place and walking around in a new environment? Or do you find it difficult to know where you are in an unfamiliar environment and easily get lost even with the help of a map? Are you confident that you have a good sense of direction, or are you aware of a person who has severe difficulties with orientation in the environment? Why and how do you think people differ in the ability to stay oriented in space, which is indispensable for daily functioning in the environment? The task of spatial orientation seems straightforward on the surface, but it involves elaborate cognitive processing, namely acquiring, representing, and coordinating information about space inside and outside the traveler. To know where you are, you need to identify your position and orientation in the environment by understanding the relationship between yourself, the environment, and spatial representations in your head (spatial knowledge) and in your hand (a map or a smartphone). What is the structure of a mentally represented environment? Why does the accuracy of "inner maps" differ from person to person? How do people understand and use externally provided navigational information in the form of maps, satellite navigation tools, or verbal descriptions? What is the basic ability that relates to spatial cognition in the environment, and how does it develop from infancy through childhood to adolescence? We will consider these questions in Chapters 3–8.

Lastly, consider our feelings and emotions in and about the environment. We daily use expressions such as: "I love my neighborhood," "I feel at home in this town," "I am attached to this place," or "the living environment is getting worse." What do they mean and how are the feelings of good and bad, pleasant and unpleasant, and active and inactive determined? Do you prefer urban environments or rural scenes? Are there any general tendencies in the pattern of landscape preference? What does it mean when you say that you are emotionally attached to a particular place or identify yourself with the place? We will discuss these *affective* components of environmental perception in Chapter 9.

Now we are ready to look at how we perceive and interact with the environment and what we think and how we feel about the environment. We set out on a journey of space, or, more specifically, the space in our minds.

Review box for Chapter 1

1. Write in a diary what you did today and think about the "whereness" of your activities. How is your behavior spatial?
2. Identify a set of places important to you on a local map. Are the places located as you thought they were? Are there any differences between their actual locations and your estimated (imagined) locations?
3. Find some examples in your daily life of the statement: "Everything is related to everything else, but near things are more related than distant things."
4. Give the characteristics of major approaches in geography and psychology in relation to the study of spatial cognition.
5. Discuss various ways in which people interact with the environment. Think about human–environment interaction in terms of "representation."
6. How do you answer the question, "What is space?" What do you know about space and what do you want to know about space?

Chapter 2

Brief history of research into spatial cognition

Interest in space and cognition of space

Knowing about surrounding environments and representing the knowledge of learned (and sometimes imagined) environments are major activities that have characterized humans since antiquity. The earth has a history of 4.5 billion years, and modern humans (Homo sapiens) appeared on its surface in Africa around 200,000 to 300,000 years ago. Since then, humans have been living in interaction with the environment: they perceive and react to it, and it affords them functions for their survival. Successfully living in the environment requires an understanding of orientation, particularly knowing where you are and where other people, objects, and places are in your surroundings. Think about, for example, hunter-gatherers in ancient societies. They would set out to hunt animals and birds, sometimes involving long journeys in the wilderness. On successfully killing animals at a series of outbound hunting locations, they would need to collect the prey and bring it back home, keeping track both of the location of the prey (where they found individual animals) and of their homes. To do that, they may have relied on celestial cues, such as the position of the sun or the stars, and landscape features including hills, valleys, mountains, trees, or shorelines; they did not have digital maps or satellite navigation tools available as we do now—they needed to be self-navigating. As well as hunting, humans also foraged, collecting vegetables and harvesting fruits and seeds, which depends on the memory of the location of these foods in the tracts of farms and within complex vegetation.

Also, think about marine navigators. They explore the wild sea during the day or night, relying on their instinct for space (or "sense of direction") and reading celestial cues that are too subtle for laypersons, particularly for people who are accustomed to satellite navigation guidance. Being at sea in a small canoe may be likened to an ant moving in a featureless desert. Keeping track of one's position and orientation in a vast sea or a wide plain seems an enormous feat to us; but humans, animals, birds, and insects are in fact all capable of this (Etienne and Jeffery, 2004; Mittelstaedt and Mittelstaedt, 1980).

Polynesian and Micronesian navigators have developed and inherited from generation to generation a system of marine navigation, called the Etak System. In the system, the navigators secure the course of a canoe to a destination island by dead reckoning, supported by piloting in reference to the location of a series of stars along the lines of sight from the canoe to an imaginary reference island (Figure 2.1). Lewis (1976) described that in the navigation process envisioned in this system, "the canoe is visualised as stationary, while islands are regarded as 'moving' from one horizon star position to the next as the voyage proceeds" (p. 263). Similarly, Lewis reported with respect to terrestrial navigation the remarkable spatial orientation ability for Aboriginal people in the Western Desert region of Central Australia. They navigate at will on a land that appears simply featureless to laypersons and possess surprisingly accurate and precise knowledge of various locations in it (e.g., water and food resources, compass points, and sacred sites). Lewis stated,

> It would appear then, that the essential psychophysical mechanism was some kind of dynamic image or mental "map", which was continually updated in terms of time, distance and bearing, and more radically realigned at each change of direction, so that the hunters remained at all times aware of the precise direction of their base and/or objective.
>
> (p. 262)

Humans learned of the environment through exploration, and represented and communicated this knowledge to others in map form. Wall paintings in the Lascaux cave,

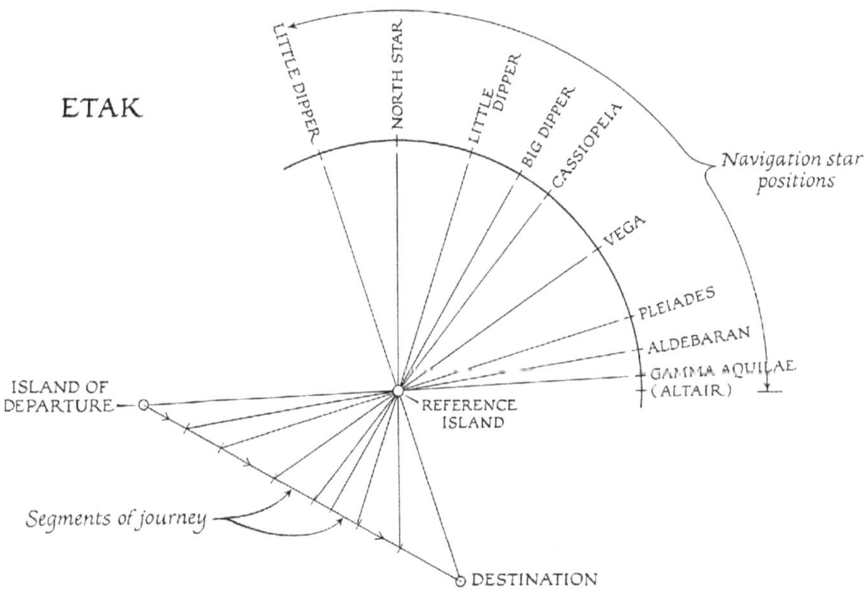

Figure 2.1 Etak System used by marine navigators in Polynesia and Micronesia. They set the course of a canoe in reference to a series of stars located along the lines of sight to an imaginary reference island. Navigation with celestial cues in the midst of an ocean seems remarkable, especially for people who are accustomed to satellite navigation. Reprinted with permission of Harvard University Press from Gladwin, 1970, p. 185. © 1970 by the President and Fellows of Harvard College.

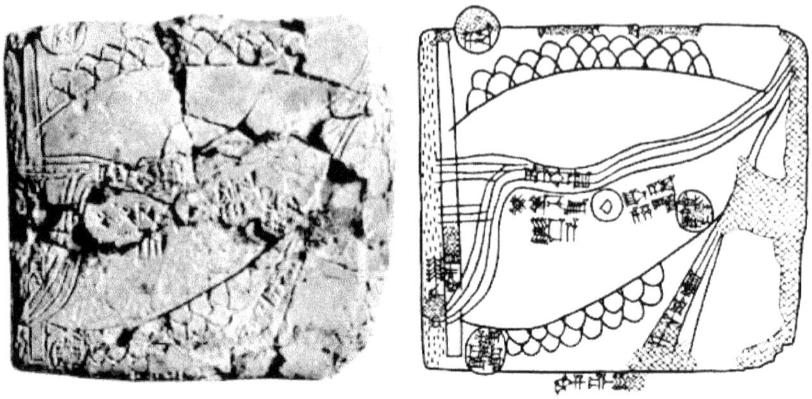

Figure 2.2 **Nuzi map (Nuzi was an ancient Mesopotamian city), inscribed on a clay tablet about 2300 BC. It depicts settlements, streams, and mountains, and also specifies three cardinal points by the appropriate wind. The west wind is shown at the bottom of the map, the east wind at the top, and the north wind on the left. Reprinted from Meek, 1935, Plates I and XCIII. Materials in the public domain.**

supposed to date back 17,000 years, depict animals and humans, and some of the paintings are thought to be a map of the night sky, showing the constellations of the Pleiades (Rappenglueck, 1999). The Nuzi map, which is believed to have been inscribed on a clay tablet about 2300 BC (Figure 2.2), represents settlements, streams, and hills in the region with symbols and pictures. It shows orientation information as cardinal points for three sides of the tablet corresponding to east, west, and north (Millard, 1987; Muhly, 1978). Similarly, the Babylonian World Map, the earliest known world map dating from about 600 BC, shows on a clay tablet the world as envisioned from the perspective of Babylonians. On the map, the earth is depicted as a circle, surrounded by water, and the city of Babylon is located above the center of the circle, with the Euphrates River running through it. The map also represents nearby cities, mountains, and marshes (Ehrenberg, 2006, p. 21).

The basic desire of humans to know about the environment and externally represent the knowledge is nicely captured by the well-known painting by Vermeer, *The Geographer* (Figure 2.3). It depicts human–environment interaction: a geographer contemplating the outer world that lies beyond the readily observable surroundings, out there waiting to be discovered, and various cartographic instruments that surround him such as a marine chart, a globe, and a cross-staff (a tool used to measure the angle and movement of celestial bodies).

> The Geographer stands in a room surrounded by the knowledge of the day, ready to discover something of importance himself. The maps, globes, and cartographic instruments scattered about the room are reminiscent of any great mind seeking to understand ideas previously unsought. . . . The opening of trade, the influence of other civilizations in knowledge and technology, and the cultural mingling that was just beginning to occur all influenced both Vermeer's life as well as his depiction of The Geographer.
>
> (Borneman, 2014)

Figure 2.3 **Vermeer painting, *The Geographer*, 1669. We can of course appreciate it in an artistic way, but also from a human–environment interaction point of view. It depicts a geographer contemplating the world and representing his knowledge as a map or a globe. The geographer is surrounded by the environment and interacts with it, and his knowledge about the environment is represented internally in his head and externally as spatial representations. © Städel Museum/ARTOTHEK. Reprinted by permission.**

We have thus been interested in the environments that surround us (and oftentimes extend beyond our immediate senses) and interested in other people's thinking about those environments.

The interest that humans have in space and spatial representations is reflected in their philosophical contemplations of space since the days of Ancient Greece. Aristotle provided in his book *Physics* (Book IV, chapter 4, trans. Hardie and Gaye) a metaphysical discussion about what *place* is. He described place as "the boundary of the containing body at which it is in contact with the contained body" and "the innermost motionless boundary of what contains." He then distinguished place from things (form and matter), saying:

> The form and the matter are not separate from the thing, whereas the place can be separated. . . . Hence the place of a thing is neither a part nor a state of it, but is separable from it. For place is supposed to be something like a vessel—the vessel being a transportable place. But the vessel is no part of the thing.

Aristotle's view may be called a container view of a place: a place is the boundary or surface of a body in it, and is no part of and separable from the contained body (Couclelis, 1998). Aristotle further elaborated on the idea of place in the book: he described the concepts of void and time, interrelated place and time through the concept of motion, and identified the properties of a place such as *up* and *down*. Aristotle's conceptual discussion of space shows that the issue of space has been treated as an important topic worth contemplation since the early days of human thoughts.

Another well-known philosophical argument of space was the one developed by the philosopher Kant. In his *Critique of Pure Reason* (1781/1787), Kant explored the question of human cognition in terms of sensibility and understanding and distinguished between things and appearances. In particular, he argued that space and time are a priori conditions for human understanding of objects in the world:

> Space and time are only forms of sensible intuition, and therefore only conditions of the existence of the things as appearances. . . . We have no concepts of the understanding and hence no elements for the cognition of things except insofar as an intuition can be given corresponding to these concepts. . . . Even if we cannot *cognize* these same objects as things in themselves, we at least must be able to *think* them as things in themselves.
>
> (trans. Guyer and Wood 1998, p. 115, emphasis in original)

The capacity (receptivity) to acquire representations through the way in which we are affected by objects is called *sensibility*. Objects are therefore *given* to us by means of sensibility, and it alone affords us *intuitions*; but they are *thought* through the understanding and from it arise *concepts* (p. 172). Kant called the intuition relating to the object through sensation *cognition* (p. 172), which is based on and enabled by empirical experience, and contrasted it with pure intuition that is independent of experience:

> All our cognition begins with experience; for how else should the cognitive faculty be awakened into exercise if not through objects that stimulate our senses. . . . But although all our cognition commences *with* experience, yet is does not on that account all arise *from* experience. . . . It is therefore at least a question requiring closer investigation . . . whether there is any such cognition independent of all experience and even of all impressions of the senses. One calls such *cognitions a priori*, and distinguishes them from *empirical* ones, which have their sources a posteriori, namely in experience.
>
> (p. 136)

The two forms of sensible intuition that define a priori cognition, Kant discussed, are space and time:

> By means of outer sense (a property of our mind) we represent to ourselves objects as outside us, and all as in space. In space their shape, magnitude, and relation to one another is determined, or determinable. Inner sense . . . gives . . . no intuition of the soul itself . . . yet it is still a determinate form, under which the intuition of its inner state is alone possible, so that everything that belongs to the inner determinations is represented in relations of time.
>
> (p. 174)

> Space is nothing other than merely the form of all appearances of outer sense, i.e., the subjective condition of sensibility, under which alone outer intuition is possible for us.
>
> (p. 177)

> Time is an a priori condition of all appearance in general, and indeed the immediate condition of the inner intuition (of our souls), and thereby also the mediate condition of outer appearances.
>
> (p. 181)

He thus concludes that space is a form of pure intuition, by saying, "Objects in themselves are not known to us at all, and . . . what we call outer objects are nothing other than mere representations of our sensibility, whose form is space" (p. 178).

Furthermore, Kant discussed in his treatise "On the First Ground of the Distinction of Regions in Space" (1768/1991) the issue of spatial orientation as mental processing, or more specifically the alignment of an internal representation of space with the represented space (the term "region" as it is used by Kant in the following passage is interpreted as "direction"; May, 1970, p. 70):

> For the positions of the parts of space in relation to one another presuppose the region towards which they are ordered in such relation; and this region, in ultimate analysis, consists not in the relation of one thing in space to another (which is properly the concept of position) but in the relation of the system of these positions to the absolute world-space. In anything extended the position of parts relatively to one another can be adequately determined from consideration of the thing itself; but the region towards which this ordering of the parts is directed involves reference to the space outside the thing.
>
> (Kant, 1768/1991, p. 27)

He clarified the importance of spatial orientation (or mental perspective taking) by saying:

> The most complete chart of the heavens, however perfectly I might carry the plan in my mind, would not teach me . . . on which side to look for sunrise, unless, in addition to the positions of the stars in relation to one another, this region were also determined through the position of the plan relatively to my hands. Similarly, our geographical knowledge, and even our commonest knowledge of the position of places, would be of no aid to us if we could not, by reference to the sides of our bodies, assign to regions the things so ordered and the whole system of mutually relative positions.
>
> (p. 29)

Kant thus considered concepts of space and people's conceptions of space fundamental to an understanding of how people make sense of the world.

The issue of spatial orientation has also been discussed from an evolutionary perspective. Darwin (1873) discussed the ability (or instinct) of organisms to find their way and orient themselves in space. In considering the manner in which animals find their way back home after traveling a long distance, he pointed to an exceptional wayfinding performance by natives in North Siberia:

> [A report observed] the wonderful manner in which the natives kept a true course towards a particular spot, whilst passing for a long distance through hummocky ice, with incessant changes of direction, and with no guide in the heavens or on the frozen sea. . . . No one will suppose that they possessed any special sense which is quite absent in us. We must bear in mind that neither a compass, nor the north star, . . . suffices to guide a [hu]man to a particular spot through an intricate country . . ., when many deviations from a straight course are inevitable, unless the deviations are allowed for, or a sort of "dead reckoning" is kept.

Darwin suggested a specialized function in the brain that relates to the maintenance of sense of direction, saying:

> All [humans] are able to do this in a greater or less degree, and the natives of Siberia apparently to a wonderful extent, though probably in an unconscious manner. . . . The manner in which the sense of direction is sometimes suddenly disarranged . . . and the feeling of strong distress which, as I know, has been experienced by persons when they have suddenly found out that they have been proceeding in a wholly unexpected and wrong direction, leads to the suspicion that some part of the brain is specialised for the function of direction.
>
> (p. 418)

Humans, like other animals, possess specific mechanisms for processing information about space, but, as we will see below, they do not necessarily find spatial orientation and wayfinding easy all the time.

In the early periods of scientific research into spatial cognition, notes on the episodes of remarkable spatial orientation and feelings of severe spatial disorientation appeared sporadically. In the first volume of the *Psychological Review*, Binet (1894) described illusions of orientation, which he defined as

> a group of facts which have not yet been studied methodically, but which have been cited by many authors and described in different words. These facts consist of illusions or hallucinations of orientation, which arise spontaneously either when we waken in the darkness of night, or during the day when we awake.
>
> (p. 337)

Episodes that Binet reported include a sensation of the relationship between the surrounding landscape and the direction of travel being turned around when entering a specific region in a town, a feeling of going in the opposite direction to the desired one when crossing a specific street, and a sense of mismatch between the direction of travel in an environment and the memory of the environment in one's mind.

In a similar vein, Warren (1908) reported a case of a 5-year-old boy who comprehended the points of the compass correctly in any place at all times, day or night, as though he had the magnetic sense existent in migrating birds. When Warren asked the boy to point to cardinal directions several times after walking him around blindfolded, the boy answered correctly with a precision of a few degrees every time. DeSilva (1931) also reported a 12-year-old boy who had an unusual sense of orientation to the points of the compass in either a familiar or an unfamiliar environment. Interestingly, his mother told the researcher that because the boy frequently had trouble telling left from right, she used to give instructions to him in terms of cardinal directions, for example, "Get me the brush on the north side of the dresser" or "Go and sit in the chair on the east side of the porch" (p. 394). DeSilva reported that the boy had developed out of this training the exceptional ability to move about a long, complicated path and retain his orientation without paying attention to the process (p. 394). We will discuss in depth the abilities pertaining to the perception and cognition of space in Chapter 6.

Spatial orientation, maps, and imaginary maps

One of the earliest articles about spatial cognition in geography concerns spatial orientation and map use in the context of geography education. Gulliver (1908) noted the difficulties that elementary school students had in understanding and using maps, and pointed out as correlated observations that many maps were presented to students in various different orientations in a classroom. He stated that in other school subjects, such as reading and music, students learned to read words and musical notes on printed pages in standard orientations, and wondered why in the teaching of geography maps lacked a definite orientation:

> Books are printed with a definite orientation of letters and words; music is also printed with a definite relation to five lines of the staff; maps should likewise conform to the definite rule of position on the page. The child may then be taught that he [or she] would see the surface of the earth as it is represented upon a map, if he looked down upon it as a bird would do in flying through the air from the south to the north.
>
> (p. 55)

Gulliver argued that students' learning of maps would be substantially enhanced if students were instructed to orient themselves in the environment by aligning their headings with a true north–south line and practicing it with maps that are aligned with the surrounding environment in a north-up orientation.

> A child trained in our modern schools must be able to read quickly the printed page; he [or she] must also be able to read and sing musical notes; likewise he should be able to read intelligibly and quickly any map which is presented to him for use or study.
>
> (p. 58)

Freeman (1916) emphasized the role that spatial imagination plays in geography, through which "our experience may be extended beyond the confines of our own immediate surroundings" and "the location of places upon the earth becomes organized in our minds" (p. 161). One aspect of spatial imagination that Freeman pointed out in particular was spatial orientation, an understanding of either the orientation of objects in relation to one's own position or the spatial relations between different places in a fixed frame of reference. Freeman argued that children's skills of spatial imagination are fostered through the use of maps in the learning of geography, and, in line with Gulliver's (1908) discussion above, stressed the importance of the orientations of maps when using them:

> A difficulty with the understanding of maps sometimes arises from regarding a certain side of a map as representing the wrong direction in the concrete world. By convention we use the top of the map to represent north. This association must, of course, be formed in the mind of the child. . . . If the child faces the east or even if the map is on the east wall, and nothing to the contrary is said to him [or her], he will think of the top of the map as representing the east. We have no experimental facts on the matter, but observation by teachers seems to indicate that it is desirable at first, in order to form the proper association, that maps be placed so that the direction on the map corresponds to the direction on the earth.
>
> (pp. 176–177)

In 1913, Trowbridge published an article in *Science* that discussed two methods of spatial orientation that people employ in the environment, one being accurate and the other erroneous. Trowbridge described that people who orient themselves accurately in the environment are

> attached to [their] home by one very strong elastic thread of definite length [so that] all changes of position of the creatures can be referred at any moment, to definite distances and angles, forming a simple trigonometric figure which gives the direction to the home.

By contrast, people who are prone to spatial disorientation "look outward towards the horizon [rather than looking] backward toward their starting point . . . [with elastic threads] of indefinite lengths, which present no basis whatever (lines or angles) for a trigonometric figure that relates to the home." The latter method of spatial orientation, Trowbridge argued, "leads to the conception of a mental image of an orientation map that is entirely imaginary, and erroneous"; such an imaginary map brings about confusion at the exit of theaters and subway stations and difficulties with finding the way home in an unfamiliar environment (p. 890).

Trowbridge (1913) conducted a behavioral experiment to examine the accuracy of maps that people have in their heads, by asking them to estimate directions from New York City to various other locations: the North Pole, London, San Francisco, and Panama. On the basis of their responses, Trowbridge distinguished between people who had difficulty with spatial orientation and those who did not, describing them as having imaginary maps, and not, respectively. A mean absolute error of direction estimates (calculated from the data tabulated in the article) was 99.2° for the former people and 17.3° for the latter.

Trowbridge (1913) suggested, as Gulliver (1908) did, the necessity of the instruction of spatial orientation in the classroom of geography, to correct for the distortion of imaginary maps:

> If it is desirable to correct this very common defect in orientation training, it would appear necessary that children should be seated at school in a special manner when studying geography, with the cardinal points of the compass marked in the room, and the maps in the books properly oriented, and the imaginary maps systematically corrected in childhood.
>
> (p. 893)

He mentioned that his participants had a relatively high level of education, and suggested that perception of space and orientation in space taps into a specific domain of ability. In this article, Trowbridge used the term imaginary maps to point to a set of misconceptions people have about space that lead to disorientation, but, as we will see below, people have mental representations of the environment, some of which are accurate and some of which are distorted; that is, the accuracy of mentally represented environments varies greatly from person to person (see Chapter 7).

Piaget's developmental theory

One of the most influential theories in psychology in general, and in spatial cognition in particular, is the developmental theory by Piaget. Piaget and his colleagues discussed, in various domains of intellectual development, the developmental course in which people start to gain knowledge about the world as infants with direct sensorimotor actions and proceed to higher levels of understanding through childhood to adolescence in successive stages of equilibrium between the states of their knowledge and the world that they interact with (see Piaget, 1983). What particularly characterizes the Piagetian course of intellectual development is the distinction between an understanding at the perceptual level and an understanding at the representational revel (Mandler, 1983). According to Piaget, people *construct* the knowledge of the world through different stages of increasing representational quality, from concrete actions such as pure reflexes, to the manipulation of real objects, and finally to symbolic thought, exhibiting fully intelligent behavior of internally coordinating perceptions and mentally organizing representations. The development of human intelligence thus points to the direction of increasing levels of operation, transformation, and abstraction. As infants grow and mature, their concrete and simple ways of thinking become more abstract and complex; eventually they are able to think logically and systematically, and attain a rich understanding of the world to engage in hypothetical and probabilistic ways of reasoning about the world. Piaget conceptualized this intellectual development as a discrete process proceeding through the *stages* of sensorimotor, preoperational, concrete operational, and formal operational periods.

Piaget and colleagues applied the stagelike framework to the context of spatial cognition and explained the development of spatial concepts in children as qualitative shifts in three stages (Piaget and Inhelder, 1948/1967). In the earliest stage of development, spatial concepts in infants are tied to sensory perceptions and motor activities, in which various sensory spaces (particularly visual and tactile-kinesthetic spaces) are not coordinated to each other, and thus only elementary spatial relations can be grasped, such as proximity, separation, order, enclosure, and continuity (pp. 6–8). This stage signifies "the beginnings of abstraction of shapes, but curiously enough, the shapes first recognized are not euclidean but topological; thus the circle and the square cannot be distinguished because both are closed forms, although they are distinguished from open forms" (p. 21).

In the next developmental stage, children's perception becomes distinguished from their own activity, and in their spaces their perspective projections start to be coordinated. This coordination leads to "the formation of both projective and metric relationships, whose synthesis results in the constancies of shape and size" (p. 12). Thus, "there begins a crude differentiation of rectilinear from curvilinear shapes, whilst different rectilinear shapes (square, rectangle, etc.) or curvilinear shapes (circles, ellipses, etc.) are not differentiated among themselves" (p. 21). Subsequently, children progressively recognize Euclidean shapes, signifying a developmental stage that indicates the "beginning of recognition of euclidean shapes, based on the distinction between straight and curved lines, angles of different sizes, parallels, and especially on relations between equal or unequal sides of a figure" (p. 43).

In the final stage, children achieve representational understanding of shapes, and at this stage "the construction of shape is clearly quite separate from its perception and from its imitative or pictorial representation" (p. 37). The process of development that started with elementary sensorimotor perceptions and actions culminates, through symbolic images and concrete operations, in abstract operations (p. 446). In sum, infants first supposedly do not distinguish between rectilinear and curvilinear shapes, or between circles, squares, and triangles (topological space), but come to distinguish views from different perspectives (projective space), and finally achieve as adolescents a metric understanding of figures and shapes in terms of angle and distance in the frames of horizontal and vertical axes (Euclidean space).

A basic tenet behind Piaget's developmental theory is that successful organisms survive various conditions of the environment, some of which are familiar and some novel. The key to survival is *adaptation*, namely development through the reciprocal processes of understanding the environment on the basis of existing knowledge and updating the knowledge by learning new facts. Consider the concept and function of adaptation with an example. An inventory of knowledge about the types and appearances of varieties of mushrooms would allow you to choose only edible ones and avoid poisonous ones. You may identify every mushroom you find in your territory, and there is no danger of suffering from poisoning—so far so good. But what if you encounter a novel one whose appearance does not match any of the mushrooms in your "pictorial book"? You somehow need to know whether it is edible or poisonous, by eating it by yourself, seeing other people eat it, or asking an experienced and knowledgeable person. Of course you can just ignore the mushroom, which may be a clever decision; it does not, however, lead to an expansion of your knowledge base about the edibility of mushrooms. An examination of the edibility of the novel mushroom in any way would add a new item in your pictorial book. That is, knowing about the edibility of the new mushroom *restructures* your knowledge by incorporating a new piece of information, which is in the end indispensable for successful adaptation and survival. (Later in the discussion of *schemata*, we will see that an encounter with new or unexpected information initiates the cycle of information exploration and knowledge modification; Neisser, 1976.)

Piaget thus discussed the developmental process through the concepts of *assimilation* and *accommodation* and the equilibrium between them. People learn about the environment, think about the environment, and feel about the environment as they perceive, explore, and *interact with* the environment. This process of interaction is bidirectional and iterative (Piaget, 1983). Humans are not passive recipients of information from the environment, and hence external environments are not simply copied onto their minds—humans interpret and select a subset of information that fits their purposes of survival. Information from the environment is assimilated to the structure of existing knowledge; and, at the same time, the existing knowledge is modified, restructured, and enriched to accommodate the new knowledge. This sort of restructuring is at the core of development as Piaget sees it.

Such a cognitive cycle of selection, interpretation, and modification does not work for information that is completely novel, however. For example, Haken and Portugali (2003) regarded the city as a collection of information, and discussed the cognition of the city, focusing on the amount of information represented and conveyed by different urban elements. In particular, they pointed out that Shannonian and semantic information is related because of the necessity of categorization in the quantification of information. We will discuss this at length in relation to the concept of schemata in Chapter 4.

The idea of cognitive maps

We have already used in the foregoing discussions the expressions of "maps in the head," "internally (or mentally) represented environments," and "knowledge about the environment." These terms convey an idea that is central to the study of spatial cognition, as we will see in detail in the chapters that will follow. The idea was first put forth in the literature of spatial cognition by Tolman (1948) in terms of the famous metaphor of "cognitive maps."

Tolman and colleagues conducted a series of experiments with rats in a maze. In one experiment (Tolman, Ritchie, and Kalish, 1946), rats were trained in an elevated maze, shown in Figure 2.4 (*left*), for four nights with three trials per night to go from A, across an open circular table through CD, and finally to G, where they found food. Then, they were placed in a new maze (Figure 2.4, *middle*), where a series of radiating paths was added: the starting point and the circular table remained the same, but the path that they had learned was blocked at D. Tolman et al. were interested in which of the newly added paths the rats would choose, when they found that the original path was blocked. Results showed that a majority of the rats (36 percent) chose path 6 in Figure 2.4 (*right*), a path that pointed directly to the goal location. The only other path that was chosen by comparatively many rats (17 percent) was path 1, which pointed perpendicularly to the food site in the room. Tolman (1948) argued that these results suggest that the rats acquired maplike knowledge of spatial relations of the maze, or they learned the *place*, not merely the sequence of *responses*. He stated,

> These results seem to indicate that the rats in this experiment had learned not only to run rapidly down the original roundabout route but also, when this was blocked and radiating paths presented, to select one pointing rather directly towards the point where the food had been or else at least to select a path running perpendicularly to

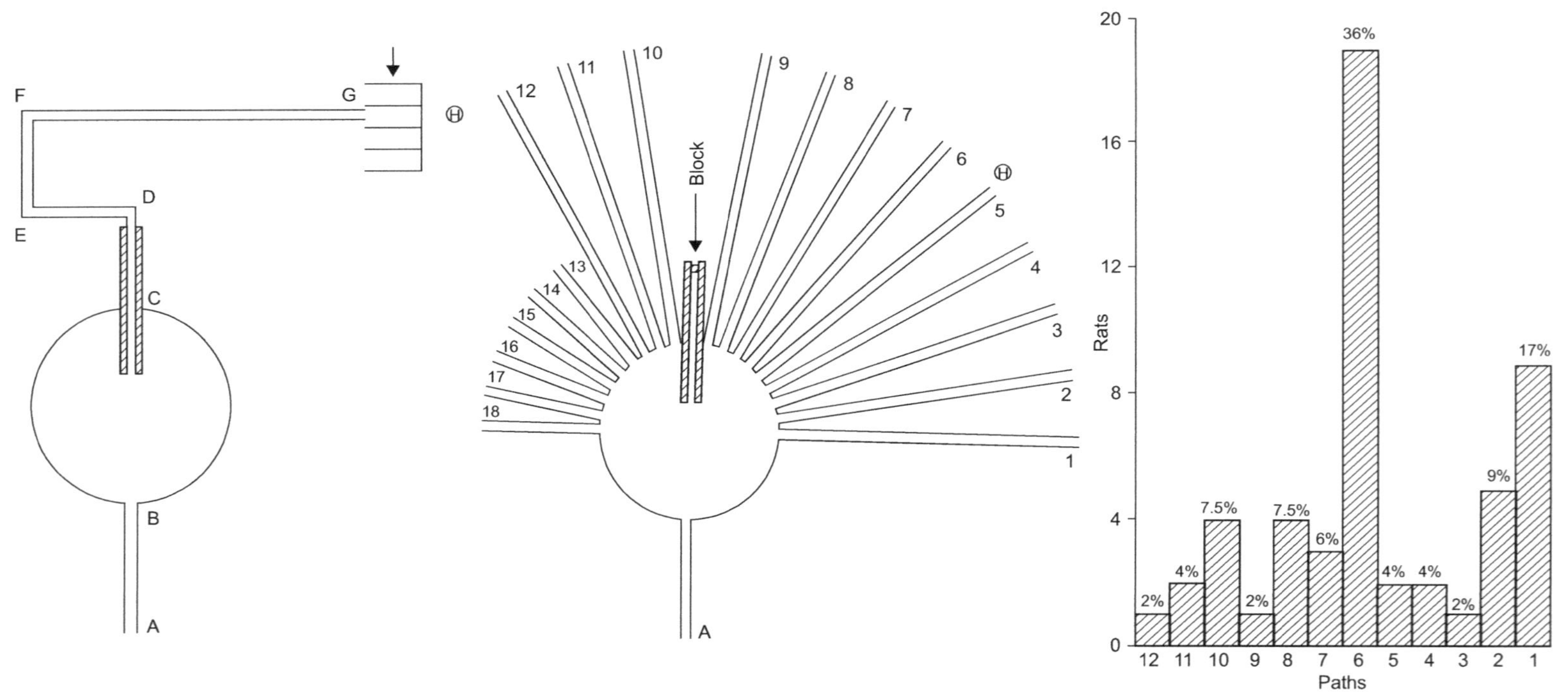

Figure 2.4 **Well-known experimental settings and results about a rat's spatial learning and behavior. After being trained to find food in a maze (*left*) and put in a new maze with a series of radiating paths (*middle*), the majority of rats chose a path that pointed directly to the goal location (path 6, *right*). Tolman described it as an indication of place learning by the rats, instead of sequence learning. Reprinted from Tolman, 1948, pp. 202, 203, and 204.**

> the food-side of the room. As a result of their original training, the rats had, it would seem, acquired not merely a strip-map to the effect that the original specifically trained-on path led to food but, rather, a wider comprehensive map to the effect that food was located in such and such a direction in the room.
>
> (Tolman, 1948, p. 204)

To argue for the acquisition of maplike cognitive representations, not the connections of stimuli and responses, Tolman (1948) pointed to the existence of *latent learning* in rats; that is, rats learned a maze during trials without rewards, which was not manifest until food was introduced. Tolman and Honzik (1930) examined the learning of a maze by three groups of rats. One group of rats (Group 1) was put at the entrance to a 12-unit alley T-maze and wandered about through the maze (both true path segments and blind alleys) to the goal without any food in it, one trial a day. Another group of rats (Group 2) similarly ran the maze, but these rats (unlike the rats in Group 1) found food at the end of the maze. A third group of rats (Group 3) was not fed for the first ten days, but found food at the end of the maze on the eleventh day onward. Tolman and Honzik observed the errors that the three groups of rats made (the number of entrances to blind alleys) over the series of trials. Figure 2.5 shows that the rats made fewer errors over trials, with the errors decreasing more rapidly for Group 2 (with food in every trial) than Group 1 (without any

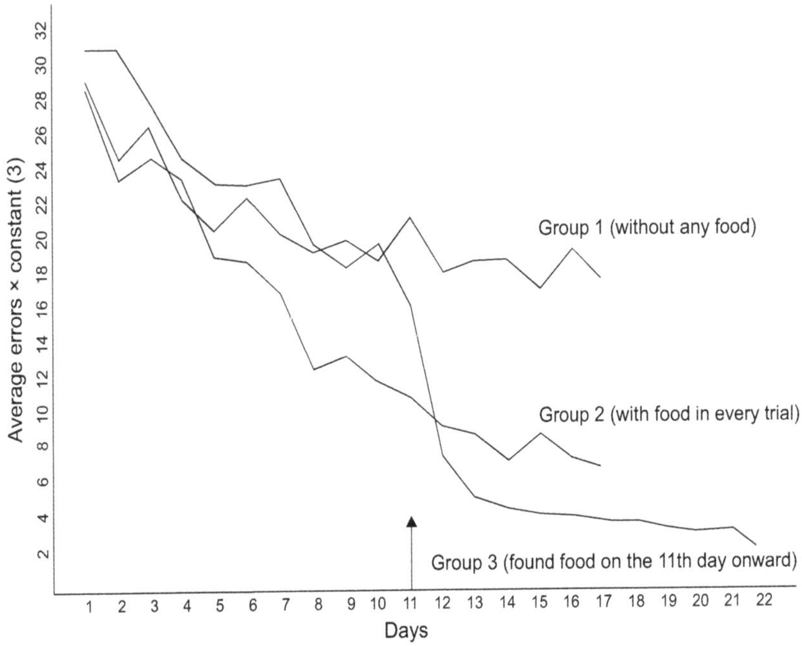

Figure 2.5 **Observations of errors for three groups of rats over a series of trials. The errors for rats that received food in every trial (Group 2) decreased more rapidly than the errors for rats without any food (Group 1). Rats that received no food until the tenth day (Group 3) showed a pattern of errors similar to Group 1, but on the eleventh day, when they received food for the first time, the errors dropped suddenly, to the level of performance comparable to or better than that for Group 2. Tolman emphasized that it indicates latent learning by the rats. Reprinted from Tolman, 1948, p. 195.**

food). Notably, although the error for Group 3 showed a pattern similar to Group 1 until the tenth day, it dropped suddenly and significantly on the eleventh day, when the food was introduced for the first time, to the level of performance comparable to or better than that for Group 2. Tolman (1948) stated:

> It will be observed that [the rats] as long as they were not finding food did not appear to learn much. . . . But on the days immediately succeeding their first finding of the food their error curves did drop astoundingly. It appeared, in short, that during the non-rewarded trials these animals had been learning much more than they had exhibited. This learning, which did not manifest itself until after the food had been introduced . . ., [is] called latent learning. . . . They had been building up a "map," and could utilize [it] as soon as they were motivated to do so.
>
> (pp. 194–195)

The significance of these contentions becomes clearer when the historical background of the discipline of psychology around that time, called behaviorism, is taken into account. Spatial cognition and behavior were then ready to be investigated from the perspective of internal representations and processing. We will turn to this in Chapter 3.

Ecological psychology and affordances

We saw in Chapter 1 that the study of spatial cognition focuses on the process of interaction between humans and the environment, which echoes the ecological approach to visual perception advocated by J. J. Gibson (1979/1986). For Gibson, neither the physical environment nor organisms mean anything when they are discussed separately, because they "make an inseparable pair" (p. 8). Organisms do not exist without the environment that surrounds them and the environment cannot be conceived without the organisms that inhabit it, a tenet of mutual dependence that Gibson refers to by the term *ecology*. Ecology signifies that organisms and the environment exist in mutual dependence; in this interdependence, organisms *perceive* the environment, or they *pick up* information from (or specified in) the environment that is critical for their survival. Gibson puts it as "every animal is . . . a perceiver *of* the environment and a behaver *in* the environment" (p. 8). That is, organisms understand (or sometimes misunderstand) what the environment can do for (or *affords*) organisms. What they perceive in the environment is the shape of the terrestrial environment (which Gibson calls the *layout* of the environment), some parts of which are invariant and some changing. Organisms discriminate, through locomotion in the environment, the properties of the environment that persist and the properties that change, or they perceive invariant properties in the background of changing properties.

In explicating the mutual dependence of the environment and organisms, Gibson first discusses the constituents of the environment: mediums, substances, and surfaces. Major examples of *mediums* for animals in the environment are air and water, which permit their locomotion and are filled with a flux of light. The flux of light or *ambient* light surrounding the viewer's point of observation has a structure that signifies the meaning of the environment and the situation of the animals. The environment also contains solid bodies that are more or less rigid and resistant to deformation, which Gibson calls *substances*; for example, rock, soil, sand, wood, minerals, plants, and animals. *Surfaces* separate the medium from the substances of the environment and play particularly important roles in the perception of the environment. They are the places where most of the actions of animals take place—any information important for animals' survival is present there as a layout specified in the ambient optic array.

Surfaces provide distinctive information for perception. For one thing, they differ in texture. Different substances such as mud, clay, and sand have different textures. Likewise, different animals have different textures, for example, having fur, feathers, or skin. Also, any surface has a characteristic shape, in the sense that it signifies the layout and structure of the environment at the larger and coarser scale compared with the fine-scale structure shown by texture. The layout of a surface concerns environmental enclosures (e.g., a cave, a tunnel, or a room) and objects with different shapes (e.g., a sheet, a stick, a cube, or a pyramid): "differently shaped enclosures afford different possibilities of inhabiting them, and differently shaped solids afford different possibilities for behavior and manipulation" (p. 29). Most importantly, ambient light reflected by a surface possesses a structure that specifies the meaning of the environment to the perceiving animal. In other words, the structure specified in ambient light is *out there* for animals to pick up directly.

Therefore, ambient light provides information about the environment for animals, and animals perceive ambient light in the form of an ambient optic array; *ambient* meaning it surrounds a position of the observer and *array* meaning it has structure or arrangement. As shown in Figure 2.6a, the optic array from the same environment (in this case, a room with a window) changes as the observer moves, and, accordingly, the point of observation moves. However, it does not change completely:

> Some features of the array do not persist and some do. The changes come from the locomotion, and the nonchanges come from the rigid layout of the environmental surfaces. Hence, the nonchanges specify the layout and count as information about it; the changes specify locomotion and count as another kind of information, about the locomotion itself.
>
> (p. 73)

Gibson stresses that an essential structure is invariant, irrespective of the change in superficial structure of an array owing to the observer's movement, and that the variation serves to reveal the nonvariation of the structure: "What is invariant does not emerge unequivocally except with a flux. The essentials become evident in the context of changing nonessentials" (p. 73). For example, a tabletop is perceived as rectangular when the image available to the eyes contains a large number of trapezoidal forms.

The changes in the ambient optic array owing to the changes in perspective specify important information about the observer's locomotion (Figure 2.6b). During locomotion in the environment, the flow of the ambient optic array, or a motion perspective, specifies the direction of movement along the axis of the displacement of the observer by the focus of expansion forward and the focus of contraction backward, superimposed on the nonchanging features of the array (the earth–sky contrast at the horizon and the texture of the earth). This optical flow provides the observer in motion with the sense of *visual kinesthesis*; it is supplemented by information about the movement of a limb (which

Figure 2.6 **The environment offers, or is equipped with, an ambient optic array that specifies the structure of the environment surrounding a position of the observer. Changes in the optic array (views from the window) also provide information about the observer's movement in the environment (a). An array flowing to one's left specifies one's own movement to the right (b). Importantly, the optic array changes as the observer and the point of observation move; invariant in the optic array is an essential structure or the layout of the environment.**

Gibson calls muscle-joint kinesthesis) and constitutes "the only reliable information about displacement" (p. 125).

These considerations form the basis of Gibson's idea of *affordances*: the environment specifies, in the composition and layout of surfaces, the values and meanings of things in the environment to animals, or describes "what the environment *affords* animals" (p. 127). An example given by Gibson is the affordance of support offered by a terrestrial surface that is nearly horizontal, flat, and sufficiently extended. It is a functional property of the environment and has a meaning for the animal that perceives it. A vertical, flat, extended, and rigid surface, Gibson says, is perceived as a barrier to locomotion, and slopes between the vertical and horizontal positions afford walking or climbing depending on the degree of steepness. The perception of the layout and the perception of affordances are thus closely interrelated: "The affordance of a certain layout is perceived if the layout is perceived" (p. 132). Furthermore, in Gibson's explanation, visual perception of the environment, or more correctly visual perception of the observer's environment, is *direct:* "the affordance, being invariant, is always there to be perceived" (p. 139) and information about the environment "is not conveyed [but] is simply there" (p. 242). Animals pick up information that is specified available in the ambient optic array in the perception of affordances.

The idea of direct perception, when it is interpreted in the strict sense of the term, contrasts with the ideas of internal representations and mental processes, because it states that everything necessary is specified in the environment. A discussion of visual perception with a focus on computational aspects of information processing has also been developed, notably by Marr (1982). We will see it in Chapter 7 in relation to the discussion of spatial frames of reference.

Image of the city

Another perspective from which researchers look at the cognition and perception of the environment concerns the mental image that people have about a city. People perceive the environment and have their own images of it, or they evaluate the environment. Some environments are tidy and coherent, some are complex but intriguing, and some are mysterious. There are environments that are confusing and defy spatial orientation inside them, and also environments that evoke the feeling of being at home and a sense of community. Some of your neighborhood areas are active, vivid, and popular, and some produce negative feelings. Thus, people think and feel about physical environments and make affective responses to them (Appleton, 1975; Kaplan and Kaplan, 1989). In general, people prefer to live in comfortable environments, and, accordingly, planners aim to develop urban environments that are comfortable, beautiful, and delightful, or to protect residential environments from excessive and irregular developments. To do so, are there any guiding principles available to planners, particularly ones that take into account the thoughts and feelings that residents have about their environments?

Lynch (1960) addressed this question by focusing on the visual quality of a city and clarified the importance of studying people's perceptions at the city scale:

> The apparent clarity or "legibility" of the city space . . . is of special importance when considering environments at the urban scale of size, time, and complexity. To understand this, we must consider not just the city as a thing in itself, but the city being perceived by its inhabitants.
>
> (pp. 2–3)

Of particular importance is the appreciation that the perceptions of the environment by its residents are a result of their interactions with the environment:

> Environmental images are the result of a two-way process between the observer and his [or her] environment. The environment suggests distinctions and relations, and the observer . . . selects, organizes, and endows with meaning what he sees. The image so developed now limits and emphasizes what is seen, while the image itself is being tested against the filtered perceptual input in a constant interacting process. Thus the image of a given reality may vary significantly between different observers.
>
> (p. 6)

While such variations may exist in individuals' environmental images, there must be a "substantial agreement among members of the same group" (p. 7). Lynch (1960) thus focused on the commonality of people's environmental images, or *public images*, seeking to reveal the common mental pictures possessed by the majority of a city's inhabitants. Questions of interest then include: what elements of a city appear on the inhabitants' minds, and what are the characteristics of such publicly recalled elements of the city? A key to answering these questions is an examination of the visual qualities of city environments that induce the residents to form a vivid, strong, and coherent image. Lynch called these qualities *imageability* and *legibility*:

> This study will look for physical qualities which relate to the attributes of identity and structure in the mental image. This leads to the definition of what might be called imageability: that quality in a physical object which gives it a high probability of evoking a strong image in any given observer. . . . A highly imageable (apparent, legible, or visible) city in this peculiar sense would seem well formed, distinct, remarkable. . . . Such a city would be one that could be apprehended over time as a pattern of high continuity with many distinctive parts clearly interconnected.
>
> (p. 10)

Understanding what makes complex urban environments imageable and legible would enable planners to design environments that are vividly remembered and produce feelings of urban beauty and delight in the minds of residents (which Lynch thought were lacking in American cities at that time).

To that end, Lynch (1960) examined the characteristics of public images held by residents in the cities of Boston, Jersey City, and Los Angeles. Lynch asked residents in each city to describe the city verbally, identify distinctive elements in the city, explain feelings about different parts of the city, and draw sketch maps of the city, as a means of "extracting" their images. Specifically, he asked the residents, for example in Boston, (a) to describe Boston in the physical sense, (b) to make a quick map or a rough sketch of central Boston, (c) to give directions for a normal trip from home to work, (d) to talk about any emotional feelings about parts of the work trip, (d) to give distinctive elements of central Boston and associated feelings, and (e) to indicate the distinctive elements and the direction of north on the sketch map drawn.

The residents' responses to these questions were summarized as a public image in map form. A map for the public image of Boston is shown in Figure 2.7. On the map, one can identify the parts of the city that stand out in the residents' minds (e.g., the Boston Common, the State House, the Charles River, Washington Street, Commonwealth Avenue, and Copley Square); in contrast, some parts of the city do not emerge on the public image (e.g., the triangular region between the Back Bay and the South End) or are represented in an ambiguous and distorted manner (e.g., the five-sided shape of the Boston Common). These results show that Boston is "a city of very distinctive districts and of crooked, confusing parts" (Lynch, 1960, p. 17), where thematic vividness is coupled with formlessness. The strengths and difficulties in imageability revealed on the public image would supply possible ideas for urban planning and design. For example, one can consider how to improve the city's imageability on the basis of the problems of isolations, discontinuities, lacks of differentiation,

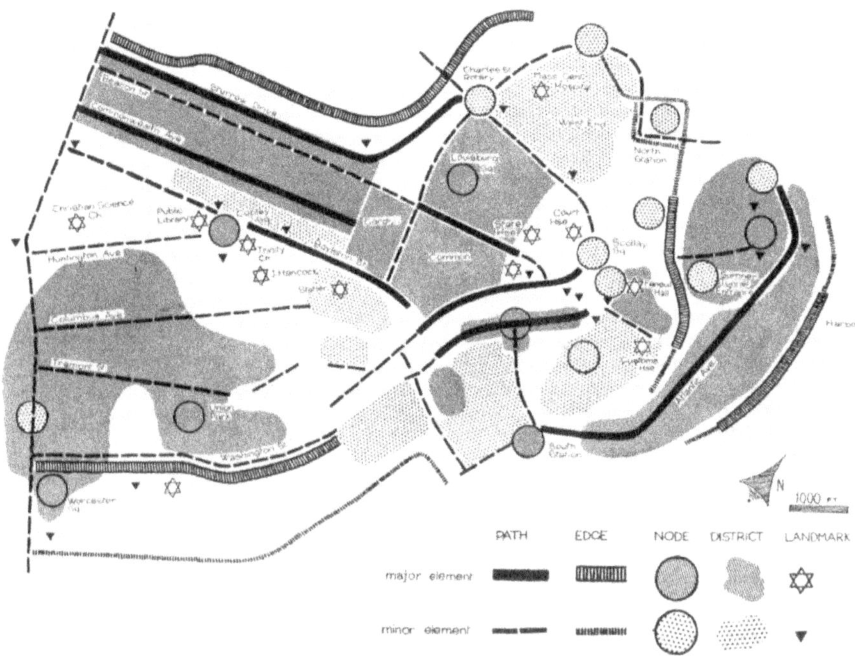

Figure 2.7 **Map of the public image held by the residents of Boston. It shows the parts of the city that stand out in the residents' minds or have higher degrees of imageability. Reprinted with permission of The MIT Press from Lynch, 1960, p. 19. © 1960 Massachusetts Institute of Technology.**

confusions about locations and spatial relations, and ambiguities about directions and shapes identified on the image.

Building on these analyses, Lynch (1960) classified the elements of physical environments that emerged in the residents' minds into five categories: paths, edges, districts, nodes, and landmarks (Figure 2.8). In Lynch's terms, paths are linear elements along which the traveler moves, such as streets, walkways, or transit lines. Edges are linear elements not used or considered as paths by the traveler, being the boundaries between two paths or linear breaks in continuity, such as shores, railroad cuts, or edges of development. Districts are two-dimensional sections of the city, recognizable as having some common character and thematic continuities in regard to facade, color, texture, or planting. Nodes are strategic spots in a city into which the traveler can enter, such as junctions or crossings. Landmarks are point-reference objects or scenes, external to the traveler and recognizable when seen, such as buildings, signs, trees, or mountains.

In addition to identifying these elements, Lynch (1960) emphasized that these individual elements are desirably interrelated in people's minds on the grounds that they provide the city with its form and structure. Importantly, Lynch acknowledged that people have difficulty in accurately interrelating individual elements with each other and

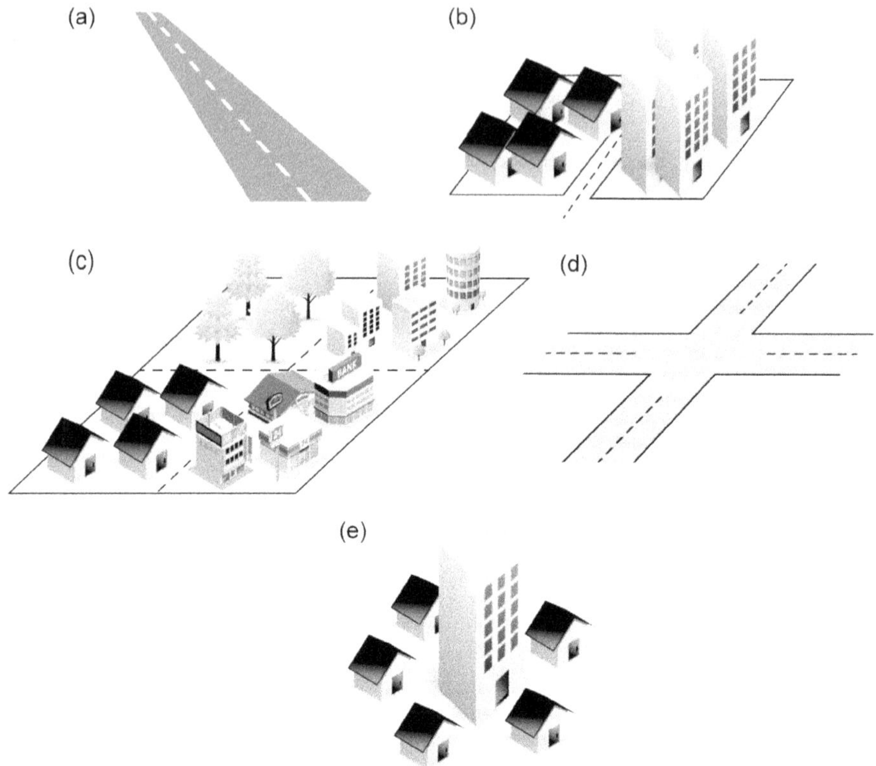

Figure 2.8 **Five elements of physical environments that Lynch identified in the public images of local residents: (a) paths, (b) edges, (c) districts, (d) nodes, and (e) landmarks. Illustration based on Lynch, 1960.**

integrating them into a whole. Such difficulty leads to the abstraction, schematization, and distortion observed in many city images.

> Rather than a single comprehensive image for the entire environment, there seemed to be sets of images, which more or less overlapped and interrelated. They were typically arranged in a series of levels. . . . This arrangement by levels . . . imposes an extra burden of organization on the observer, especially if there is little relation between levels.
>
> (p. 86)

> The image itself was not a precise, miniaturized model of reality, reduced in scale and consistently abstracted. . . . It was as if the map were drawn on an infinitely flexible rubber sheet; directions were twisted, distances stretched or compressed, large forms so changed from their accurate scale projection as to be at first unrecognizable. But the sequence was usually correct, the map was rarely torn and sewn back together in another order. This continuity is necessary if the image is to be of any value.
>
> (p. 87)

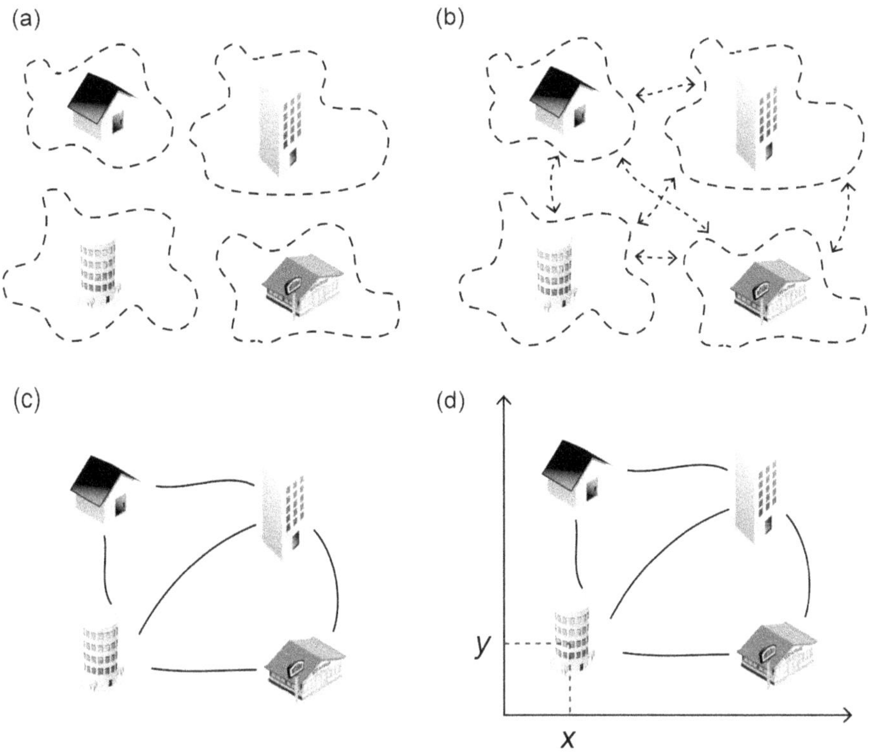

Figure 2.9 **Four stages of development of city images: (a) elements without any spatial relations, (b) elements roughly related, (c) elements connected loosely in terms of sequences, and (d) elements interrelated with each other in two dimensions. Illustration based on Lynch, 1960.**

Lynch (1960) postulated four stages of development with respect to the quality and structural accuracy of city images, in which (a) various elements are conceived individually without structure or interrelation, (b) elements are still disconnected but roughly related by general direction or relative distance, (c) elements are connected loosely in terms of sequences, and (d) elements are interconnected with rigid structure (Figure 2.9).

Lynch's (1960) systematic examinations of public images clarify that it is possible and of theoretical and practical significance to look at the images that people have of their environments. Internally represented environments are obviously not observable directly, but through the methods of verbalization, map drawing, and preference judgment, internal images can be *externalized*. Also Lynch's work points to the importance of examining people's perceptions at the urban scale of size, time, and complexity. For instance, some major concepts of urban design, such as the axis of a city and the greenery network in a city, need to be assessed from the perspective of city dwellers. For those concepts to be truly effective, the axis and the network in question need to be experienced and apprehended by people who view them horizontally in the environment, not only when a plan map is seen from above.

Spatial orientation and inner GPS

We saw earlier in this chapter marine navigators' remarkable skill of exploring the sea in a canoe using celestial cues, and likened it to an ant moving freely between feeding sites and the nest in a featureless desert. How can desert ants accomplish such accurate navigation and spatial orientation in an immense open field? Researchers have discussed two major navigation systems that desert ants rely on: path integration and landmark guidance (Hoinville and Wehner, 2018). *Path integration* refers to the updating of one's position and orientation with respect to a reference point (e.g., the origin of travel or home) on the basis of vestibular information from self-motion, as well as kinesthetic, motor efference copy, and optic flow information (Gallistel, 1990; Mittelstaedt and Mittelstaedt, 1980). In *landmark guidance*, desert ants use visual landmarks or a landmark panorama to identify the locations of, for example, the nest and a feeding site (Fleischmann et al., 2016).

The involvement of path integration in the desert ant's navigation was demonstrated by an experiment conducted by Wehner and Srinivasan (1981). In the experiment, desert ants were captured when they departed from a feeding station toward the nest, and then were released in a test area located 300–600 m away from the nest area. An observation of their return paths showed that the ants run in a straight line toward the place where the nest should have been; after moving in that direction, a distance that roughly matched the distance between the location at which they were captured and the (true) nest, they started searching for the nest, which they thought should be located there (Figure 2.10). During the search, the ants moved in loops in different directions and in increasing distances, constantly

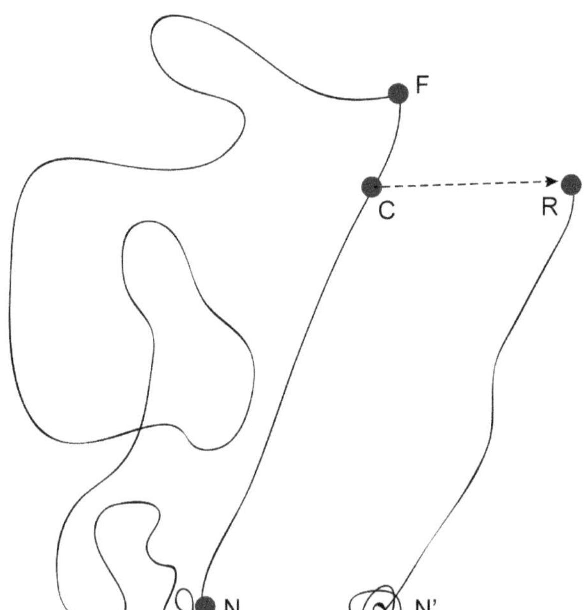

Figure 2.10 **Illustration of a path that a desert ant takes from its nest (N) to a feeding location (F). After finding food, it goes directly back to the nest. When the ant is captured on the way back to the nest (C) and released in a test area located 300–600 m away from the nest area (R), it quickly moves in a straight line toward the place where the nest should have been (N') and starts searching for the nest.**

returning to the origin of the search, probably looking for a view that would match the one around the nest stored in memory (research has indicated the use of landmarks by desert ants, particularly around the nest; Bisch-Knaden and Wehner, 2003; Fleischmann et al., 2016). Importantly, the ants did not retrace the outbound path, and thus updated the direct spatial relations between their current positions and the nest, moment by moment.

Along with behavioral observations of such remarkable navigation performances, researchers have investigated neural structures that underlie spatial orientation and wayfinding in mammals, particularly rats (Grieves and Jeffery, 2017; Figure 2.11a). One of the major early contributions in this line of investigation is the finding of *place cells* in the rat hippocampus. O'Keefe and Dostrovsky (1971) found brain cells that fired maximally in a freely exploring rat when the rat's head was in a specific region of the environment, regardless of the direction the rat was facing (Figure 2.11b). The existence of these location-specific cells suggested that some sort of spatial representation was formed in the brain (O'Keefe and Nadel, 1978), and stimulated further research into brain activities that relate to the rat's behavior of spatial orientation. In such endeavors, researchers next found cells that fired maximally when a rat's head faced a specific direction in the horizontal plane, irrespective of the rat's position (Ranck, 1984; Taube, Muller, and Ranck, 1990a, 1990b; Figure 2.11c). These cells, called *head*

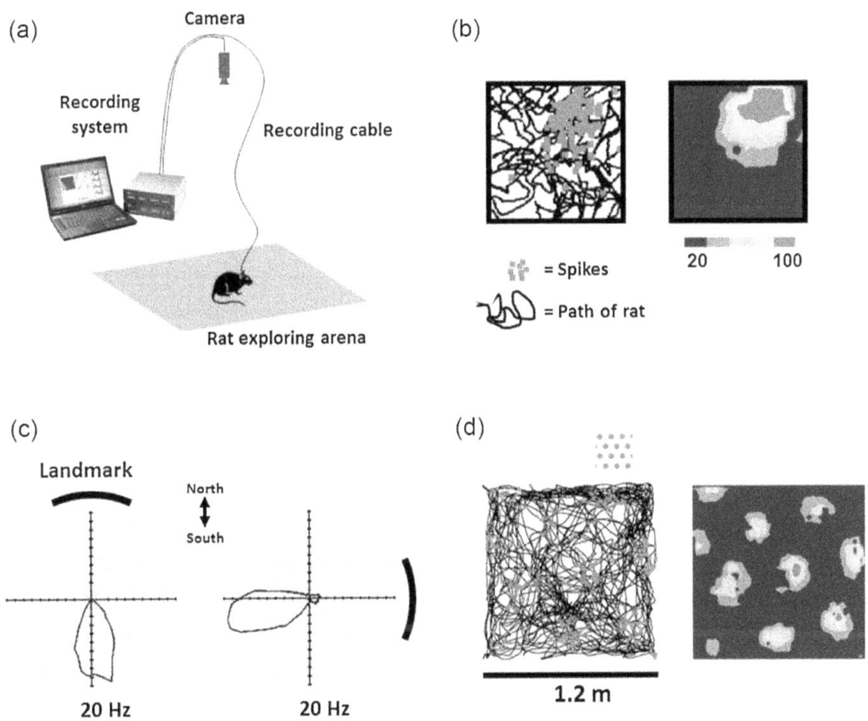

Figure 2.11 (a) Experimental setting for recording neural activities in a freely moving rat. (b) Data from a single hippocampal place cell, showing that it fires maximally when the rat is in a particular place in the environment (in this case, in the northeast part) irrespective of its heading. (c) Two trials recorded from a head direction cell, showing that it fires maximally when the rat's head faces a specific direction in relation to a local landmark. (d) Data from a grid cell, showing that it fires maximally in multiple places that form a hexagonal array. Reprinted with permission of Elsevier from Grieves and Jeffery, 2017, p. 115.

direction cells, are considered to work as a "compass" for maintaining orientation and heading, using vestibular and kinesthetic information as a major input (Dumont and Taube, 2015, p. 86). Researchers further demonstrated cells that fired when a rat was in multiple discrete locations that were regularly spaced in the environment, forming a triangular lattice or grid (Figure 2.11d). These cells, called *grid cells*, are considered to provide a spatial "metric" for locational updating (Dumont and Taube, 2015, p. 87).

Thus, there are neural mechanisms that are modulated by directional information (head direction cells) and positional and distance information (grid cells). These cells interact with each other interdependently (e.g., cells with conjunctive grid and head direction properties), but some researchers suggested a fundamental role of head direction signals generated by the vestibular system, particularly for tasks that require environment-centric representations (Chadwick et al., 2015, p. 90) and cognitive map-like strategies (Dumont and Taube, 2015, p. 89). In recognition of the contribution of this line of neuroscientific research into spatial orientation, the Nobel Prize in Physiology or Medicine 2014 was awarded to J. O'Keefe, M.-B. Moser, and E. I. Moser. The press release by the Nobel Foundation described the research as an investigation of a "positioning system in the brain" or an "inner GPS." The workings of the brain for understanding space have thus attracted much interest, and new findings have constantly been accumulated. For example, researchers suggested the existence of brain cells that encode information about environmental boundaries, spatial locations of objects, the location and direction of a goal, and movement through space. The interested reader is referred to more specialized discussions, for example, textbooks by Dudchenko (2010) and Wiener and Taube (2005).

Insights into the workings of the brain in spatial learning have been provided through an examination of neural activities associated with the encoding and processing of different types of spatial knowledge. Aguirre and D'Esposito (1997) asked four adults (three male and one female) to learn a virtual town by freely navigating between 16 places in it, and examined their regional brain activities by functional magnetic resonance imaging (fMRI) while they conducted spatial tasks about visual appearance and spatial relations of the places learned in the town. In a visual appearance task, participants were shown a series of pictures of places with place names, and asked to judge whether the places and names matched. In a position (or spatial relations) task, participants were shown a picture of a place with its correct name and given the name of another place (a current place and a target place, respectively), and were asked to indicate the direction from the current place to the target place. An analysis of fMRI data showed that when the appearance task was compared with the position task, a significantly greater activity was observed in the lingual gyrus and inferior fusiform gyrus. By contrast, when the position task was compared with the appearance task, a significantly greater activity was observed in the premotor cortex and superior parietal lobule. Aguirre and D'Esposito pointed out that the dorsal–ventral dissociation for spatial (relational or configurational) and visual (appearance) processing suggests the involvement of separate brain areas in the processing of different types of spatial information, and that it runs counter to a unitary view on spatial knowledge as a "cognitive map."

Hartley et al. (2003) asked 16 adults (all male) to learn about two virtual towns in two different manners: by freely exploring between nine landmarks in one town (Town 1) and by following predetermined routes between nine landmarks in another town (Town 2). They then examined the participants' brain activities using fMRI while they navigated between pairs of the learned places in three conditions. In one condition, participants were asked to find the shortest paths between the nine places learned in Town 1, starting from each of the nine places and navigating toward the remaining eight places successively (a wayfinding condition). In another condition, participants were asked to navigate between the nine places learned in Town 2, going from one place to the remaining eight places along the same routes that they had followed when they learned about the town (a route following condition). In a third condition, participants were asked to follow a set of trails visually identified by green markers in a virtual town (which constituted simple trail following and did not require any spatial knowledge about the environment to navigate in). An fMRI data analysis showed that the accuracy of wayfinding correlated with a greater activation of the hippocampus (relative to trail following) and a lesser activation of the right head caudate (relative to route following). Good navigators flexibly chose the caudate representation when the task required route following instead of wayfinding, the latter of which entails relational and configurational spatial reasoning. In fact, a greater activation of the right head of caudate was observed among the poorest navigators during wayfinding compared with route following. Furthermore, participants (good and poor navigators alike) activated the posterior right hippocampus when navigating accurately as opposed to when navigating inaccurately; it suggests, Hartley et al. argued, that posterior hippocampal activity was related to within-subject variations in the accuracy of navigation. The distinction of hippocampal-dependent place learning and caudate-dependent response learning has widely been observed in rats (e.g., Packard and McGaugh, 1996).

The contribution of the hippocampus to spatial memory has been observed for humans who have professional experience and expertise in wayfinding. Maguire, Woollett, and Spiers (2006) measured gray matter volume in the hippocampus by MRI scanning for 16 licensed London taxi drivers and 17 London bus drivers, and examined the relationship of the volume to their performances on various memory and cognition tasks. The two groups of participants were equivalent in age, handedness, education, abstract reasoning ability, driving experience in London, and the level of stress and anxiety. Maguire et al. found that taxi drivers had greater gray matter volume in the posterior hippocampus and less gray matter volume in the anterior hippocampus than bus drivers. Also, among taxi drivers, gray matter volume in the right posterior hippocampus increased, but that in the anterior hippocampus decreased, as a function of the length of time of taxi driving. Concerning their performances on cognitive tasks, taxi drivers did better than bus drivers on the recognition of major landmarks in London and the understanding of spatial relations between the landmarks assessed by triadic proximity judgment. Taxi drivers, however, did worse than bus drivers on delayed recall of visually presented materials (reproducing complex line drawings from memory). London taxi drivers undergo extensive training before obtaining an operating license and navigate widely and flexibly around the city, while bus drivers operate along a constrained set of routes, and Maguire et al. argued that the better spatial memory for places in London by taxi drivers was associated with their greater posterior hippocampal gray matter volume (consistent with the findings by Hartley et al., 2003, discussed above). On the other hand, Maguire et al. discussed that the poor performance on retrieving new visual-spatial information by taxi drivers was related to their reduced anterior hippocampal gray matter volume, which they assumed reflected a tradeoff between acquiring and using knowledge about large city environments and acquiring and retrieving new visual-spatial information. Similarly, Schinazi et al. (2013) observed that the size of the hippocampus (gray matter volume in the right hippocampus) correlated positively with survey learning of the environment, assessed by the accuracy of direction estimates between landmarks on two different routes. We will see this in the discussion of individual differences in cognitive mapping in Chapter 6.

Review box for Chapter 2

1. Explain the remarkable navigation skills of Polynesian and Micronesian marine navigators from the viewpoint of spatial orientation.
2. Discuss the Vermeer painting *The Geographer* from the perspective of spatial cognition. Also discuss it from the perspective of spatial representations.
3. Give your thoughts about the early discussions of space (mostly metaphysical or anecdotal) that we saw in this chapter.
4. Tell us about your experiences of spatial disorientation.
5. Summarize the studies of map orientation and imaginary maps. Why do you think these studies are important in the literature of spatial cognition?
6. Summarize Piaget's developmental theory. What do you think about his ideas of development?
7. Discuss the importance of Tolman's study. Describe why the idea of "cognitive maps" was (and still is) significant for psychology in general and spatial cognition in particular.
8. Discuss the concept of affordances in Gibson's treatment of visual perception. What do you think about his idea of direct perception, when it is contrasted with the idea of mental representations?
9. Ask a handful of people around you to draw sketch maps of your neighborhood and discuss the maps in terms of the five elements proposed by Lynch (1960).
10. If you are interested, dig into the literature of neuroscience and find more studies about neural activities pertaining to the perception and cognition of space.

Chapter 3

Maps in the head

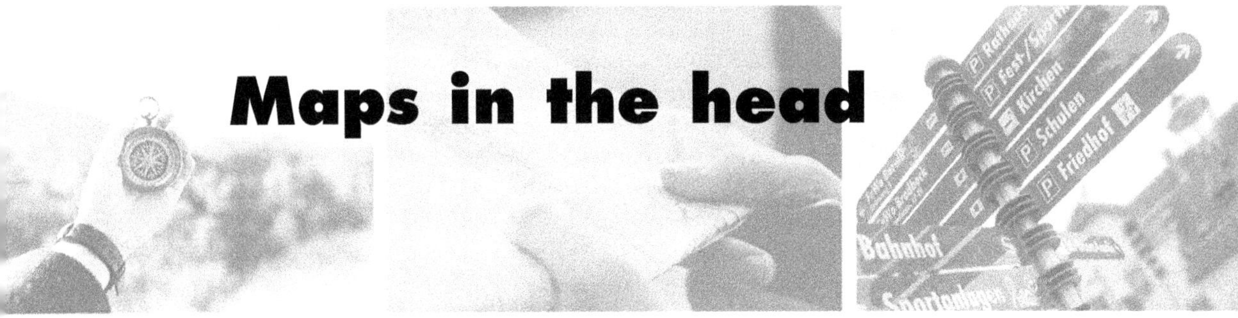

Spatiality for human behavior

As the preceding chapters clarified and as the following chapters will elaborate, space matters for us, in that our behavior takes place at a specific location in space, and our cognition, thinking, and decision making involve spatial components. That is, we daily engage in spatial behavior. As we saw in Chapter 2, geographers, who have always been interested in space, namely the structures and processes of physical environments, the distributions of human activities in social, cultural, or economic environments, and the interactions between humans and their environments, have long recognized the importance of spatiality for our behavior. In the systematic approach, which originated in the late 1950s, the geographer's interest in space was directed toward an investigation of the effect of *distance* on human spatial behavior. Distance, in the sense relevant to our everyday life, signifies the amount of separation between two places that is to be overcome by displacement in traveling or migration. The farther away the two places are, the more effortful the traveling is. If there are two grocery stores in your neighborhood that sell the same variety and quality of goods at comparable prices, you probably go to the nearer store; or if there are two mailboxes in your neighborhood, you probably go drop a letter to the nearer mailbox. These patterns of travel behavior reveal that you intend to minimize the distance to travel.

The tendency to minimize distance is conceptualized by a spatial interaction model generally called the *gravity model*. A basic and primitive idea of the model is given in the form of

$$I_{ij} = k \bullet \frac{m_i m_j}{d_{ij}^2},$$

where I_{ij} = the degree of spatial interaction between places i and j; m_i, m_j = the size of places i and j; d_{ij} = the distance between places i and j, and k = a constant. That is, the degree of interaction (e.g., the number of trips, the frequency of visits, or the amount of information flow) is greater between larger places (in terms of population or attractiveness), and it decreases with an increase in the distance that separates two places. The gravity model specifies that the degree of interaction is inversely proportional to the square of the distance between the two places.

The observation of *friction of distance*, or the concept of *distance decay*, is succinctly summarized by Tobler (1970) as "everything is related to everything else, but near things are more related than distant things" (p. 236), a statement that is often referred to as the first law of geography (Tobler, 2004). The gravity model and the concept of distance decay can be applied to various phenomena over space, including regional migration, journey to work, social interaction, and consumer behavior. Much of the empirical work of spatial modeling particularly in the 1960s and 1970s focused on the specification of the form of the model (e.g., a linear, power, or exponential function) and the estimation of parameter values, as well as discussing the theoretical meanings and implications of distance decay (Figure 3.1).

The concept of distance decay has also been incorporated into various normative models of spatial distributions, notably Christaller's (1933/1966) central place theory. Christaller noted some regularities in the pattern of urban settlements and discussed the hierarchical distribution of cities of different sizes and functions. In doing so, he conceptualized a system of central places, which provide economic goods and services

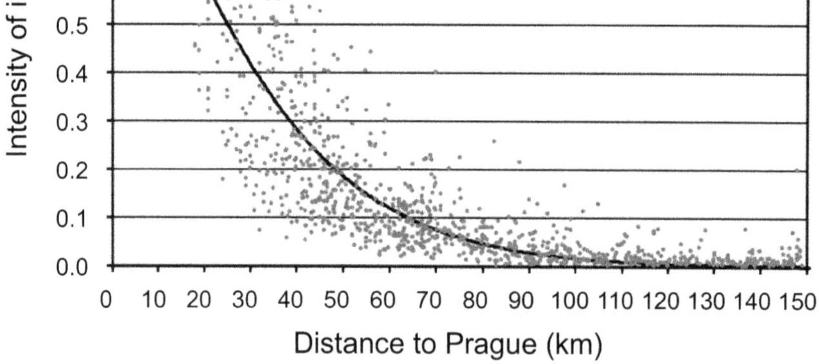

Figure 3.1 **Example of distance decay. The number of commuters decreases as the distance to travel increases (distances from various municipalities to a regional center, in this case, Prague). The decreasing function illustrates the concept of friction of distance: The degree of spatial interaction is greater between places that are located near to each other than places located far away. Reprinted with permission of Elsevier from Halás, Klapka, and Kladivo, 2014, p. 113.**

to the surrounding areas, a larger central place offering a larger number and variety of goods and services to a greater surrounding area (i.e., having a larger market area). These central places are hierarchically ordered with respect to their market areas; places of the same size are spaced evenly and larger places are located farther away than smaller ones (Figure 3.2). Importantly, Christaller based his theory on some simplifying assumptions; for example, the population is evenly distributed on an isotropic plane, the tastes and demands of people are uniform, and people minimize the distance to travel to shop. On the basis of these assumptions, he derived hierarchically organized hexagonal patterns of market areas of different sizes; at the center of each market area is located a central place of a corresponding size.

One thing we should note about these models is that they are applied to data aggregated and averaged over a specific area: regional migration is analyzed in terms of the flow of people between pairs of cities (e.g., between West Coast cities and Los Angeles); commuting behavior is analyzed in terms of the numbers of commuters to a central business district from neighboring areas (towns, cities, or counties); and consumer behavior is analyzed in terms of the numbers of customers at a certain supermarket from different neighborhoods. These analyses concern the aggregate flow of people summed over

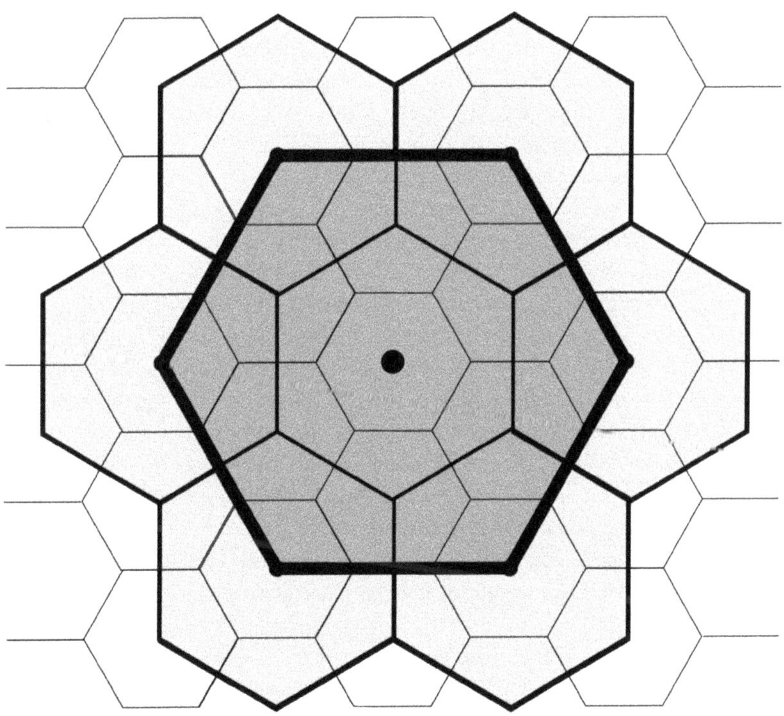

Figure 3.2 **Christaller's (1933/1966) central place theory. In this conceptualization, places are hierarchically distributed with respect to their sizes and functions, with larger central places offering a larger number and variety of goods to a greater surrounding area. Places of the same size are spaced evenly and larger places are located farther away than smaller ones.**

districts in question, and not individual behavior. Normative models make assumptions that simplify complex spatial phenomena so that they are tractable in a thoughtful and meaningful way; however, these assumptions are often at odds with the characteristics of individual spatial behavior that have been revealed by studies from the cognitive and information processing perspective.

The body of research conducted through the spatial modeling approach contributed substantially to a systematic understanding of the distributions of human activities and the prediction of travel behavior. Acknowledging its value, we should carefully consider the validity of resorting to aggregate data analysis when our aim is to reason about human spatial behavior at the individual level. Researchers caution against a fallacy of this sort, called the *ecological fallacy*, an error in reasoning when you draw a conclusion about individuals based on aggregate data, which leads to possible misinterpretations. Of course, there is similarly the danger of making a wrong conclusion about group behavior on the basis of models of individual behavior, for example in the analysis of organizational behavior or health-related variables across different countries. An important thing to note is that these fallacies arise from a mismatch between the method and the data, namely when conceptual models are conceived at one level and data are collected at another level (Diez Roux, 2015, p. 419).

To address the issue, researchers applied various modifications to spatial interaction models, for example, taking the probabilistic nature of location choice into account instead of the dichotomous determination in the gravity model (e.g., Huff's, 1963, probabilistic gravity model) or considering the component characteristics of each choice alternative in terms of utility (e.g., discrete choice models; see McFadden, 2001). These modified models aim at describing spatial behavior in more detail, but are still not specific enough to explain human behavior at the individual level. In particular, the normative approaches are based on the assumptions of rationality, complete knowledge, and utility maximization. They succeed in describing the patterns of aggregate human behavior to a considerable degree, but research has shown that human behavior, reasoning, and information processing are not perfectly rational. In the idealized case of utility maximization, people have complete knowledge of all grocery stores in the neighborhood and decide on which store to shop by identifying the one that gives the highest utility in a function of costs and benefits (e.g., distance to travel, varieties, quality, and prices). In reality, however, people know only a subset of the stores in the neighborhood, and oftentimes the knowledge is limited and inaccurate. Humans' so-called rational behavior is thus in fact crude and approximate, because of the incomplete information available and the limited computational capacity (Simon, 1956). That is, the choice they make, the solution they find, and the decision they take are somehow satisfactory and acceptable, but not strictly optimal. Such a realistic strategy of decision making by humans is called by Simon *satisficing*.

Simon (1990) viewed the human mind as a physical symbol system and discussed how it exhibits intelligent behavior. Specifically, Simon argued that the human mind, like a computer, is a physical symbol system in the sense that it can apply the operations of input, output, storing, and modification to symbol structures, and these symbolic operations enable intelligent behavior. A major characteristic of the intelligent human mind is its adaptivity to the "shaping forces of an environment" (p. 2) for survival. In the context of symbolic operations and information processing by the human mind, it needs to adapt to the limitation of its computational speed and power; in Simon's terms, the intelligent system works with *bounded rationality*:

> Since we can rarely solve our problems exactly, the optimizing strategy suggested by rational analysis is seldom available. We must find techniques for solving our problems approximately, and we arrive at different solutions depending on what approximations we hit upon. Hence, to describe, predict and explain the behavior of a system of bounded rationality, we must both construct a theory of the system's processes and describe the environments to which it is adapting.
>
> (pp. 6–7)

One of the methods that Simon (1990) discussed for coping with real-life complexity with bounded rationality is *heuristic search*. When the set of choice alternatives, or the space of possibilities, is too large to search completely (which is usually the case in our real-world behavior), the search necessarily is selective. The psychological process that underlies such a selective and heuristic information search is satisficing, "using experience to construct an expectation of how good a solution we might reasonably achieve, and halting search as soon as a solution is reached that meets the expectation" (p. 9). In other words, humans "look for good-enough solutions, instead of hopelessly searching for the best" (p. 17). Such a satisficing view on human spatial behavior is reflected in behavioral research into cognitive maps.

Cognitive-behavioral approaches to spatial cognition

An impetus for scientific research into spatial cognition, particularly in the literature of psychology, is the interest in the examination of the human mind from an information processing perspective and the appreciation of the significance of studying human behavior in an ecologically meaningful setting. The rise of the information processing approach is often identified with the emergence of cognitivism in psychology, or the emergence of the field of cognitive science, in opposition to behaviorism. Researchers in the doctrine of behaviorism intended psychology to be a "truly" scientific discipline and hence focused their analysis on (or restricted it to) the observable. The major variable of the behaviorist's choice for the observable was the modification of an animal's behavior due to the experience of reward and punishment, or operant conditioning. For example, if a rat, while moving around in a box that contains a lever, finds that the behavior of "pressing the lever" is rewarded by the provision of food, the rat becomes conditioned to press the lever (positive reinforcement). If a rat in the box receives an electric shock and finds that the pressing of the lever is followed by the removal of the electric shock, it is similarly conditioned to press the lever (negative reinforcement). Animals' behavior is thus explained by the formation of a link between a stimulus and a response, or a stimulus–response chain. As a consequence of the focus on the observable in this tradition, the interest in the mind or mental processes was basically excluded from psychology.

By the mid-1950s, however, researchers noted the inadequacy of such a behavioristic view for the understanding and explanation of complex human behavior. Notably, Chomsky (1959) criticized Skinner's (1957) application of the functional concepts of stimulus and reinforcement to the analysis of human verbal behavior (acknowledging the significance of the concepts for the investigation of animal behavior), and defended "mentalistic" approaches to human cognition. Chomsky argued that an organism's behavior is complex enough to call for an understanding of internal structure and information processing. In particular, he pointed to the existence of latent learning in various kinds of animal behavior and the capacity of children in language acquisition to generalize and hypothesize on the basis of casual observations and limited (and often ambiguous) information.

On the other hand, Skinner (1985) criticized cognitive science in defense of behaviorism. He mentioned that cognitive science considers behavior to start within the organism—people think and then act. People are the initiator of the internal process of information processing; the product of the internal processing, called a knowledge representation, is then retrieved and used in subsequent behavior. By contrast, in behaviorism, the environment is given the initiating role: the environmental histories affect the animal's behavior in terms of sequential contingencies of reinforcement. Behaviorists observe that a rat is changed by contingencies of reinforcement (the pressing of a lever is followed by the provision of food or the removal of an electric shock), whereas cognitive scientists, Skinner states, want to say that "the rat has learned, and now *knows* that pressing the lever brings food" (p. 293).

The above discussions by Chomsky (1959) and Skinner (1985) clearly show the distinction of the two disciplinary perspectives of behaviorism and cognitive science. Whereas Skinner (or behaviorism) focuses on changes in overt behavior in response to external stimuli, as long as they are amenable to causal explanations based on observed variables, Chomsky (or cognitive science) emphasizes mental faculties and processing that reveal themselves in the course of maturation in interaction with past experience (not merely in response to external stimulation), which he assumed to be innate and genetically determined.

A second impetus for scientific research into spatial cognition concerns interest in ecological validity. The approach of ecological optics advocated by Gibson is a major example in this regard (see Chapter 2). Gibson's argument for visual perception in an ecologically meaningful environment and direct perception of information specified in the ambient optic array (which he called "natural vision"; Gibson, 1979/1986, p. 303) is echoed in the interdisciplinary interest in spatial cognition and behavior. Neisser (1976) similarly developed an argument for the investigation of cognition "in the ordinary environments and in the context of natural purposeful activity" (p. 7). For example, Neisser stated that the research on memory by the 1960s "dealt primarily with the learning of 'nonsense syllables,' in tightly defined laboratory procedures of little generality" (p. 5).

Controlled laboratory tasks surely contribute to the understanding of human memory. For example, experiments with nonsense syllables revealed important characteristics of human learning and memory, such as the forgetting curve and the serial position effect (Ebbinghaus, 1885). Similarly, the experimental tasks of learning sequences of words or locations of digits within a matrix, while conducting a concurrent task of counting from one to eight or tracking a light moving along a circular path, contributed to the identification of different subcomponents of working memory (Baddeley, 1986, 2003). The point the ecological approach makes may be appreciated if you consider ecological validity in relation to internal validity (the degree to which causal statements about the relationships observed in a study is warranted) and external validity (the degree to which the results of a study can be generalized to other cases and settings). The ecological approach advocates the study of perception and cognition in a realistic setting or under "natural" conditions, but does not disregard the soundness of research design that enables valid causal statements and generalization (see Mook, 1983). Neisser (1978) stated,

> I am not recommending an aimless accumulation of ecological minutiae. . . . The challenge will be to shift from testing hypotheses for their own sake to using them as tools for the exploration of reality. . . . It is a challenge that will not be easy to meet. The realistic study of memory is much harder than the work we have been accustomed to.
>
> (p. 19)

The disciplinary inclination toward the human mind was likewise observed in geography, particularly in the 1960s. The recognition of the limited applicability of aggregate analysis to the explanation of the variation (or "subjectivity," if one prefers) and the non-rationality of human behavior led geographers to engage in process-oriented research into human spatial behavior. This approach is called behavioral geography (or cognitive geography). The term "behavioral" inadvertently caused misunderstandings among other geographers, but behavioral geographers sought an explanation of spatial behavior on the basis of an understanding of individual people's information processing and decision making (i.e., they do not confine their analysis to the observable, as behaviorism in psychology did). Geographers have long been interested in space as they set out on a journey of discovery and represented the knowledge as maps; geographers are now interested in exploring the terra incognita of the mind or internally represented environments (Wright, 1947).

Thus a major tenet of behavioral approaches in geography is the focus on individuals as an active information gather and interpreter. We, sensate beings in the environment, actively perceive and interpret information from it in a flexible and adaptive manner. This "interpretation" forms our understanding and knowledge of the environment, or internal representations of the environment. In other words, people's spatial behavior is based on internally represented environments, as well as physical environments. These internal representations do not necessarily accurately mimic the represented environments: they are in fact prone to errors and distortions. One should note, however, that these errors and distortions are not random; they show systematic errors reflecting the structures and processes of internal representations. Also notable is the fact that our knowledge about the environment is acquired through interactions with the environment, and in the process of human–environment interaction the causality is bidirectional. The structure of the environment guides our attention and behavior (Hillier, 2012; Penn, 2003) and our knowledge directs explorations in the environment; the information newly acquired from the explorations then restructures our knowledge. Our learning about the environment is thus conceived of as an iterative process of assimilation and accommodation (see Chapter 2).

An acknowledgement of the fact that people's behavior is based on their perceptions of the environment clarifies that a solid understanding (or explanation) of observed spatial phenomena requires an analysis at the individual level. This is why cognitive-behavioral approaches have significance in the study of human spatial cognition (Golledge and

Timmermans, 1990). In the context of cognitive mapping, the process of learning about the environment (or spatial knowledge acquisition) is considered an individual difference trait, rather than a universal developmental progression (Ishikawa and Montello, 2006; see Chapter 5). That is, not only internally represented environments differ from physical environments, but also the accuracy of internally represented environments differs from person to person. Research on spatial perception and cognition has theoretical and practical significance to the extent that the ability to internally represent the environment in an accurate and configurational manner shows large individual differences and the ability relates to the fundamental human behavior of spatial orientation and wayfinding in the environment. Why do some people find their way in a new environment easily, and why do some people have difficulty with spatial orientation and feel anxious in an unfamiliar environment? Why do people patronize a specific grocery store or migrate to a specific city, not other stores or cities? How can we design a residential environment that fosters a sense of community among residents? What consequences could the convenience of a spatially enabled society potentially have on our spatial awareness and sense of place? Insights about these questions will be provided by efforts of research and practice from the cognitive-behavioral perspective. We discuss them in detail in the following chapters.

Cognitive maps and cognitive mapping

Studies in the spatial modeling approach examined how spatiality affects human spatial behavior and clarified the concept of friction of distance. Building on and extending the findings from these aggregate analyses, cognitive-behavioral approaches looked at individuals' spatial behavior on the premise that it is affected by perceived (psychological or subjective) environments as well as physical (objective) environments (see Figure 1.3). The discrepancy between psychological and physical environments is revealed most readily in the studies of cognitive distance (see Chapter 4); that is, distances perceived by people are not the same as distances measured objectively in the environment. Importantly, the discrepancy is not due to random variation—the errors or misconceptions observed in people's responses show systematic patterns of distortions, simplifications, and schematizations. The degree of distortion, simplification, and schematization, or the accuracy of internally represented environments, also shows variations among people depending on their spatial aptitudes (see Chapter 6). These variations are of interest to researchers insofar as our behavior depends on our thinking and feeling about the environment that we interact with.

Now look at the interaction between people and the environment as it relates to spatial orientation and wayfinding. As our spatial behavior is based on and guided by the knowledge that we have about the environment, accurate spatial orientation and successful wayfinding are possible when our knowledge is accurate and flexible. Accurate "inner maps" or "maps in the head" enable us to stay oriented in space and know where we are in the environment. In the examples of "going to a mailbox that we think is closest to home" and "pointing to a direction in which we think the destination is located" that we saw in Chapter 1, accurate inner maps ensure that the mailbox is actually the one closest to home and the destination is actually in that direction. When inner maps are inaccurate, however, we have difficulty knowing where we are. We may go to a mailbox that is farther away than others or we may fail to reach our destination. Likewise, when inner maps are inflexible, we cannot orient ourselves or find the way in the environment adaptively and creatively, thus, for example, we are unable to go to work when a usual route is blocked due to construction.

These exactly exemplify what is meant by the premise that our spatial behavior is based on our knowledge about the environment (Figure 1.3). As the expressions "inner maps" and "maps in the head" suggest, the knowledge that we acquire through interactions with the environment is stored and represented in our minds. It is hence called *internal* (or *mental*) *representations*, and often referred to by the famous metaphor of "cognitive maps." Contrary to the implication of the term cognitive maps, some would say that you could not find a map if you were able to look into your head. That may be right, but the important thing here is that you act as if you have maps in your head (Downs, 1981). The consequences of accurate and inaccurate inner maps discussed above show that our cognition and behavior are in mutual dependence. One can say that spatial behavior consists of an iterative process of knowing about the environment, representing the knowledge in our head, and using the knowledge to act in the environment. Downs and Stea (1973) called it *cognitive mapping* and defined it as "a process composed of a series of psychological transformations by which an individual acquires, codes, stores, recalls, and decodes information about the relative locations and attributes of phenomena in his [or her] everyday spatial environment" (p. 9).

Downs and Stea (1973) clarified that the rationale behind the definition of cognitive mapping is a contention that "human spatial behavior is dependent on the individual's cognitive map of the spatial environment" (p. 9). One may consider cognitive mapping and cognitive maps as a process and structure of spatial knowledge, respectively: the process of acquiring, representing, using, and updating knowledge about the environment is called cognitive mapping, and a product of the process at a specific time in the individual's mind is called a cognitive map. Downs and Stea used the term "map" to mean a functional analog, in the sense that "a cognitive map exists if an individual behaves as if a cognitive map exists" (p. 10). They also pointed out that because of the complexity of the environment and the large spatial-temporal scale over which sensory information needs to be integrated, cognitive maps tend to show various degrees of incompleteness, distortion, and schematization, in terms of systematic errors across people and individual variations between people (remember the discussion in Chapter 1 about the existence of distortion, generalization, and simplification in cartographic maps). We turn to these issues in Chapters 4 and 5 below.

Large-scale space and environment perception

As Downs and Stea (1973) pointed out, in the discussion of cognitive mapping, the environment that people learn and explore is complex. This is closely related to the issue of spatial *scale*. For example, what do you envision when you hear the word "space"? Spare room in your suitcase, a bedroom in the house, a college campus, your neighborhood, or outer space? These are all space but they differ in scale, a matter that affects our perception and cognition in important ways—scale matters.

Montello (1993) distinguished four classes of psychological spaces in terms of scale: figural, vista, environmental, and geographical spaces. The distinctions are made with respect to size relative to the human body and how the space is thus interacted with, such as whether locomotion is needed to apprehend the space. Figural space is smaller than the body and can be perceived from a single viewpoint without locomotion (e.g., small objects and pictures). In contrast, geographical space is much larger than the body and cannot be apprehended directly through locomotion alone (e.g., counties, states, and countries). Particularly important for the present discussion of cognitive mapping is the distinction between vista and environmental spaces. Vista space is larger than the body, but can be apprehended from a single viewpoint without locomotion (e.g., rooms and town squares), typically by vision. Environmental space, also larger than the body (e.g., buildings and neighborhoods), can still be learned directly, but requires the observer to locomote around to comprehend its layout (Figure 3.3).

The characteristics of environmental space and their consequences on people's perception and cognition were cogently argued by Ittelson (1973), who presented a discussion of *environment perception* by contrasting it with *object perception*. First, Ittelson compared the perceptions of objects and environments by distinguishing between the roles of the viewer as an observer and an explorer. People manipulate and observe objects, whereas they explore the environment to learn it (Figure 3.4). In environment perception,

Figure 3.3 **Scale matters in the perception of space. This figure compares vista space (*left*) and environmental space (*right*). People can apprehend vista space from a single viewpoint, whereas in environmental space they need to move around to comprehend its layout. Photographs from https://pixabay.com/**

Figure 3.4 **Comparison of object perception and environment perception. In object perception, one manipulates an object as an observer (*left*), whereas in environment perception, one navigates in the environment as an explorer (*right*). Photographs from https://pixabay.com/**

therefore, people are explorers enclosed by the environment, not outside observers isolated from the environment. This is a major property of the environment, of particular importance for spatial cognition: environments surround people. The mutual relationship of "surrounding" and "being surrounded" between the environment and people provides unique characteristics for environment perception that object perception does not always possess, most importantly for the development of spatial knowledge in a new environment. As we discussed above, the sophistication and flexibility of spatial knowledge hinge on the degree to which it supports a maplike, configurational understanding of the environment. The necessity of moving around requires the traveler to integrate separate pieces of information at individual places into a whole to acquire an accurate configurational understanding of the environment. It poses extra cognitive effort and raises interesting and important questions about the structures and processes of spatial knowledge, which we will see in Chapters 4 and 5.

Another characteristic of environment perception is its multimodality; that is, people acquire information about the environment through multiple senses. The majority of information comes from vision: we view an array of objects on a table and learn its configuration; we notice a house with a flashy color next to a gas station; we see a grocery store in a distance, farther away than a pharmacy; or we locate an Italian restaurant on Main Street on a town map. Gibson's (1979/1986) ambient optic array (which specifies the layout of surfaces) and optical flow (which specifies changing perspectives with the observer's locomotion) are an example of visually provided environmental information (see Figures 2.7 and 2.8).

Another piece of information about the environment is auditory: we can identify the location of a sound, how far it is, and which direction it comes from. For example, in a navigation system designed for people with visual impairments, locational information about an object in the surrounding space is transmitted to the traveler through earphones as speech that appears to be coming from the object. The traveler turns in the direction of the utterance, and as he or she faces and approaches the object, the sound level of the utterance increases accordingly (Loomis, Golledge, and Klatzky, 1998). Tactile information is also important. We identify the texture of a surface by touching it, sensing whether it is smooth, coarse, rough, or curved. Touch particularly provides important navigational information for people with visual impairments in the use of tactile maps. Tactile maps convey information to the user about the locations of places and the shapes of streets through the sense of touch, and show information about mapped features with braille labels and tactile symbols (Lobben, 2015; Miele, Landau, and Gilden, 2006; Subryan, 2010). Olfactory information can also serve as a spatial cue (Jacobs et al., 2015): you may associate a certain smell with a particular place. For example, the smell and taste of a piece of madeleine brought back the whole memory of Combray to the Narrator in Proust's *In Search of Lost Time*.

Finally, information from the body plays an important role in environment perception. Let us look at it by conducting a small experiment (Klatzky et al., 1998). Stand with your eyes closed and imagine that you are walking along a two-leg pathway, without physically moving. The experimenter says, "Go forward 3 m, turn clockwise 90°, then go forward 3 m; and now turn yourself to face the origin (the point you started from)." OK open your eyes—how many degrees did you turn? Now, one more trial with the same instructions; but this time, you physically walk the instructed pathway in the room with your eyes closed, and face the origin. Did you turn the same number of degrees as you did in the imagined condition? Which walking, imagined or physical, did you find easier? In the latter case, you had body-based information (vestibular senses and kinesthesis from muscles, tendons, and joints) as well as optical flow, neither of which was available in the former case. Research has shown that people tend to make errors in the imagined condition, systematically overturning by the magnitude of the turn between the path segments (in the present case, 90°), a tendency that is not observed for people who physically walk. Thus, people fail to update heading accordingly without the experience of physical walking, which points to the important role that proprioceptive information plays in spatial updating. It should also be noted that Gibson (1979/1986) discussed that optical flow supplemented by information about the movement of a limb (visual kinesthesis and muscle–joint kinesthesis) constitutes the primal source of information about displacement (see Chapter 2). In the sense that the environment necessitates locomotion inside for its comprehension (Ittelson, 1973), information from the body associated with self-motion signifies the process of cognitive mapping.

Ittelson (1973) added to the list of the characteristics of environment perception that environments have an ambience or atmosphere. In our everyday behavior and perception, we interact with surrounding environments in specific situations and social contexts, and we make aesthetic evaluations of, and attach meanings to, the environments. In other words, environments to us are not mere spaces, but particular *places*. We will look at these aesthetic and affective components of environment perception in Chapter 9.

Internal and external representations of space

We have thus far clarified that people learn the environment through interactions with it and represent the acquired knowledge in their heads as internal representations, which in turn guide their behavior, hopefully adaptively and flexibly.

Because of the necessity of locomoting around in the environment and mentally integrating separately learned information, constructing accurate internal representations is not always easy, especially not for people with a poor sense of direction (Hegarty et al., 2002; Ishikawa and Montello, 2006; Kozlowski and Bryant, 1977). Being in the same environment, some people have accurate maps in their heads easily and quickly, while others have difficulties and acquire only fragmented and distorted knowledge (Figure 3.5). The fragmentation and distortion lead to inflexibility in reasoning about the environment globally, in a way not tied to specific landmarks or routes, and a failure to navigate in the environment freely. It illustrates the importance of providing carefully designed navigational information and wayfinding assistance tailored to the recipient's spatial aptitudes (see Chapter 7).

People acquire knowledge about the environment from various sources of information. Probably the most popular is direct navigation; that is, walking or driving in the environment, which is particularly relevant to the learning of an environmental space (Montello, 1993). Another major source of information is a map. Nowadays maps are available digitally, as well as printed on paper, on mobile devices equipped with navigation functions and often supplemented by photographs or panoramic street views. As a

Figure 3.5 **Spatial knowledge is acquired from various information sources, including maps, navigation applications, compass directions, text, verbal descriptions, signage, and direct navigation in the environment. Being in the same environment, people differ in the accuracy of their knowledge about the environment. Spatial orientation, or how we know where we are, is a fundamental question about human spatial behavior. Photographs from https://pixabay.com/**

representation of spatial information, language should not be forgotten. We daily communicate information about space verbally, for example, giving navigational directions such as "the bathroom is located down the hall," "go straight for three blocks and make a right," "you will see the hotel on your left," and even "you cannot miss it." Maps, navigation tools, and verbal directions are presented externally to the traveler and are called *external representations*, in contrast to internal representations that are stored in the traveler's head as knowledge about the environment.

In real-world navigation, people coordinate three components of information about the environment: internal representations, external representations, and the environment (Figure 3.6). In short, you need to know where you are by relating your knowledge about the surrounding environment with a view available at your current position and orientation. In an unfamiliar environment, you consult a map, but merely having a map in the hand is not enough: you need to align the map (or any other externally provided assistance, including navigation tools and verbal descriptions) with your position and orientation in the represented space. Accurate spatial orientation requires the coordination of these three components of information, and it is not as straightforward as many people may assume (Liben and Downs, 1993).

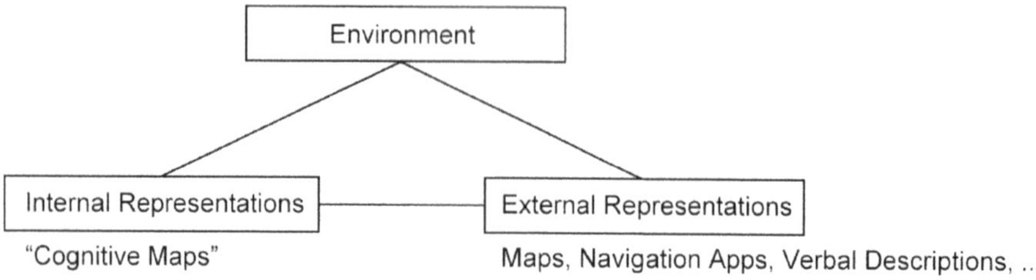

Figure 3.6 **Coordination of three components of information in navigation: the environment, internal representations of the environment, and external representations of the environment. In accurate wayfinding and spatial orientation, you need to interrelate your knowledge about the environment and externally available navigational information (representations in your head and in your hand) with the represented space.**

Interaction between internal and external representations

To discuss the coordination of the self, map, and environment, particularly the interaction between internal and external representations, it is worthwhile to look at a model of map communication proposed in the literature of cognitive cartography (Koláčný, 1969; Figure 3.7). The model describes map use as a process of communication between the mapmaker and the map-reader. Mapmakers select particular information that they want to convey about a place, on the basis of their knowledge of the world, and present the information in map form following the principles of map design and symbols. Map-readers, in turn, view the map and extract information as they understand it, on the basis of their knowledge and perceptions of the world. This process is often iterative, and the design of the map and mapmakers' and map-readers' conceptions are modified with feedback.

In this model, the success of communication depends on how accurately the intended information is conveyed from the mapmaker to the map-reader. The unidirectional and static view of communication in the model has been criticized, particularly with the advances in visualization techniques that allow map-readers to interact with maps actively and mapmakers to show large amounts of data in various formats. Moreover, in the recent practice of scientific visualization and big data analysis, researchers do not know beforehand which information to visualize or what variables to analyze (i.e., mapmakers do not decide on which information to communicate to begin with). To emphasize such an interactive and exploratory nature of information visualization, MacEachren (1995) proposed a more realistic model of *geovisualization*.

If such limitations are taken into consideration, the communication model shown in Figure 3.7 offers an insightful view on spatial orientation and navigation assistance, and clarifies the importance of an information processing approach to spatial cognition. In the practice of navigation assistance, emphasis is often placed on the development of navigation tools. But as the argument above suggests, the effectiveness of navigation tools needs to be evaluated from the perspective of the user, or more specifically interaction between the user, tool, and environment. In particular, the existence of large individual differences in the accuracy and developmental pattern of cognitive maps

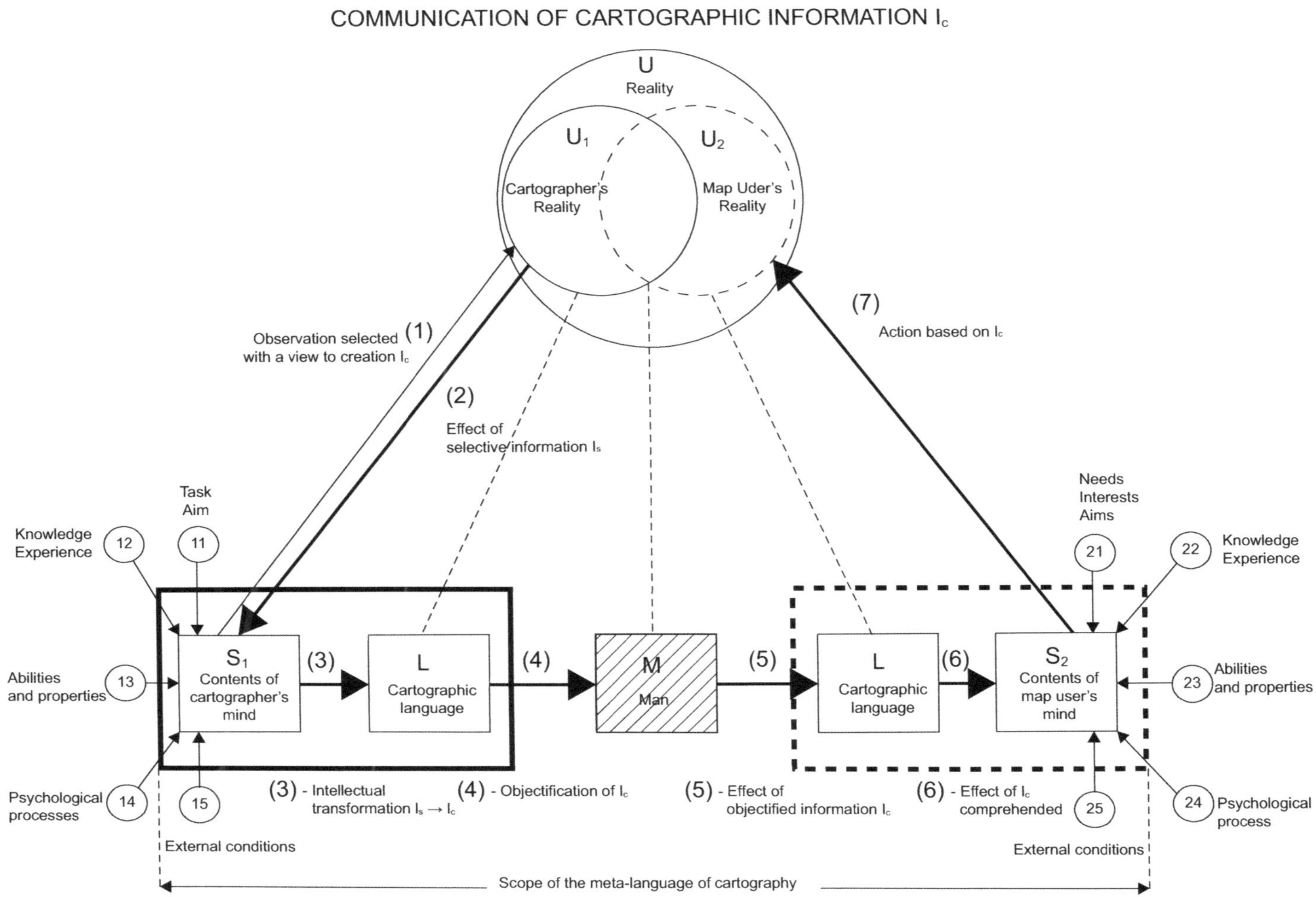

Figure 3.7 **Model of map communication. Information is communicated between the mapmaker and the map-reader. Mapmakers select information that they want to convey and represent it as a map; map-readers view the map and extract information as they understand it. Reprinted with permission of Taylor & Francis from Koláčný, 1969, p. 48.**

(details in Chapter 5) makes it unlikely that there is a single format of navigation assistance that works best at any time; rather, it needs to be adjusted to the attributes of users and the context of use.

Cognitive distance

The identification of spatiality as an important factor of human behavior in the spatial modeling approach was succeeded by cognitive-behavioral approaches in terms of the recognition of the influence of perceived space on behavior. In particular, research in the early periods noted the discrepancy between psychological and physical spaces in an analysis of people's perceptions and conceptions of environmental distance, or cognitive distance. Its novelty at the time was suggested by the title of Thompson's (1963) article, "New Concept: Subjective Distance." Thompson, with an aim to increase the precision of the prediction and description of consumer behavior, noted the possible difference between physical distance that is used in the gravity model and subjective distance that customers actually perceive, saying, "One can then question the usefulness of this factor [distance] as a simple input variable" (p. 1). He surveyed four San Francisco Bay Area communities that were located approximately equidistant from a department store and a discount store, and asked the residents who patronized both stores to estimate distances in miles from their residences to the two stores. Thompson found that their distance estimates were shorter for stores that they thought were more attractive. On the basis of these findings, Thompson suggested that subjective distance is more appropriate as an input for spatial models than objectively measured distance. Similarly, MacKay and Olshavsky (1975) reported that the frequency of visits to supermarkets was more strongly correlated with subjective distance than physical distance. Also, Cadwallader (1981) showed that a gravity model based on subjective distance significantly contributed to the prediction of store choice.

Research into cognitive distance was extensively conducted in the 1960s and 1970s, to yield a variety of empirical observations of systematic patterns in psychological judgments of distances. Lee (1970) asked students in the city of Dundee, Scotland, to estimate walking distances from a single location (the entrance to a university building) to 22 locations in the city. Test stimuli (route distances) were composed of 11 pairs of equal distances; one of each pair was in the direction inward to the city center and the other was in the direction outward from the city center. An analysis of the students' estimates showed that inward distances were estimated as relatively shorter than outward distances, the former being overestimated by 14 percent and the latter by 36 percent, on average. Golledge, Briggs, and Demko (1969) asked students in the city of Columbus, Ohio, to judge interpoint distances between 11 locations in the city. Estimates were made by paired comparison and analyzed through the method of multidimensional scaling (see Chapter 4). Golledge et al. found that distances away from the city center tended to be underestimated and distances toward the city center overestimated, a pattern of results contrary to that observed by Lee. With participants from the West Los Angeles area, Cadwallader (1976) likewise found a relative overestimation in the judgment of distances toward the center of the metropolitan area.

Byrne (1979) asked students in the town of St Andrews, Scotland, to estimate walking distances between 12 pairs of places in the town. Half of the places were in the central shopping district and the other half were in the residential outskirts; half the distances were along straight routes and the other half were along routes that contained two to four major changes of direction. Byrne found that distances within the town center were estimated as longer than distances in the outskirts, and that distances along straight routes were estimated as shorter than distances along routes with turns.

Sadalla and Magel (1980) and Sadalla and Staplin (1980) examined the effects of the numbers of turns and intersections on the perception of distances, respectively, and found that the paths that contained more turns or intersections were estimated to be longer, although they were of the same physical length. Specifically, Sadalla and Magel asked participants to walk two paths laid out on a hallway in a building with masking tape and estimate the distances traveled. The paths were of the same length (200 feet), one path containing two right-angle turns and the other path seven right-angle turns (Figure 3.8). Sadalla and Magel found that the two-turn path was estimated as shorter than the seven-turn path, a mean estimate for the latter being 1.3 times longer than that for the former. Sadalla and Staplin, using a pathway grid laid out on a floor of a building with masking tape (Figure 3.9), constructed nine linear paths that were of the same length (28 feet) and had varying numbers of intersections (one, four, or seven). After walking these paths, participants estimated the distances traveled. Their estimates were found to be longer for paths with more intersections, in the ratio of 1:1.2:1.3 for the paths with one, four, and seven intersections, respectively.

Montello (1997) examined existing studies of cognitive distance and identified three variables that affect the knowledge of environmental distance. One is the number of features in the environment (e.g., turns or intersections). As seen above, routes with more turns or intersections tend to be perceived as longer than routes of the same length with fewer turns or intersections. Another variable is travel time. For example, MacEachren (1980) asked residents in the city of Lawrence, Kansas, to estimate distances to a set of familiar supermarkets along routes that they usually took to go there. Their estimates were found to correlate 0.85 with travel time and 0.80 with physical distance, indicating a close relationship between cognitive distance and time. A third variable that affects cognitive distance is travel effort. Okabe, Aoki, and Hamamoto (1986) examined the effects of slopes on cognitive distance by using as experimental stimuli three paths of the same length (98 m) in a botanical garden. One path was flat and the other two were sloped with an average gradient of 8.5°,

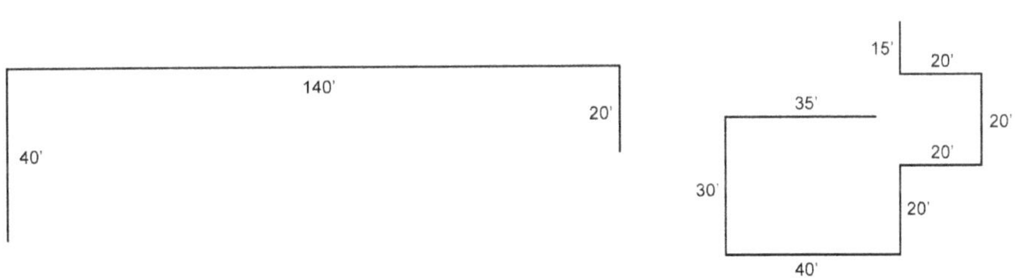

Figure 3.8 **Effects of the number of turns on the perception of distance. Of two paths that were of the same length, a path with two right-angle turns was perceived as shorter than a path with seven turns. Reprinted from Sadalla and Magel, 1980, p. 69.**

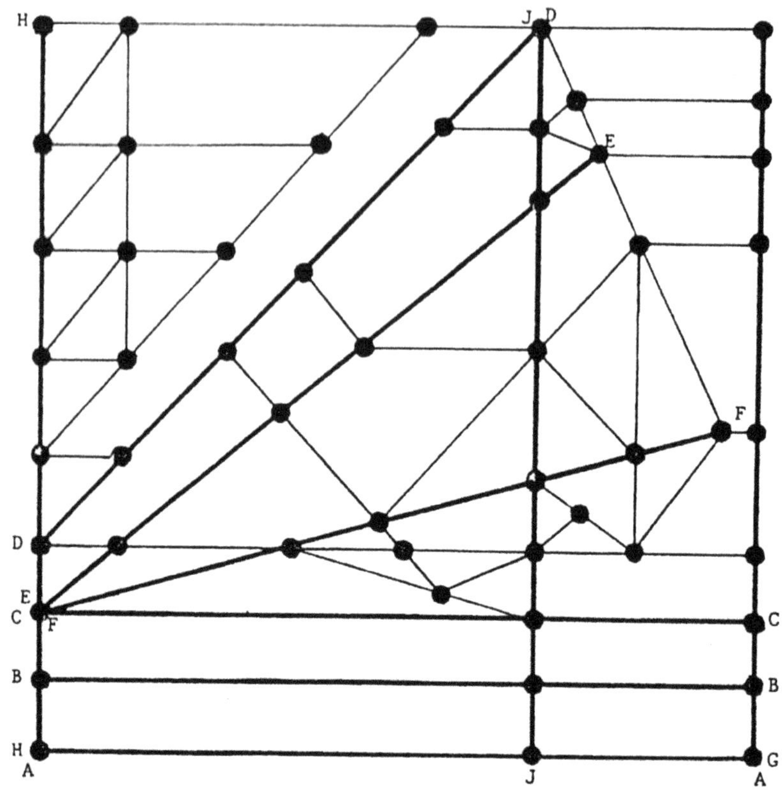

Figure 3.9 **Effects of the number of intersections on the perception of distance. In a pathway grid laid out on a floor of a building, nine linear paths that were of the same length and had varying numbers of intersections (one, four, or seven) were constructed. Paths with more intersections were perceived as longer than paths with fewer intersections. Reprinted from Sadalla and Staplin, 1980, p. 172.**

one upward and the other downward. Participants walked each path and then estimated the distance traveled. Their estimates were longer for the sloped paths: participants perceived the upward and downward slopes as 1.1 and 1.2 times as long as the flat path, respectively, on average. Caution should be used in interpreting these results, because travel time and travel distance are inherently related closely to each other and the number of features was not controlled for in the studies. With such a caveat, however, one can interpret these findings as an indication of the effects of travel time and travel effort on cognitive distance.

Thus, early studies of spatial cognition demonstrated a mismatch between the physical "reality" in the environment and the "subjective" knowledge of the environment that people possess in terms of cognitive distance. Such a mismatch has theoretical and practical significance when considered in the context of Simon's (1990) contentions of bounded rationality and satisficing. Physical environments exist out there and surround us; psychological environments exist inside us as represented in the mind with possible distortions, schematizations, and ambiguities. We will now look in more depth at the nature of distortions in spatial knowledge (or the structure of cognitive maps) in the next chapter.

Review box for Chapter 3

1. Examine the applicability of the concept of friction of distance to your daily behavior (e.g., shopping, commuting, or recreation).
2. Compare the idea of utility maximization (based on complete rationality and optimization) with the idea of satisficing (based on bounded rationality).
3. Discuss the characteristics of cognitive-behavioral approaches to spatial cognition.
4. One professor says that you cannot objectively study the unobservable. Another professor says that it is important to examine the process that underlies the observable. How would you moderate a debate between the two professors?
5. Discuss internal and external validity. What do you need to consider when you want to conduct a study of spatial cognition with a sufficient level of internal and external validity?
6. Explain the idea of cognitive mapping and cognitive maps.
7. Compare object perception and environment perception. Discuss the effect of scale on spatial cognition.
8. Describe different sources of knowledge about the environment.
9. Discuss the communication model of map reading and understanding.
10. Describe your experience with the cognition of environmental distance. Are there cases where you think a distance in one direction is longer or shorter than the distance traveled in the opposite direction?

Chapter 4

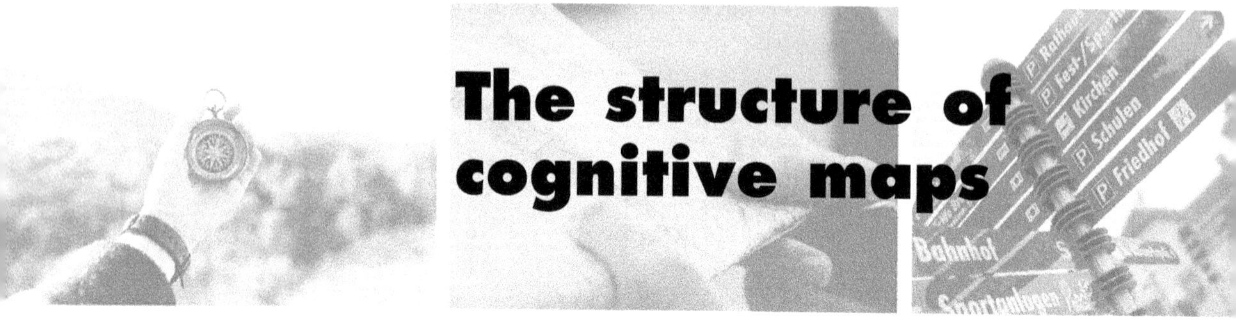

The structure of cognitive maps

Mental representations

In the field of cognitive psychology, after its rise with the surge of cognitivism, early studies examined the structures and processes of the human mind, including memory, attention, and imagery, on the premise that complex human behavior is the result of mental processing and hence an understanding of perception and cognition requires a systematic look into internal representations. In particular, researchers looked at mental imagery and psychological operations applied to images envisioned in the mind. One of the characteristics of mental images and operations that has been extensively studied concerns *mental rotation*. For example, people physically rotate an object, such as a die or a cube, by manipulating it by hand. Or people can see a cube being rotated in another person's hand. People can also mentally rotate a cube in their minds; that is, they do not physically rotate it but imagine it being rotated. To describe it, one may say that people envision a cube or construct a *mental representation* of the cube and rotate it in the mind. Is such a statement viable? If so, in what sense? What is a mental representation like, and how is it manipulated in the mind?

To examine mental rotation, Shepard and Metzler (1971) developed experimental stimuli of two-dimensional drawings of objects that are composed of a series of solid cubes: ten cubes are assembled in different shapes with three right-angled elbows (Figure 4.1). Shepard and Metzler presented pairs of such drawings to participants and asked them to judge for each pair whether the objects were the same. Half the pairs were congruent, such that they could be matched after rotation (or one object could be rotated into the other object's orientation) either in the picture plane or in depth. The other half of the pairs were different, being mirror images of each other. The stimuli were presented in various orientations; the angle of rotation that separated the drawings of each pair was varied from 0° to 180° in 20° increments. Shepard and Metzler examined participants' response times for judgments, being particularly interested in whether the response times were proportional to the angular disparity between the two drawings, or the degrees of mental rotation required.

Figure 4.2 shows results of mean response times for congruent pairs, plotted against angular disparity between the two drawings of a pair (shown only for correct responses; an

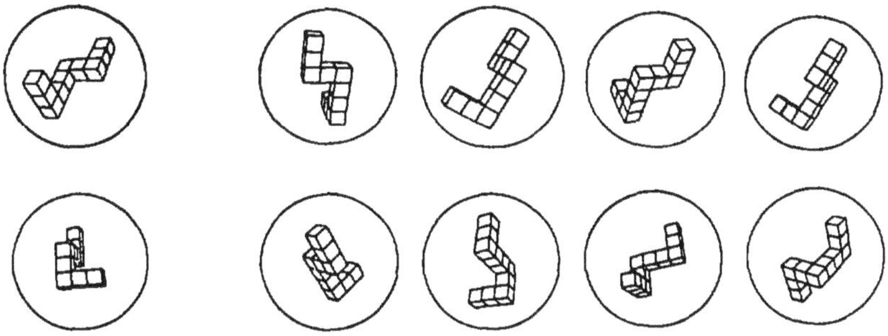

Figure 4.1 **Stimuli of three-dimensional mental rotation tasks. Participants view pairs of objects and judge whether the objects are the same (one object can be rotated into the other object's orientation) or different (mirror images of each other). The stimuli were presented in various orientations, drawings of each pair being separated by 0° to 180°, in 20° increments. Reprinted from Vandenberg and Kuse, 1978, p. 600.**

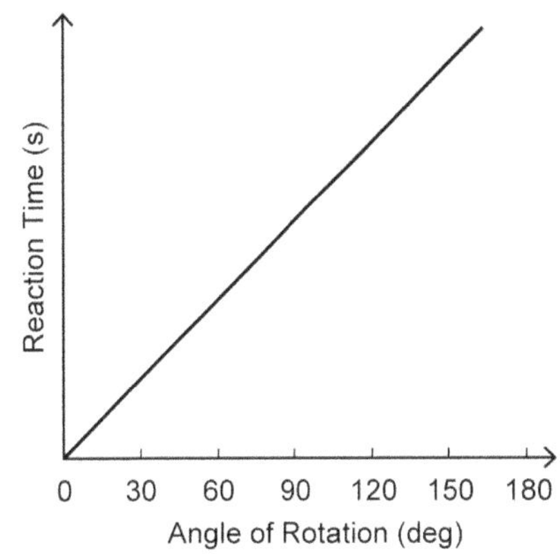

Figure 4.2 **Schematic illustration of results of mean response times, plotted against angular disparity between the two drawings of a pair. Response times increased linearly with angular disparity between paired drawings, or the degrees of mental rotation required to match one drawing to the other.**

error rate was as low as 3.2 percent). The graph shows that the response times increased linearly with angular disparity between paired drawings, for both rotation in the picture plane and rotation in depth. Shepard and Metzler (1971) reasoned that the linear functions indicated that the respondents "mentally rotated" the objects in three dimensions, at a rate of about 60° per second.

The results suggest that people apply mental operations and transformations to images of objects that they envision in their minds. Importantly, there seems to be a functional correspondence between manipulations of mental representations and manipulations of physical objects in the real world; that is, people rotate images mentally as they rotate objects physically. Cooper and Shepard (1973) examined the characteristics of mental

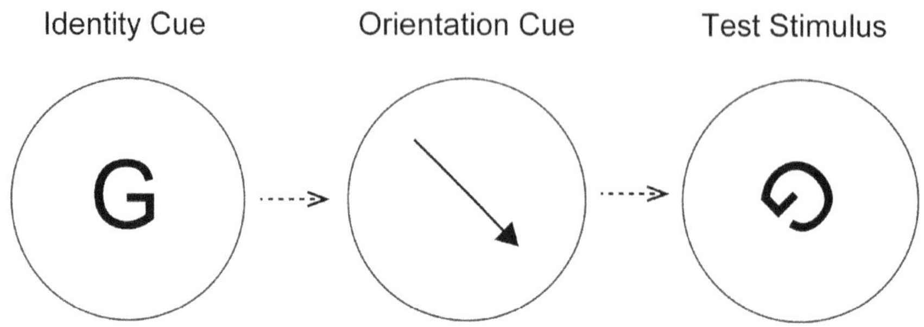

Figure 4.3 **Experimental stimuli of mental rotation tasks. Participants viewed an alphanumeric character in its normal, upright orientation (an identity cue), followed by an arrow indicating the direction at which the top of a test stimulus would appear (an orientation cue); participants were then presented with a test stimulus and asked to judge whether it was a normal character or a flipped image. In other conditions, participants were presented with (a) no advance information, (b) only the identity cue, (c) only the orientation cue, or (d) the identity and orientation cues combined (i.e., an alphanumeric character shown in the same orientation as a test stimulus). Illustration based on Cooper and Shepard, 1973.**

rotations further using asymmetric alphanumeric characters (upper-case letters G, J, and R, and numerals 2, 5, and 7). In a basic condition of the experiment, participants viewed an alphanumeric character in its normal, upright orientation (an identity cue) for a duration of 2,000 milliseconds, and then an arrow indicating the direction at which the top of a test stimulus would appear (an orientation cue) for varying durations of 100, 400, 700, and 1,000 milliseconds; after that, they viewed a test stimulus (Figure 4.3). Thus, in this condition, both identity and orientation cues were presented in sequence (denoted by B in Figure 4.4), and participants judged whether a test stimulus was a normal character or a mirror image (flipped over). There were four additional conditions with respect to the manner in which advance information was presented: (a) no advance information was given (denoted by N), (b) only the identity cue was given (denoted by I), (c) only the orientation cue was given (denoted by O), and (d) the identity and orientation cues were given simultaneously such that an alphanumeric character was shown in an orientation in which a test stimulus would be presented later (denoted by C).

Figure 4.4 shows reaction times for the five conditions. It reveals, as in Shepard and Metzler (1971), a monotonic increase in reaction time as a function of deviation of the orientation of a test stimulus from its upright orientation. (The departure of the shape of the function from strict linearity was, Cooper and Shepard (1973) discussed, due to the familiarity of the alphanumeric characters to participants.) Longer response times for larger angular deviations show that participants mentally rotated a tilted character (presented stimulus) so that it would be aligned with a mental representation of the character stored in their minds in its normal, upright orientation. Results for the four different durations of time for viewing the orientation cue (100, 400, 700, and 1,000 milliseconds) showed that for duration times shorter than 1,000 milliseconds (or 1 second), participants took time to rotate a mental representation of a character into the orientation of a test stimulus (showing a pattern of reaction time similar to the ones for conditions N, I, and O in Figure 4.4). When a duration of 1,000 milliseconds was given, participants had a mental

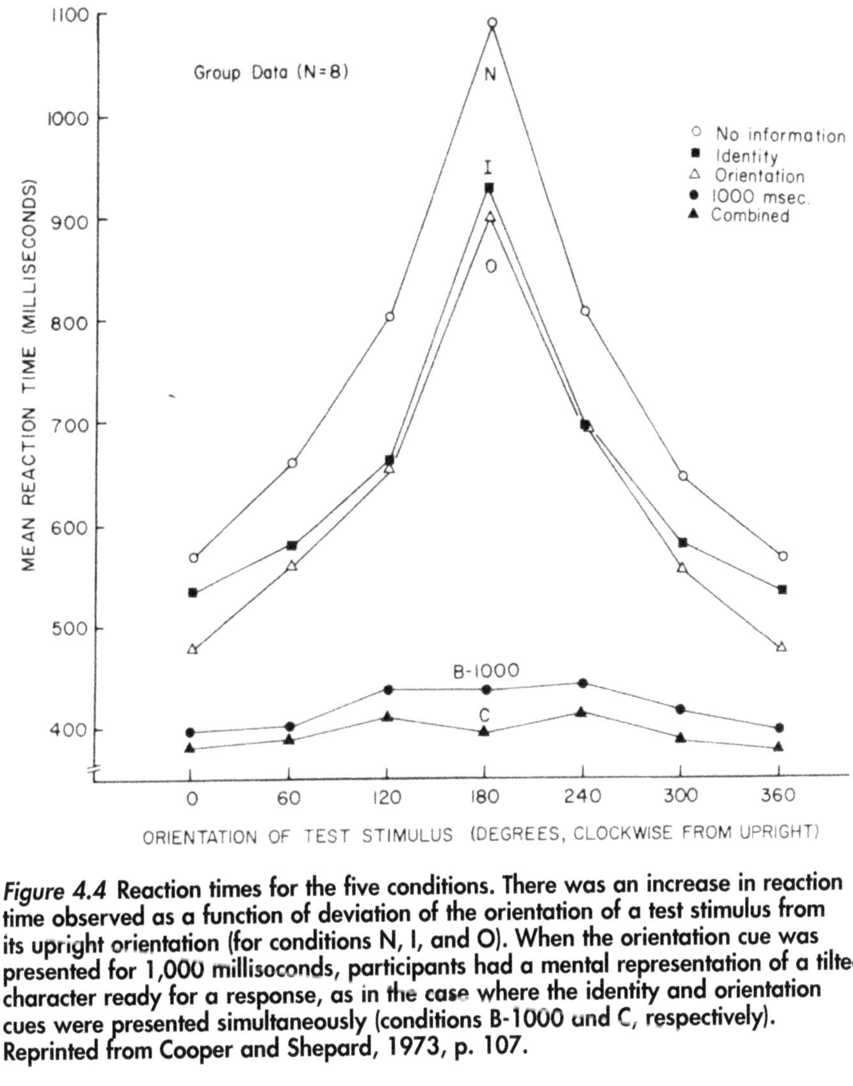

Figure 4.4 **Reaction times for the five conditions. There was an increase in reaction time observed as a function of deviation of the orientation of a test stimulus from its upright orientation (for conditions N, I, and O). When the orientation cue was presented for 1,000 millisoconds, participants had a mental representation of a tilted character ready for a response, as in the case where the identity and orientation cues were presented simultaneously (conditions B-1000 and C, respectively). Reprinted from Cooper and Shepard, 1973, p. 107.**

representation of a tilted character (a character in a to-be-presented orientation) ready for a response (see B-1000 in Figure 4.4, a pattern of reaction time similar to the one for condition C).

Cooper and Shepard (1973) conducted a further experiment to corroborate the idea that people mentally rotate internal representations: they asked participants to imagine an alphanumeric character rotating clockwise at a specific rate in harmony with the verbal commands, "up," "tip," "tip," "down," "tip," "tip," which set a pace of 60° rotation per half second. When the tape-recorded command "up" was heard, participants imagined the character being in its upright orientation, and three commands later when the word "down" was heard, they imagined it being rotated into the 180° orientation or upside down. Participants were then presented with a test stimulus either in the orientation that the participant should be imagining at that moment according to the verbal commands or in another orientation different from that; they were asked to judge whether the stimulus was a normal or reflected version of the imagined character. Cooper and Shepard found that reaction times increased linearly as a function of the angular disparity of the orientation of a test stimulus from an imagined (or expected) orientation. When the stimulus's orientation coincided with the imagined orientation (i.e., 0° of disparity), the reaction time was around 600 milliseconds, and when the stimulus's orientation departed 180° from the imagined orientation, the reaction time was around 900 milliseconds, indicating a rate of mental rotation similar to the one observed above. These results provide further support for the idea of mental rotation in people's minds.

Kosslyn, Ball, and Reiser (1978) examined the idea that mental representations preserve a functional relationship with represented objects, in the context of mental scanning of an image. Kosslyn et al. showed a map of a fictitious island to participants, on which seven locations were identified by illustrations of objects: a hut, tree, rock, well, lake, sand, and grass (Figure 4.5). Participants studied the island until they could draw it and indicate the seven locations accurately without viewing them. They were then given the name of one object, and after 5 seconds the name of another object, and asked to "scan" a map envisioned in their minds to enable them to mentally trace a walk in the shortest straight line from the first object to the second object on the imagined island. An analysis of response time for the mental scanning showed that the time to scan mental images

Figure 4.5 **Map used as a stimulus for mental scanning. Participants were asked to scan a map envisioned in their minds so that they mentally trace a walk between two objects on the imagined island. Their reaction times were proportional to the distance to be traveled. Reprinted from Kosslyn, Ball, and Reiser, 1978, p. 51.**

increased linearly with the distance between the two locations to be traveled. The results showed, according to Kosslyn et al., that metric distances were preserved in mental images in the functional sense as in a percept of a picture. Shepard and Chipman (1970) described such a functional correspondence between mental images and objects as a second-order isomorphism, a correspondence in the sense that the relations among external objects map onto the relations among their internal representations. They stated, "although the internal representation for a square need not itself be square, it should (whatever it is) at least have a closer functional relation to the internal representation for a rectangle than to that, say, for a green flash or the taste of persimmon" (p. 2). Put differently, mental representations are functionally related to physical objects that they represent, but need not necessarily be picture-like images.

Schemata and the understanding of the world

The preceding section looked at mental representations from the perspective of psychological transformations that humans apply to internal images and revealed an isomorphism between mental and physical rotations. Another perspective on mental representations concerns how they guide our behavior and direct our explorations, so that we make sense of our experiences and understand the world. People understand the world on the basis of their knowledge, and at the same time they restructure the knowledge by incorporating new information. Human learning and intellectual development are thus considered as an iterative process of assimilation and accommodation. In the context of cognitive mapping, internal representations of the environment (maps in the head or cognitive maps) guide people's behavior in the environment, and at the same time are updated by new information acquired through interactions with the environment.

This aspect of internal representations as a guiding principle or knowledge structure has been discussed in cognitive psychology as the concept of *schemas* or *schemata*. The concept is often traced back to Bartlett (1932), who conducted a series of experimental studies to examine people's perception and memory. In the studies, groups of people were presented with pictures and verbal passages (folk stories or newspaper reports), which they were asked to recall after a delay. Bartlett observed how the participants' drawings and descriptions of the presented materials changed in the course of a series of reproductions by the same person on different occasions (repeated reproduction) or by a number of different people (serial reproduction). A well-known material used in the study was a North American folk tale called *The War of the Ghosts*. The story belonged to a culture that was unfamiliar to the participants, and described incidents without clear order or interconnection, when seen from the participants' perspective. Bartlett observed that the structure and elements of the story were simplified and transformed over the chain of reproductions into those that looked more familiar, coherent, and rational (or stereotyped) to the participants and to their social environments. Bartlett stated:

> It is now perfectly clear that serial reproduction normally brings about startling and radical alterations in the material dealt with. . . . At the same time the subjects [participants] may be very well satisfied with their efforts, believing themselves to

> have passed on all important features with little or no change, and merely, perhaps, to have omitted unessential matters. . . . In fact, . . . human remembering is normally exceedingly subject to error. It looks as if what is said to be reproduced is . . . really a construction.
>
> (pp. 175–176)

These findings led Bartlett (1932) to argue that remembering is a process of *restructuring* based on past experiences, with schemata playing active roles of affecting our behavior and being affected by our behavior. He mentioned: "'Schema' refers to an active organisation of past reactions, or of past experiences, which must always be supported to be operating in any well-adapted organic response" (p. 201). To emphasize the active nature and continuous engagement of schemata, Bartlett suggested alternative terms such as "active developing patterns" (p. 201) or "active organised settings" (p. 212). In essence, schemata direct us to the past experiences that are most relevant to our current needs and allow us to react to the incoming information correctly and adaptively; and, at the same time, we have the capacity to act on the schemata and restructure them by accommodating the new information (p. 206). Put differently, schemata offer a general framework (or a script) that works for a family of related instances (e.g., ordering at a restaurant). The framework is skeletal with several open slots, which are to be filled in (or instantiated) according to each corresponding instance applicable (e.g., checking at the table or with the cashier).

The concept of schemata was elaborated by Neisser (1976), particularly in the context of cognition in the real world and behavior in the ordinary environment. Specifically, Neisser discussed that perception is a selective process being guided by and dependent on the cognitive structure internal to the perceiver (i.e., schemata). The schemata prepare the perceiver to attend to specific kinds of information, and not others. The perceiver explores the stimulus (or the ambient optic array in Gibson's term) for the information anticipated; and the schemata now are modified by the information that is picked up and direct the perceiver for further exploration (pp. 20–21).

> Because schemata are anticipations, they are the medium by which the past affects the future; information already acquired determines what will be picked up next. . . . The schema is just one phase of an ongoing activity which relates the perceiver to his [or her] environment. . . . Normally, however, the function of an unexpected stimulus is to initiate the cycle of perception proper; there is usually enough continuing information to support the cycle once it has begun.
>
> (pp. 22–23)

Neisser clarified that perception is a constructive process, in which the perceiver actively interacts with the environment, gathers information about the environment, and interprets it on the basis of existing knowledge; and then the knowledge structure is changed by the newly acquired knowledge (p. 57). Specifically in the context of cognitive mapping, Neisser described the perceptual cycle of "direction–exploration–modification," such that people explore the environment being guided by cognitive maps and cognitive maps are modified by new information acquired through locomotion in the environment (Figure 4.6).

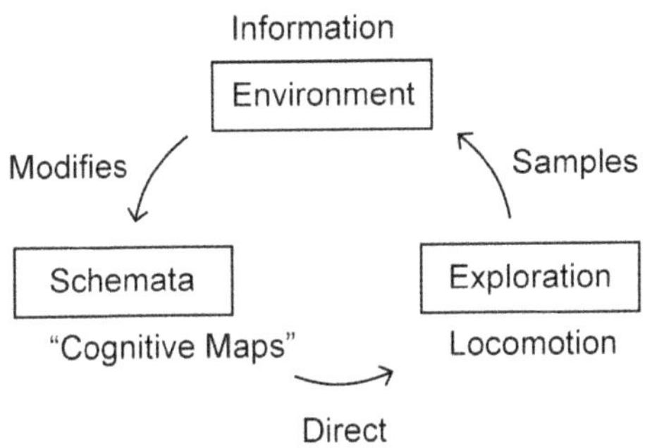

Figure 4.6 **Neisser described the perceptual cycle of direction, exploration, and modification of information in human spatial behavior. It signifies the constructive process of perception and the dynamic nature of cognition with schemata. Illustration based on Neisser, 1976.**

Schemata and bodily experience in space

The concept of schemata as a mental structure that guides people's behavior and interpretations has been discussed by a number of writers, in various contexts and from various perspectives. The arguments by Kant and Chomsky (which we saw in Chapters 2 and 3) and an argument from an evolutionary perspective (which we will see in Chapters 6 and 9) emphasize the innateness and genetic inheritance of our mental facilities (a "nature" view). Schemata have also been discussed from a "nurture" point of view, focusing on the influence of experience on our reasoning and thinking. Research into the influence of language on spatial and nonspatial reasoning is one such example (linguistic relativism, which we will see in Chapter 7). Another example of a nurture view on spatial cognition is the theory of *experientialism* put forth by Johnson (1987) and Lakoff (1987). Johnson discussed that *meaning* (or broadly *meaningfulness* to us) is closely related to the structures of mind through which we experience and understand the world. He clarified that schemata provide a generic, skeletal structure for people, so that they understand many different situations by filling in open "slots" of the structure with specific values pertaining to the instance being experienced at the moment. In other words, specific situations are understood when they are fitted (or assimilated) into the framework or script of a schema. Importantly, the theory of experientialism argues that schemata are formed by our experiences in the world; namely, the manner in which we organize incoming information and interpret the world emerges from our perceptual interactions with, and bodily movements in, our spatial environments. Schemata are in constant operation in our perception, manipulation of objects, and exploration of the environment, to organize our experiences and provide meanings for them. To emphasize the constant operation, generality, and malleability of schemata, Johnson used the term "image schemata" or "embodied schemata."

For example, Johnson (1987) pointed out a schema of "in–out orientation" and discussed how it organizes our experience, understanding, and language. The schema is based on our bodily experience of moving an object or ourselves, or seeing an object moving, out of a container along a path (an OUT schema; Figure 4.7a). As a basic instance of the OUT schema, Johnson gave the expression "John went out of the room" (p. 32).

The general structure of the schema is modified in each particular situation, so that various instances can be represented, such as "Mary got out of the car," "He squeezed out some toothpaste," and "Get out of bed" (p. 33). He stated,

> We couldn't begin to understand our experience without them [image-schematic meaning structures]. They give comprehensible structure and definiteness to our experience and connect up different experiential domains to establish a measure of coherence and unity in our understanding of our world. . . . On my view, so-called linguistic meaning is treated as a specific case of meaning in a broader sense. . . . Meaning structures in this broader sense are the basis for human reasoning and inference.
>
> (p. 37)

The influence of embodied schemata on our perception and understanding of the world was further discussed by Lakoff and Johnson (1980) with a focus on the metaphorical nature of our thinking. To clarify the roles that metaphorical concepts play in our thinking and reasoning, Lakoff and Johnson discussed orientational metaphors. These metaphors allow us to understand various related concepts in terms of spatial orientation, such as up–down, in–out, and on–off, based on our experiences of acting in and interacting with the physical environment with our bodies. They are rooted in our physical and cultural experiences, and the manner in which they structure our conceptual system is called *spatialization*. They are metaphorical in that they lead to an understanding of one kind of thing in terms of another, or more specifically a conceptualization of less delineated experience with more sharply delineated experience. In this metaphorical relationship, the most sharply delineated one is our bodily experience in the physical environment (e.g., standing up); that is, our sensory-motor experiences form the basis of our conceptualization of nonphysical experiences. From the basic experience of the normal functioning of our body in the upright posture arise orientational metaphors of emotions and well-being (pp. 15–16), such as HAPPY IS UP ("I'm feeling up" or "I'm feeling down"), HEALTH IS UP ("He's in top shape" or "He fell ill"), HAVING CONTROL IS UP ("She has control over me" or "His power is on the decline"), and MORE IS UP ("The number of books keeps going up" or "My income rose last year"; Figure 4.7b).

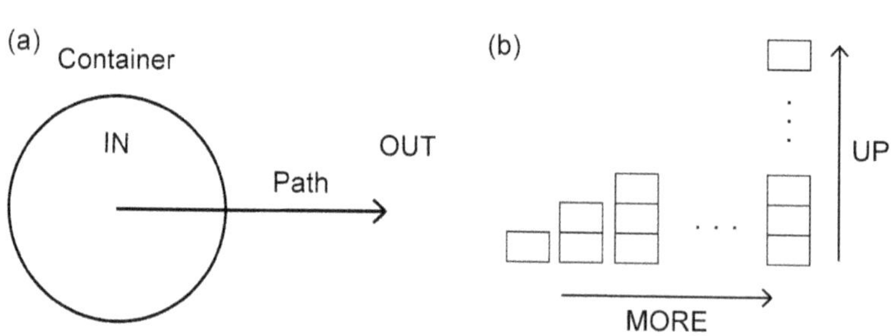

Figure 4.7 **Importance of interactions with, and bodily movements in, the environment for our cognition and thinking. (a) Schema of "in–out orientation," based on the bodily experience of moving an object or ourselves, or seeing an object moving, out of a container along a path. (b) Orientational metaphor of "more is up," based on the experience of normal functioning of our body in the upright posture. Illustrations based on Johnson, 1987, and Lakoff and Johnson, 1980.**

Lakoff and Johnson (1980) also discussed the metaphorical conceptual system in reference to the cognitive system of *categorization*. Categorization lies at the heart of human understanding of things and phenomena in the world. We would be baffled when encountering an animal that cannot be readily classified into any known category—it looks like a cat, like a dog . . . and like a fox, but not quite. What is it? At the same time, we find it difficult to give every single attribute that defines the cat, dog, or fox. Rosch (1978) discussed that categories are defined by prototypical exemplars and the degree of resemblance to them. In the orientational metaphors, the upright posture may be considered a prototype and other UP metaphors possess family resemblance to the prototype. Most importantly, the perspective of experientialism points to the significance of human experiences in and with space, and delineates how space matters for human cognition.

Hierarchical structure of spatial knowledge

In Chapter 3, we saw that mental representations of the environment tend to be distorted, fragmented, and schematized (Downs and Stea, 1973; Golledge and Spector, 1978; Lynch, 1960). Now we look in detail at the structure of cognitive maps, namely the manner in which mentally represented environments are systematically distorted, fragmented, and schematized. One of the early studies of distortions in spatial memory was conducted by Stevens and Coupe (1978). They examined people's knowledge of spatial relations between geographic locations, by asking participants (students in San Diego, California) to indicate the directions between five pairs of places in North and Central America: (a) from Portland, Oregon, to Toronto; (b) from San Diego, California, to Reno, Nevada; (c) from Montreal to Seattle, Washington; (d) from the Atlantic entrance of the Panama Canal to the Pacific entrance; and (e) from Portland, Oregon, to Portland, Maine. Figure 4.8 shows participants' responses and correct directions for the five pairs of locations, an analysis of which reveals a systematic pattern of errors. Importantly, Stevens and Coupe pointed out that there is a strong tendency for estimated directions to be distorted "toward the direction of the superordinate relationships" (p. 424) and ascribed it to the *hierarchical* structure of spatial memory.

Take the pair of San Diego and Reno as an example. Saying that spatial memory is represented hierarchically means that knowledge about places is ordered from higher to lower levels, the latter being encompassed by the former; and in the hierarchy, places at the same level are not completely linked to each other. In people's spatial memory, therefore, the direct spatial relation between the states of California and Nevada is stored, but the spatial relations between cities in the two states are not specified (see the schematic diagram in Figure 4.9). To judge the spatial relation between the cities of San Diego and Reno, people need to access (a) the stored spatial relations between San Diego and California (San Diego is part of California) and between Reno and Nevada (Reno is part of Nevada) and (b) the stored superordinate relationship of California and Nevada (California is generally to the west of Nevada). From these pieces of stored information, people *infer* that San Diego is west of Reno. That is, the spatial relation between San Diego and Reno is judged on the basis of the superordinate relationship of the states in which the cities are located. Similar directional biases toward superordinate relations are identified for other pairs of

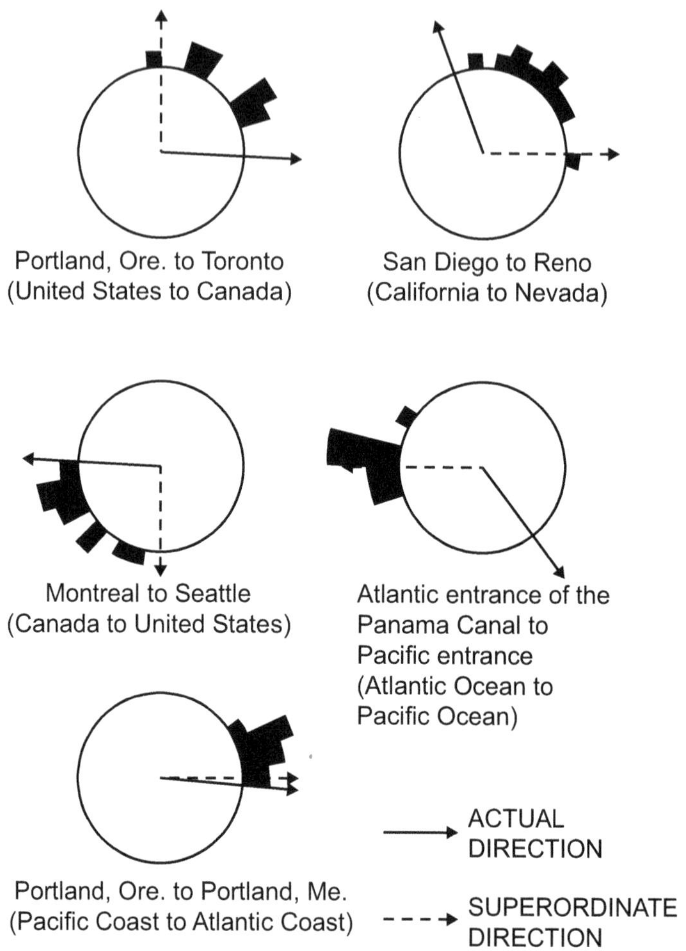

Figure 4.8 **Participants were asked to indicate the directions between five pairs of places: (a) from Portland, Oregon, to Toronto; (b) from San Diego, California, to Reno, Nevada; (c) from Montreal to Seattle, Washington; (d) from the Atlantic entrance of the Panama Canal to the Pacific entrance; and (e) from Portland, Oregon, to Portland, Maine. The graphs show systematic patterns of errors: judged directions tend to be distorted toward the direction of the superordinate relationships, suggesting a hierarchical structure of spatial memory. Reprinted with permission of Elsevier from Stevens and Coupe, 1978, p. 424.**

locations in Figure 4.8 (Canada on the north of the United States, and the Pacific Ocean toward the west of the Atlantic Ocean).

Hierarchical organization of spatial knowledge has also been identified in the context of categorization of spatial knowledge. Maki (1981) asked students in North Dakota to recall cities in North Dakota and Minnesota and locate them in their correct positions on outline state maps, respectively. Then they were asked to judge spatial relations of cities within the same state (both cities from North Dakota or from Minnesota) and cities

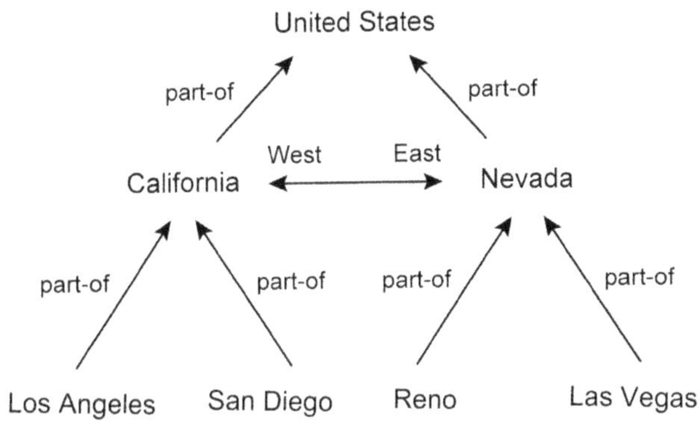

Figure 4.9 **Hierarchical structure of spatial memory. The spatial relation between the cities of San Diego and Reno is judged based on the stored superordinate relationship of the states of California and Nevada. The stored knowledge says that California is generally to the west of Nevada, so people (incorrectly) infer that San Diego is west of Reno. Illustration based on Stevens and Coupe, 1978.**

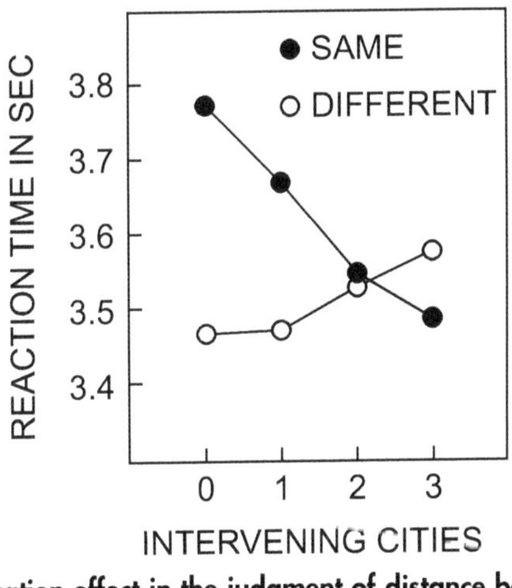

Figure 4.10 **Categorization effect in the judgment of distance between two cities. When the two cities were from the same state, mean response times decreased with an increase in the distance between them, but not when the two cities were from different states. Reprinted from Maki, 1981, p. 18.**

between the two states (one city from North Dakota and the other from Minnesota). To make spatial judgments, participants read a statement given in the form of "City A is west (or east) of City B," and answered whether it was correct. Maki observed that participants' response times decreased as the distance between two cities increased for within-state judgments, but not for between-state judgments (Figure 4.10). Maki identified in the pattern of response times the influence of categorization and hierarchical ordering. For two cities from different states, category information can be used to verify the statement of spatial relations (one city is from Minnesota and one is from North Dakota, and Minnesota

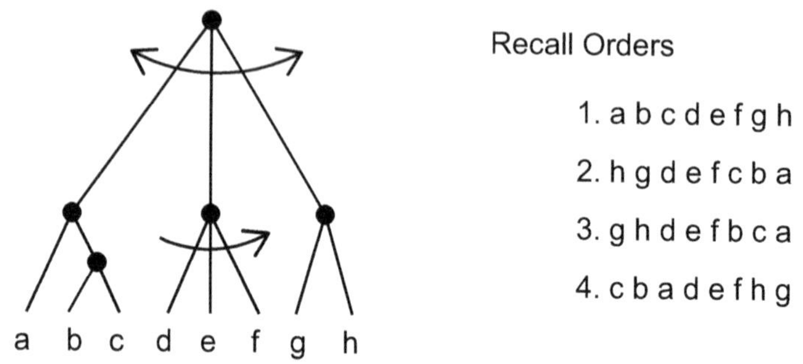

Figure 4.11 **Hierarchically ordered tree diagram. The ordered-tree algorithm examines participants' protocols of free recall and identifies sets of items that are recalled contiguously. The diagram represents the regularities in the recalls as a hierarchical tree structure, assuming that items that are mentioned closely together in free recall are stored and represented as a group in memory. Reprinted with permission of Springer Nature from Hirtle and Jonides, 1985, p. 210.**

is east of North Dakota). For two cities within the same state, such category information is not enough for the judgment of their spatial relations. That is, spatial relations of cities in different states are judged on the basis of spatial relations of the states. But when two cities are in the same state, one needs to scrutinize the knowledge of the state (i.e., zoom in on the mental representation of the state), which takes longer as the cities are located closer. These results, Maki discussed, are consistent with the symbolic distance effect observed in the literature for a variety of stimuli—the time required for a comparison of two objects in memory increases as the distance between the two objects decreases on the dimension relevant for the judgment (Moyer and Bayer, 1976: Paivio, 1975; Van Opstal et al., 2008).

Hierarchical structures have been observed for memory of spaces at various scales, for example, for the memory of geographic locations, landmarks in a neighborhood, and an array of objects in a room. To examine the structure of spatial knowledge, an ordered-tree algorithm (Reitman and Rueter, 1980) is often used, which is a variation of cluster analysis and identifies regularities in the order of recalls of places or objects from memory. Participants are asked to memorize a set of items, and recall them repeatedly, starting with any item and proceeding in any order. The ordered-tree algorithm examines the protocols of free recall to identify sets of items that are recalled contiguously, and represents the regularities in the recalls as a hierarchically ordered tree diagram (Figure 4.11). The rationale behind the analysis is that items that are mentioned closely together in free recall are stored and represented as a group in memory.

Using this method, Hirtle and Jonides (1985) and McNamara, Hardy, and Hirtle (1989) showed that the memory for landmarks in a city and objects laid out on a floor was represented hierarchically. Importantly, cluster membership predicted response time and distance judgments better than did the extent of physical separation. Hirtle and Jonides asked students in Ann Arbor, Michigan, to memorize 32 major landmarks in the city, the names of which were written on index cards. The students were then asked to recall the landmarks, and estimate distances between pairs of landmarks. From an analysis of free recalls, an ordered tree diagram was constructed for each participant, which shows hierarchical clustering of landmarks in their memory. For the participant in Figure 4.12, for example,

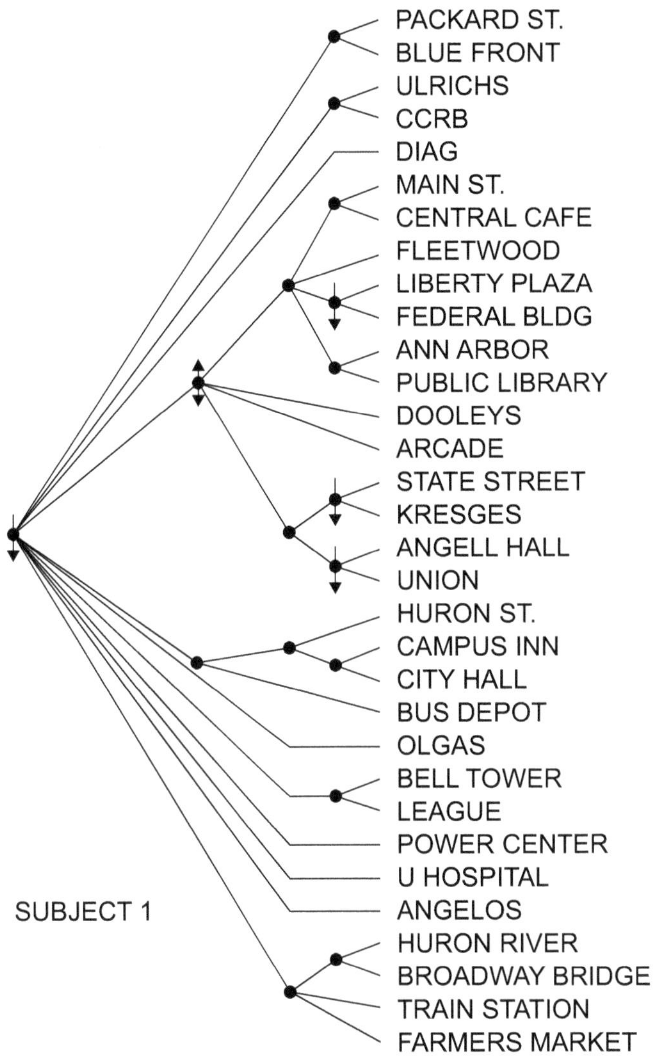

Figure 4.12 **Ordered tree diagram for free recalls of landmarks by a student in Ann Arbor, Michigan. For this participant, the landmark pair of Huron St. and Campus Inn is classified as *across cluster* relative to the pair of Campus Inn and City Hall; whereas it is *within cluster* relative to the pair of Huron St. and Bus Depot. Reprinted with permission of Springer Nature from Hirtle and Jonides, 1985, p. 213.**

the landmark pair of Huron St. and Campus Inn is classified as *across cluster* relative to the pair of Campus Inn and City Hall; whereas it is considered *within cluster* relative to the pair of Huron St. and Bus Depot. Hirtle and Jonides found that across-cluster distances tended to be overestimated and within-cluster distances were underestimated, relative to each other (Figure 4.13). McNamara et al. conducted a similar experiment with objects laid out on a floor and likewise observed the influence of objects' cluster membership on response time

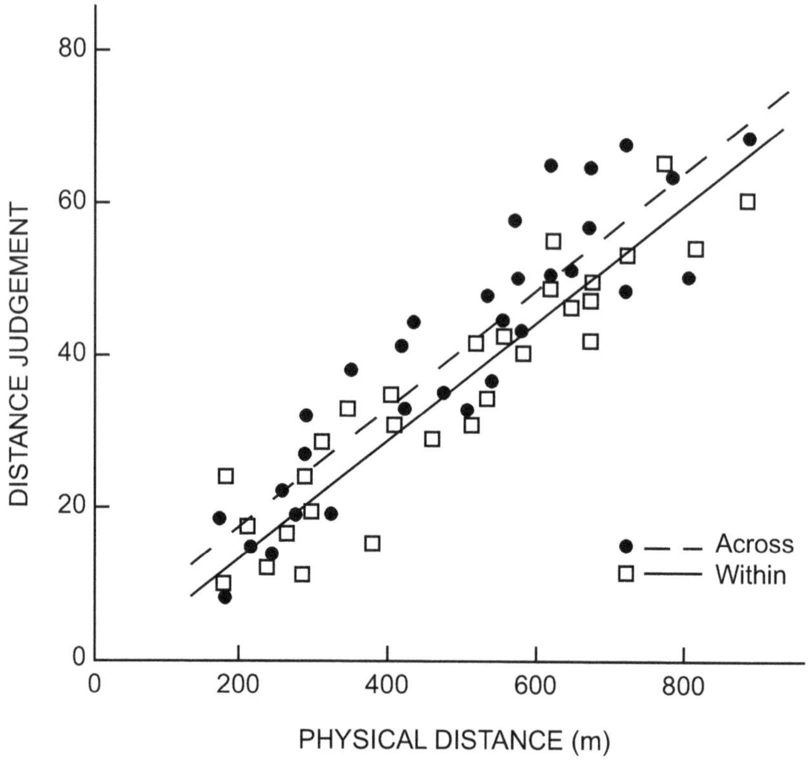

Figure 4.13 **Cluster membership and distance judgment. An analysis of distance estimates for cities in the same cluster and in different clusters showed that across-cluster distances tended to be overestimated and within-cluster distances were underestimated, relative to each other. It is another indication of hierarchical structure of spatial memory: items in the same cluster are interrelated with each other more closely than items in different clusters. Reprinted with permission of Springer Nature from Hirtle and Jonides, 1985, p. 215.**

and distance judgment. In a recall task, participants responded to a target object faster when it was primed by an object in the same cluster than when it was primed by an object in a different cluster (in a ratio of 1:1.1 on average). They also judged distances to be shorter for within-cluster pairs than for across-cluster pairs (in a ratio of 1:1.3).

Spatial knowledge acquired from maps and direct experience

As we discussed in Chapter 3, one of the major characteristics of the environment—it cannot be viewed in its entirety from a single viewpoint—makes it necessary for the explorer to integrate and interrelate separate pieces of information observed at individual locations. Because of this, accurate and flexible spatial behavior in the environment is not as easy as many people think. Importantly, a significant portion of people have difficulty in understanding the spatial layout of the environment metrically and configurationally (we will see it in detail in

Chapter 5). So, in the context of spatial orientation and navigation assistance, information about the environment that is externally represented and given to us plays an important role. In particular, maps allow us to comprehend spatial relations and configurations as seen from above. With such a potential benefit, on the other hand, maps affect the knowledge acquired in important ways, differently from direct navigation in the environment; namely, the manner in which the environment is learned or the sources of information about the environment influence the structure of spatial knowledge and the flexibility of spatial reasoning.

Thorndyke and Hayes-Roth (1982) examined spatial knowledge learned about a floor of a company building by two groups of participants. One group of participants consisted of 24 employees at the company, who had learned the layout of the building by directly navigating in it while they were working for various lengths of time, ranging from one month to two years. The employees had never viewed a map of the floor plan of the building. Another group of participants was 24 college students who had never navigated in the building; they studied a map of the building until they were able to draw the map from memory without errors. The participants were individually taken to a series of locations in the building and asked to estimate directions and distances to other locations. For distances, they estimated two kinds: distances along the hallways (route distances) and distances in a straight line. The locations used as home and target locations were landmark places in the building that participants had learned either from their experience of direct navigation (the employees, or navigation participants) or from viewing the map (the students, or map-learning participants).

The results of distance and direction estimates are shown in Figure 4.14. Navigation participants estimated route distances more accurately than straight-line distances (a distance correlation around 0.88 vs. 0.80), while map-learning participants estimated both distances equally accurately (a distance correlation of 0.82). For route distance estimates, the accuracy was equivalent among participants with different lengths of employment; but for straight-line distance estimates, the accuracy improved with the amount of experience. Participants with 6–24 months of experience estimated straight-line distances more accurately than did participants with 1–2 months of experience. That is, direct experience of navigation in the building facilitated the acquisition of accurate knowledge of route distance, compared with straight-line distance. By contrast, studying a map helped people to acquire accurate knowledge of straight-line distance; map-learning participants estimated straight-line distances as accurately as did navigation participants with 6–24 months of experience. On the other hand, navigation participants estimated directions more accurately than did map-learning participants (an angular error around 20° vs. 40°), and their estimates became more accurate as the amount of experience increased.

Thus, mental representations of environments differ in their sensitivity to metric and configurational information depending on whether they are acquired from direct navigation or maps. Knowledge from maps is good for configurational understanding in terms of straight-line distance, but not for estimating directional spatial relations in the represented space. This points to the distinction between survey learning of distance and direction. Straight-line distances can be conceived of from a fixed perspective, but flexible estimation of relative directions entails taking (or imagining) different perspectives in the environment, which extends beyond the perceptual horizon available at a specific viewpoint (see the discussion of mental rotation and perspective taking in Chapter 6). This kind of directional inflexibility of internally represented environments offers important implications about orientation dependence of spatial knowledge, which we turn to in the

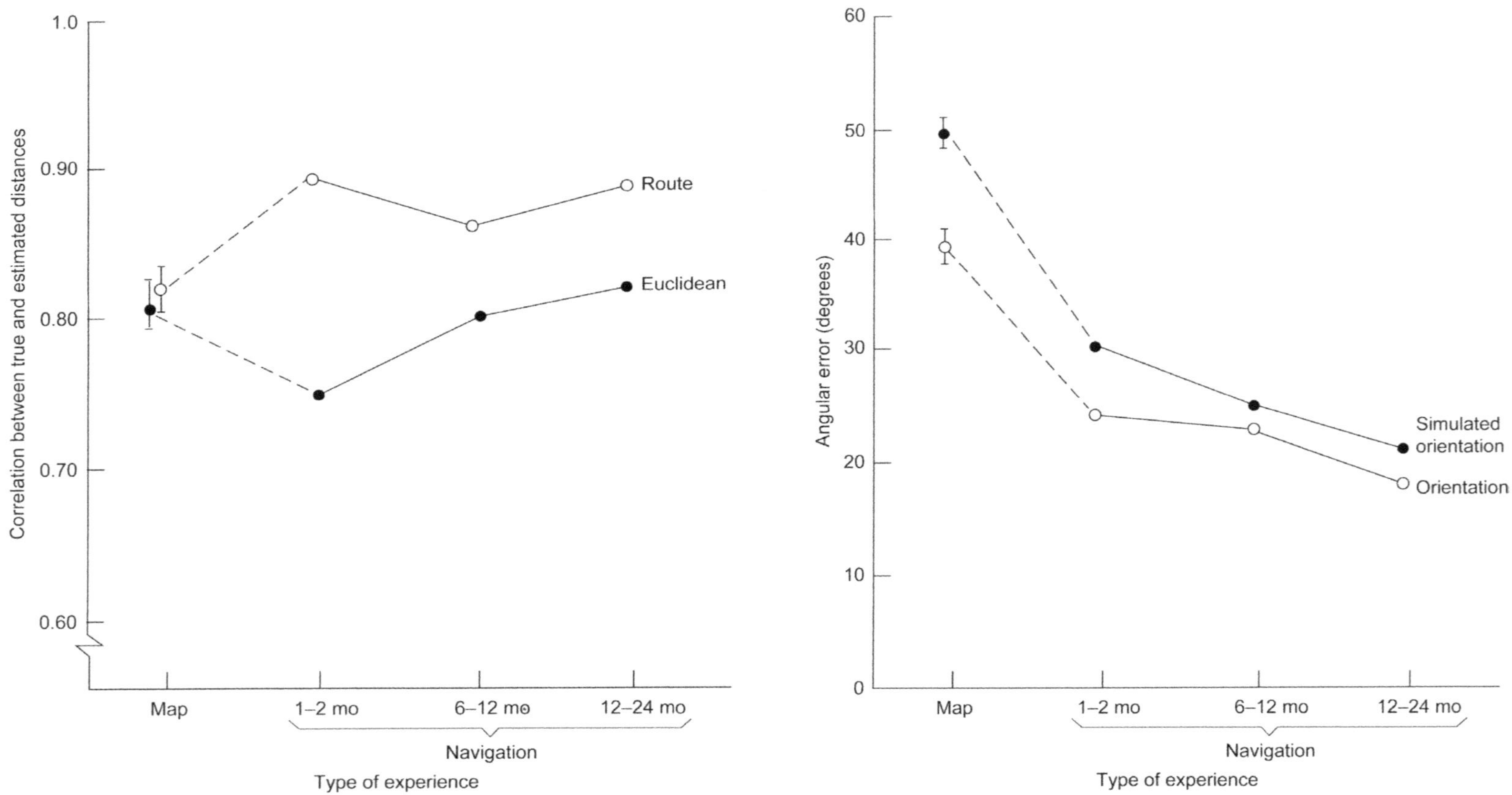

Figure 4.14 **Spatial knowledge acquired from direct navigation and a map. Navigation participants estimated route distances more accurately than straight-line distances, whereas map-learning participants estimated both distances equally accurately. On the other hand, navigation participants estimated directions more accurately than map-learning participants, and their estimates became more accurate as the amount of experience increased. That is, direct experience of navigation in the building facilitated the acquisition of accurate knowledge of route distance and relative direction; studying a map helped people to acquire accurate knowledge of straight-line distance. Reprinted with permission of Elsevier from Thorndyke and Hayes-Roth, 1982, pp. 576 and 580.**

next section. We will also consider it in the context of an "alignment effect" in map use in Chapter 7.

Orientation dependence of spatial knowledge

When you use a map in navigation, do you hold it in a fixed orientation (e.g., in a north-up orientation) or rotate it so that the direction of travel always coincides with the upward direction on the map? The ease with which a map is used depends on its orientation, particularly for the latter type of people. That is, (cartographic) maps have a privileged orientation or preferred direction (we will discuss this in relation to spatial skill in mental rotation in Chapter 7). What about maps in the head? Is there a specific direction from which you can infer spatial relations or recall an environmental scene more easily and accurately than other directions? In other words, do your internal representations of the environment have a preferred direction?

An insight into this question was provided by an experiment conducted by Rieser (1989). Rieser laid out nine objects on the floor of a room, at 40° intervals in a circle with a 2.5-m diameter (Figure 4.15). Participants viewed the array, standing at the center of the circle and facing one of the objects. After studying the layout with a fixed facing direction, participants were instructed to point, with their eyes closed, to a specific object with a new facing direction (facing a different object from the one that they faced while studying the layout). In one condition, participants were physically rotated by the experimenter to face the new object (being aligned with the new facing direction). In the

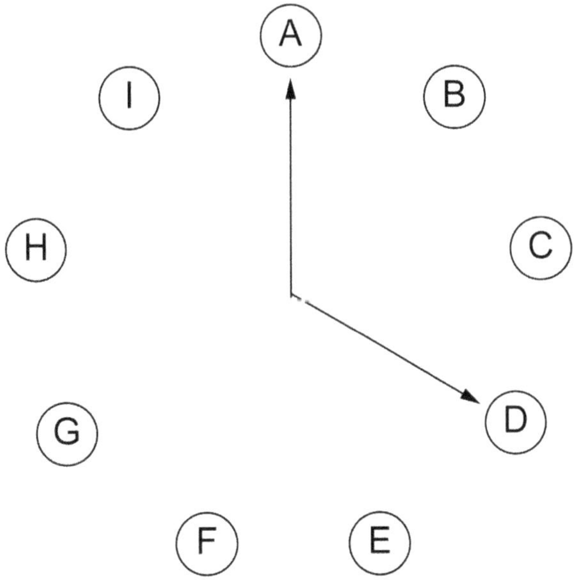

Figure 4.15 **Nine objects laid out on the floor in a circle. Participants viewed the layout with a fixed facing direction, and then were instructed to point, with their eyes closed, to a specific object with a new facing direction. Arrows indicate two possible different facing directions. Reprinted from Rieser, 1989, p. 1159.**

other condition, participants merely imagined facing the new direction without physically rotating (they were actually still facing the original direction). An analysis of mean response times revealed that participants in the imagined condition took a longer time to respond as the degrees of imaginary rotation increased. When they were physically rotated, participants responded faster, irrespective of the new facing directions. Also, participants responded less accurately in the imagined condition than when they were physically rotated, with mean absolute errors of 8°–16° and 13°–24°, respectively (depending on the degrees of rotation). These results show that people's mental representations of spatial layouts are orientation dependent, in this case tied to the original viewing perspective, and making spatial judgments from an imagined different perspective is effortful. Information of rotation from the body facilitates spatial inference based on stored spatial knowledge from a new perspective.

In a follow-up experiment, Rieser (1989) asked participants to point to a specific object by imagining standing at one of the objects (being at a new position with the same facing direction). That is, participants imagined that they moved to the new position by linear displacement or translation without rotation (they imagined it but made no physical movement). For comparison, Rieser also instructed the participants to point to a specific object by imagining that they faced another object, as people in the imagined rotation condition did above. Results showed that participants responded faster in the imagined translation condition than in the imagined rotation condition. They also responded more accurately in the imagined translation condition, although the difference was marginally significant. Thus, imagining translation does not pose extra cognitive effort, as imagining rotation does (you may remember the small experiment that you did in Chapter 3).

Orientation dependence of spatial memory has been examined in a variety of settings and conditions; for example, participants viewed arrays of objects of a small or large size, learned spatial layouts through direct navigation in the real world or virtual environments, viewed maps of object arrays or spatial layouts, learned a spatial layout from a single viewpoint or multiple viewpoints, were seated or standing in the learning and testing phases, learned a spatial layout and conducted spatial memory tasks in the same room or different rooms, and conducted spatial tasks in an in-situ condition or imagined condition (e.g., Evans and Pezdek, 1980; Levine, Jankovic, and Palij, 1982; McNamara, Rump, and Werner, 2003; Mou et al., 2008; Presson, DeLange, and Hazelrigg, 1989; Presson and Hazelrigg, 1984; Presson and Montello, 1994; Richardson, Montello, and Hegarty, 1999; Rieser, 1989; Rieser, Guth, and Hill, 1986; Roskos-Ewoldsen, McNamara, and Shelton, 1998; Shelton and McNamara, 1997; Sholl and Nolin, 1997). Acknowledging the existence of some inconsistent results, McNamara (2003) argued that research evidence overall suggests that people's memories of spatial layouts are orientation dependent and have privileged directions.

Importantly, McNamara (2003) clarified that the privileged directions are not necessarily a viewing perspective: he pointed to the influence of a reference system that is defined by the environment or intrinsic to the spatial layout. In some cases, a mental representation is tied to a perspective from which a layout was viewed, and in other cases it may be aligned with the symmetric axis of a layout, the rectangular shape of a room, the first or longest leg of a traveled path, grid-like street patterns in the neighborhood, and terrain features or salient landmarks in the environment (Mou et al., 2008; Shelton and McNamara, 2001; Werner and Schmidt, 1999). McNamara et al. (2003) demonstrated orientation dependence of spatial memory based on environment-centered frames of reference in an experiment conducted in a large city park. In the experiment, 24 people walked along one of two paths, either a path aligned with a large rectangular building at the center of the park, running parallel to its exterior walls, or a path misaligned with the building being rotated by 45° (Figure 4.16). Participants learned the locations of eight objects along the paths, and estimated directions between the objects by imagining standing at one of the objects and facing another object (e.g., they were instructed, "Imagine you are at the bench and facing the historical sign, and point to the frame"). Results showed that participants estimated directions more accurately when they walked along the aligned path than the misaligned path. For the aligned path, the accuracy of direction estimates was significantly higher when the imagined headings were parallel to the legs of the path (0°, 90°, 180°, and 270°) or oriented toward a nearby lake (135°). For the misaligned path, the accuracy was highest when the imagined heading was oriented toward the lake. McNamara et al. discussed that

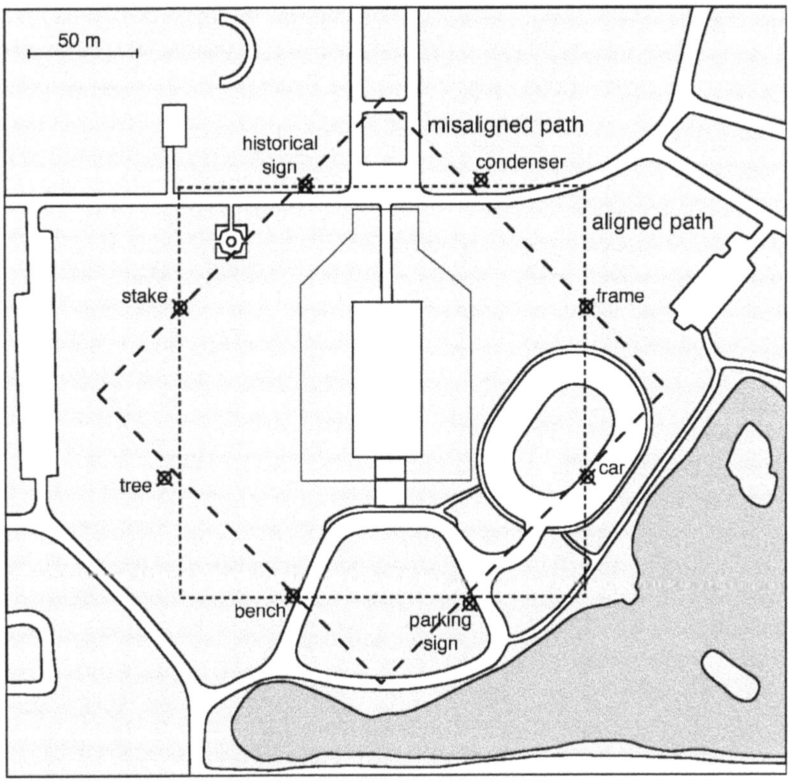

Figure 4.16 **Participants walked along a path running parallel to the exterior walls of a building at the center of a park (dotted lines) or a path misaligned with the building walls, rotated by 45° (dashed lines). Participants estimated directions between objects along the paths by imagining standing at one of the objects and facing another object. Participants estimated directions more accurately when they walked along the aligned path than the misaligned path. Reprinted with permission of Springer Nature from McNamara, Rump, and Werner, 2003, p. 592.**

participants in the aligned condition mentally represented the layout of the environment in intrinsic reference systems that are defined by the paths, the rectangular building, and the direction toward a very salient landmark (the lake). Participants in the misaligned condition mentally represented the environment in a single reference system defined by the salient landmark. The orientational incongruence of the path and the rectangular building posed difficulty for participants in the misaligned condition; the egocentric experience of walking the path did not allow them to mentally represent the learned environment in a framework aligned with the traveled path (e.g., along the 45°–225° axis).

Externalization of spatial knowledge

We have now looked at some major characteristics of internal representations of the environment—hierarchical structure, categorization, and orientation dependence. In so doing, we of course did not focus on the structure of spatial knowledge directly. Rather, we looked at people's responses to spatial tasks, and from their responses we reasoned about possible structures of internal representations. We examined people's estimates of environmental distances, estimates of directions between geographic locations, and free recalls of landmarks in a city—systematic patterns in their estimates and recalls allow us to discuss the structures of internally represented environments.

Researchers thus observe overt behavior to address the questions of "how" and "why." They describe the distributions of human activities from naturalistic observations of the gatherings and movements of people. They describe the patterns of human behavior based on the analysis of aggregate data on shopping, commuting, and migration. Current studies tend to explore the structures and processes of spatial knowledge through examinations of people's performances on experimental spatial tasks such as distance and direction judgments. Historically, the fact that the analytical step of going from overt behavior to internal representations involves some sort of *inference* caused misunderstandings among some researchers, who criticized the "black box" nature of these "positivistic" investigations (Bunting and Guelke, 1979; Graham, 1976). On the contrary to these unfortunate misunderstandings, scientific research from the cognitive-behavioral perspective aims to *explain* the reasons or underlying processes for complex and deliberate (even though unconscious) spatial behavior through systematic and controlled observations (Golledge, 1981, 1996). As we discussed in Chapter 3, in relation to the debate between behaviorism and cognitivism, research into spatial cognition strives for high internal, as well as external and ecological, validity. To achieve this, researchers need valid and reliable methods of *externalizing* people's internal representations.

By the early twentieth century, along with anecdotal reports of remarkable sense of orientation and experiences of spatial disorientation (e.g., Binet, 1894; DeSilva, 1931; Peterson, 1916; Warren, 1908), a few examples of experimental studies of spatial orientation appeared in the literature (for example, Brown, 1932, 1937, conducted experiments to study human behavior in a maze). Trowbridge's (1913) discussion of imaginary maps was one of them: he conducted a behavioral experiment and examined participants' performances on direction estimation. In the experiment, people in New York City were asked to estimate directions to the North Pole, London, San Francisco, and Panama. To do that, participants were given a circular piece of paper and instructed to mark on the circle, assuming that the center represented New York City, to indicate the directions to the four places.

Woodring (1939) suggested that deriving a good method of direction estimation was a requisite for systematic investigation of human spatial orientation. He clarified the rationale of a scientific study of spatial cognition and explained how to collect and analyze direction estimates, as done by many researchers today. He stated,

> Our first task is to develop a technique for reducing to quantitative terms the individual error in indicating the direction of an object which is not at the moment acting directly on the sense organs of the subject [participant], but with which he [or she] has had previous experience. . . . A survey of the literature reveals few attacks on this specific problem. . . . [In the pointing technique Woodring proposed here] there is placed before the subject a paper on which has been drawn a five-inch circle with the center marked by a dot. He is then given a pencil and a straightedge and instructed to draw from the center of the circle a line which points directly toward a designated point, known to him but not visible from his location.
>
> (pp. 147–148)

Another spatial task intended to examine people's spatial knowledge is distance estimation. The method basically applies the technique of psychophysical scaling (psychophysics examines the relationship between psychological and physical magnitudes) to an analysis of the perception and cognition of environmental distance. In estimating distances, participants are typically provided with a yardstick, or a standard distance, which they experience by viewing or traveling it, and estimate the lengths of experimental routes in comparison with the standard distance in terms of a ratio or a magnitude. In ratio estimation, participants were presented with a set of lines drawn on paper, one of which represents the length of the standard distance (Figure 4.17). Participants make a response

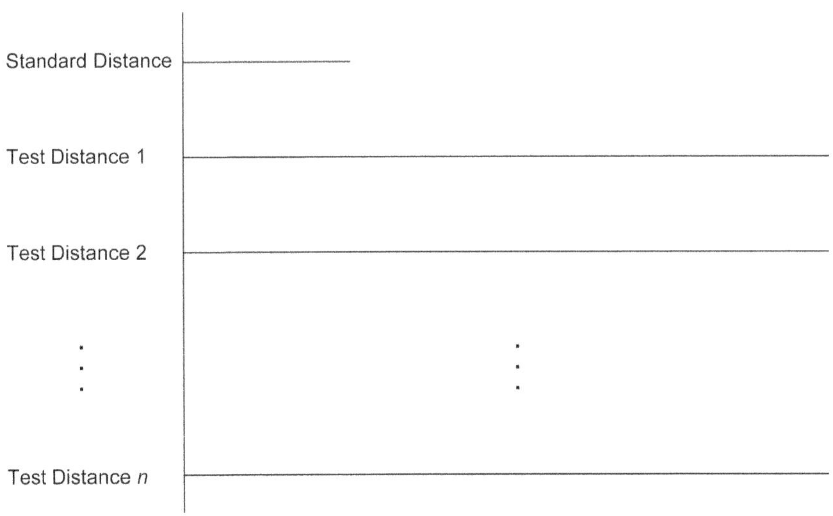

Figure 4.17 **In ratio estimation, participants were presented with a set of lines drawn on paper, one of which represents the length of a standard distance. Participants marked each line below the standard line to indicate the ratio of a test distance to the standard distance.**

on each line below the standard line, by placing a mark on the line to indicate the ratio of the distance to be estimated to the standard distance. For example, if participants think that the experimental route is twice as long as the standard distance, they place a mark on the response line so that the length is twice as long as the standard line. In magnitude estimation, participants are given a specific value (e.g., 10 or 100) that corresponds to the length of the standard distance, and estimate the lengths of experimental routes in comparison with it. If participants think that the experimental route is twice as long as the standard distance, they give a value twice as large (e.g., 20 or 200) as an estimate.

Psychophysical scaling in general examines the relationship between the magnitude of sensation and the intensity of a stimulus, such as brightness, loudness, or acceleration (Gescheider, 1997). Across a variety of stimuli, Stevens (1957) discussed that the relationship is written as a power function in the form of

$$\psi = k\varphi^a,$$

where ψ = sensation magnitude, φ = stimulus intensity, k = a constant, and a = an exponent. In the case of the cognition of environmental distance, ψ is a cognitive distance and φ is a physical distance. The value of the exponent, which depends on the type of stimulus or sensory modality, determines the shape of a graph for the relationship between the two magnitudes. An exponent of $a = 1$ shows a linear relationship, such that sensation magnitude is proportional to stimulus intensity (falls along a straight line). When $a > 1$, the relationship is positively accelerated, and when $0 < a < 1$, the relationship is negatively accelerated (Figure 4.18). For example, for the estimation of visual length, a value of a close to 1 is typically observed, indicating that apparent length is proportional to actual length. An exponent greater than 1 has been observed for heaviness, angular acceleration,

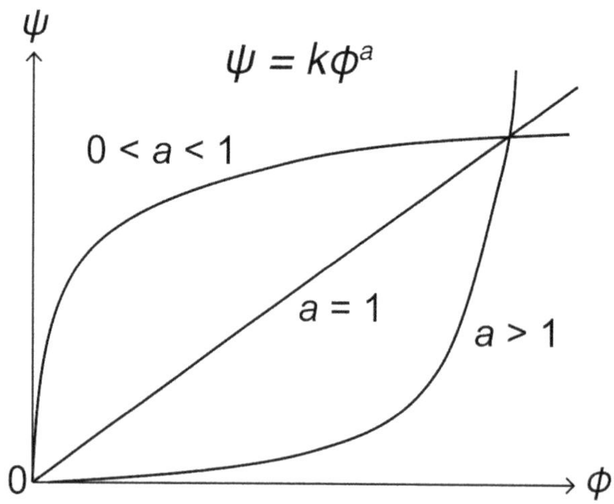

Figure 4.18 **Stevens's power law, $\psi = k\varphi^a$, for the relationship between sensation magnitude and stimulus intensity. An exponent of $a = 1$ shows a linear relationship, such that sensation magnitude is proportional to stimulus intensity. When $a > 1$, the relationship is positively accelerated, and when $0 < a < 1$, the relationship is negatively accelerated.**

or electric shock, showing a rapid increase in sensation magnitude with an increase in stimulus intensity, while an exponent less than 1 has been observed for loudness, brightness, or vibration (Gescheider, 1997, p. 303). For the conception of distance in the broad sense of the term, a negatively accelerated function has generally been found. For example, the difference between 10 m and 11 m or $10 and $11 is perceived as larger than the difference between 100 m and 101 m or $100 and $101. That is, the increase in perceived distance becomes smaller as physical values become greater (e.g., Liberman, Trope, and Stephan, 2007; Zauberman et al., 2009).

As a method of analyzing distance estimates, multidimensional scaling (MDS) produces a cognitive configuration that underlies the estimates of distances between pairs of places. MDS attempts to find a two-dimensional (or higher) configuration that fits a matrix of interpoint distances. For instance, suppose that you locate four places on a cartographic map (A–D in Figure 4.19). You can obtain the distances between the four places by measuring the lengths of corresponding line segments on the map with a ruler and multiplying them by the reciprocal of the representative fraction (e.g., by the value of 25,000 for a map with a scale of 1/25,000). You thus come up with an interpoint distance matrix in a straightforward manner (shown in Figure 4.19). MDS goes the other way around—it starts with an interpoint distance matrix and finds a two-dimensional configuration that maximally fits it. Usually in the study of cognitive maps, the distance matrix contains participants' estimates of straight-line distances between pairs of places (in this case, a 4 × 4 matrix of cognitive distances). To find a planar configuration whose interpoint distances match the distances in the input distance matrix, MDS basically starts with a guess of the coordinates of four points. It measures interpoint distances on the output configuration and compares them with the corresponding distances in the

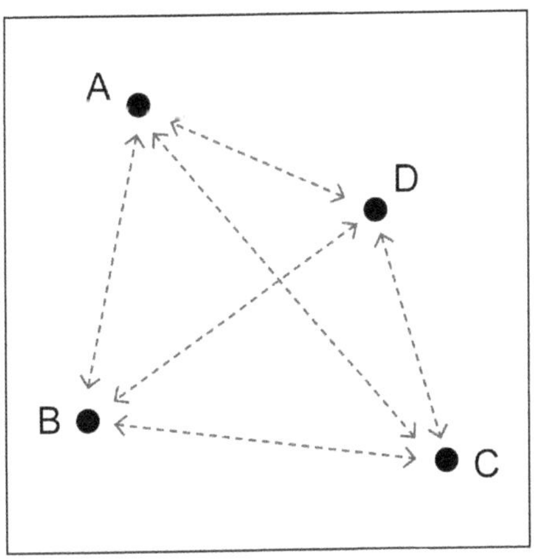

Map

	A	B	C	D
A	0			
B	3.3	0		
C	5.0	3.8	0	
D	2.8	3.7	2.8	0

Distance Matrix

Figure 4.19 **Multidimensional scaling (MDS) and interpoint distances. You can obtain the distances between places on a cartographic map (A–D in this figure) by measuring the lengths of interpoint distances on it (*left*). MDS goes the other way around—it starts with an interpoint distance matrix (*right*) and finds a two-dimensional configuration that maximally fits it.**

input matrix. It iteratively modifies the configuration to make the difference smaller, and stops when the difference becomes lower than a predetermined threshold value. (There is not a predetermined number of dimensions in an MDS analysis. But in this case, descriptions are confined to two dimensions because we are interested in configurations of places on a plane, or on a map.) The difference between input and output distance matrices is called *stress*, and is used as an index of the degree of discrepancy between the two sets of distances. A smaller stress value indicates a better fit of a two-dimensional configuration that approximates the input distance matrix. Figure 4.20 shows two

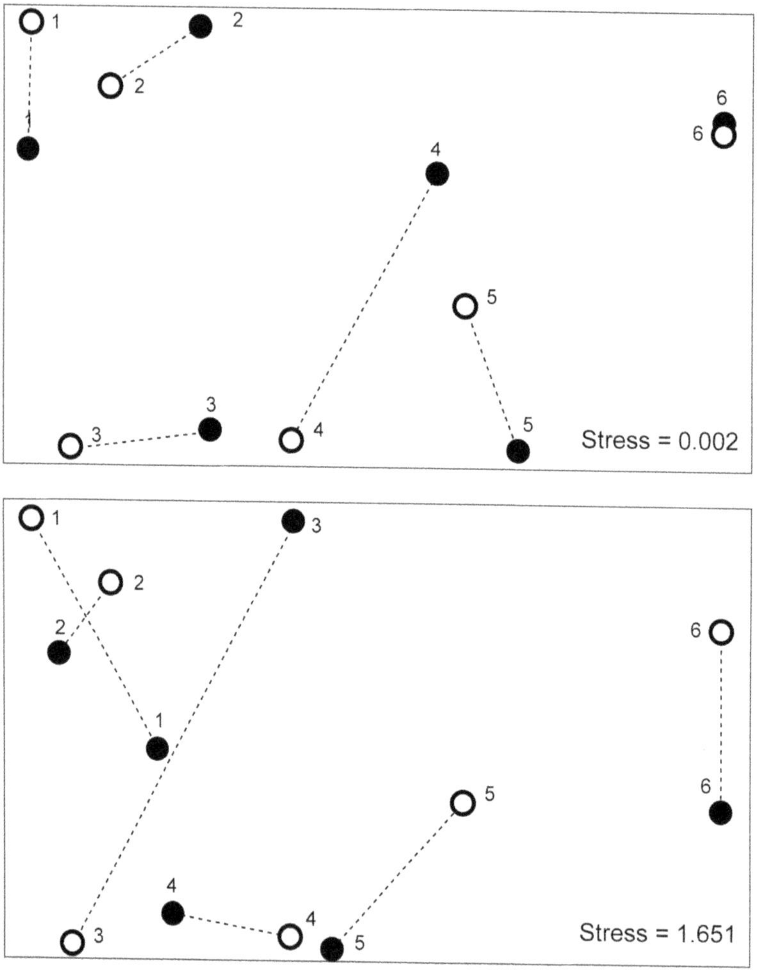

Figure 4.20 **Two configurations generated by MDS for two people's distance estimates (open circles represent actual locations and solid circles estimated locations). The configuration on the top has a smaller stress value and matches the actual locations of places more closely than does the configuration on the bottom (stress values of 0.002 and 1.651, respectively).**

configurations generated by MDS for two people's distance estimates. The configuration on the top has a smaller stress value and matches the actual locations of places more closely than does the configuration on the bottom.

Finally, we look at the method of map drawing. It has been extensively used, particularly since Lynch (1960) systematically used it in his analysis of people's environmental images. For example, Appleyard (1970) discussed local residents' knowledge of surrounding environments through an examination of their sketch maps. Appleyard asked groups of inhabitants in Ciudad Guayana (a city in Eastern Venezuela) to draw maps of the city. He then examined the structures of the drawn maps. He classified the drawn maps on the basis of the types of elements predominantly depicted into two categories: sequential maps and spatial maps. Within each category, Appleyard identified four subtypes corresponding to their levels of sophistication. Within the *sequential* type of maps, he identified a gradation from the most primitive-looking ones that contained fragments of sequences, through maps that depicted chains, branches, and loops, to more detailed and accurate network maps. Within the *spatial* type of maps, he identified a progression from scattered and cluttered maps of dots, points, or names, through maps that were mosaic or linked in form, to more sophisticated and assured representations of patterned maps (p. 103).

Examples of the eight types of sketch maps identified by Appleyard (1970) are shown in Figure 4.21. In the progression of sequential maps, individual elements were first not connected with each other (fragmented), and then gradually became connected so that they formed a route and represented changes in directions (chains) or the shapes of straight lines and circles (branches and loops), and were finally shown as complete and accurate road systems (networks). On sequential maps, spatial layouts, such as networks of roads or rivers, were rarely shown. In the progression of spatial maps, individual elements, such as buildings, were depicted at more or less correct positions as groups, but not in accurate spatial relations with each other first (scattered and cluttered). The groups of elements gradually became roughly interrelated in terms of boundaries (mosaic) and road connections (linked), and finally were configured accurately in relation to each other (patterned). Pocock (1976) examined sketch maps drawn by residents in the city of Durham, and similarly identified sequential and spatial types of maps according to the types of elements depicted and the degree of accuracy of spatial orientation.

The accuracy of sketch maps may be quantified by counting the numbers of (a) landmarks and paths depicted on the maps, (b) landmarks sequenced correctly, and (c) turns specified correctly in terms of left or right. Sketch maps may also be analyzed in terms of overall metric accuracy of relative spatial relations through bidimensional regression (Friedman and Kohler, 2003; Tobler, 1965, 1994; see Chapter 5), which is an extension of ordinary regression analysis to two dimensions. In the analysis, a sketch map and a correct map are overlaid so that a set of anchor points (e.g., landmarks, turns, or decision points) on the two maps matched as closely as possible, after translation, scaling, and rotation. The degree of correspondence between the two maps is given as a bidimensional correlation coefficient, ranging in value from 0 to 1 (the larger, the better correspondence).

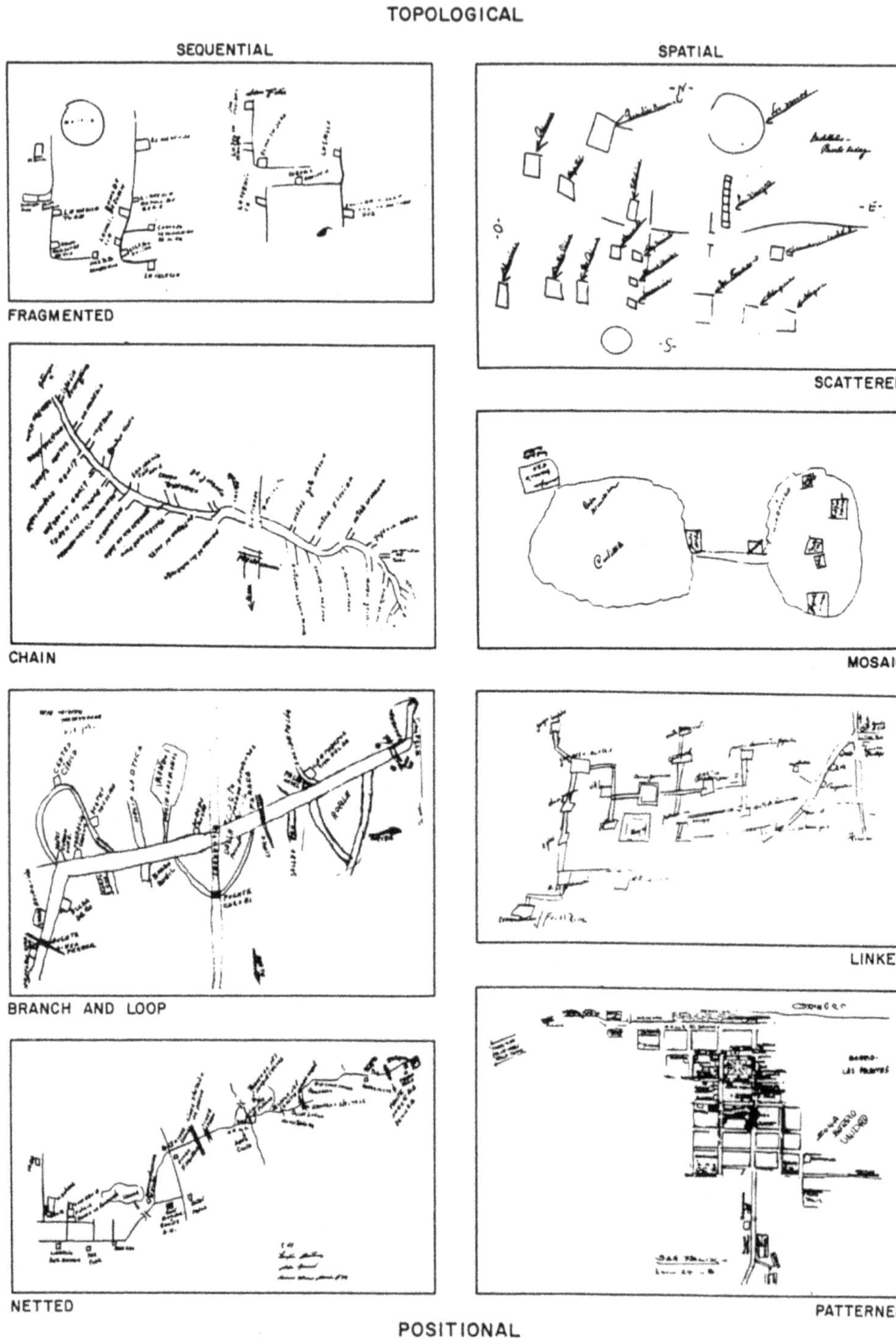

Figure 4.21 Sequential and spatial types of sketch maps. In the progression of sequential maps, fragmented elements became connected as chains, branches, or loops, and were finally represented as complete road networks. In the progression of spatial maps, individual elements were first scattered and cluttered, and gradually became roughly interrelated as a mosaic or links, and finally were configured accurately in relation to each other or patterned. Reprinted from Appleyard, 1970, pp. 105 and 106.

The geometry of cognitive maps

Researchers in spatial cognition are interested in the structures and processes of internal representations of the environment, and thus they systematically observe and examine overt behavior. In other words, they strive for valid and reliable methods of externalizing people's spatial knowledge, so that they can provide objectively grounded discussions about cognitive maps and cognitive mapping. To clarify the research objective along these lines and the distinction between internal and external representations, Liben (1981) used the term *spatial products*: the results of externalization, or responses to systematically designed spatial tasks, provide insights about the structures and processes of internal representations. A good deal of research has been accumulated in the literature through analysis of various kinds of spatial products, as partly seen in this chapter. Some of the research findings, interestingly, pointed out inconsistencies in regard to the geometry of cognitive maps. That is, past research provided findings that are at odds with the (implicit) assumption that internal representations of the environment follow Euclidean geometry, or cognitive maps are maplike.

Now look at the properties that define a space that is called *metric* (Cullinane, 2011). In metric spaces, the quantity of distance (denoted by d) between any two points (m and n) can be measured systematically and consistently, so that it satisfies the following four axioms:

(i) $d(m, n) = 0$ if and only if m and n coincide;
(ii) $d(m, n) \geq 0$;
(iii) $d(m, n) = d(n, m)$; and
(iv) $d(m, n) + d(n, o) \geq d(m, o)$ for another point o in space.

You probably do not find these axioms difficult to figure out in reference to your knowledge of Euclidean geometry in high-school math (Euclidean space is an instance of metric space; see Figure 4.22). A point on a plane does not give distance (the first axiom, nondegeneracy) and any distance should not be negative (the second axiom, nonnegativity). Distance does not depend on the direction of measurement; that is, the distance from point m to point n is equal to the distance from point n to point m (the third axiom, symmetry). And finally, for any triangle on a plane, the sum of the lengths of two sides is greater than or equal to the length of the remaining side (when they are equal, they form a straight line; the fourth axiom, triangle inequality).

Empirical observations have shown, however, that people's spatial judgments do not satisfy the metric axioms. In particular, the axiom of symmetry often does not hold in people's estimates of environmental distances. The findings by Cadwallader (1976), Golledge et al. (1969), and Lee (1970) discussed in Chapter 3 point to the asymmetry of cognitive distance. Distances toward the city center were estimated to be relatively shorter than distances away from the city center in Lee's study, and vice versa in Golledge et al.'s study, even though they were of the same physical length.

Burroughs and Sadalla (1979) discussed the asymmetry in distance judgments in relation to the cognitive salience of reference points. Places and buildings that serve as landmarks in a region are cognitively salient, in the sense that they are salient enough to serve as reference points such that the locations of other places and buildings are identified

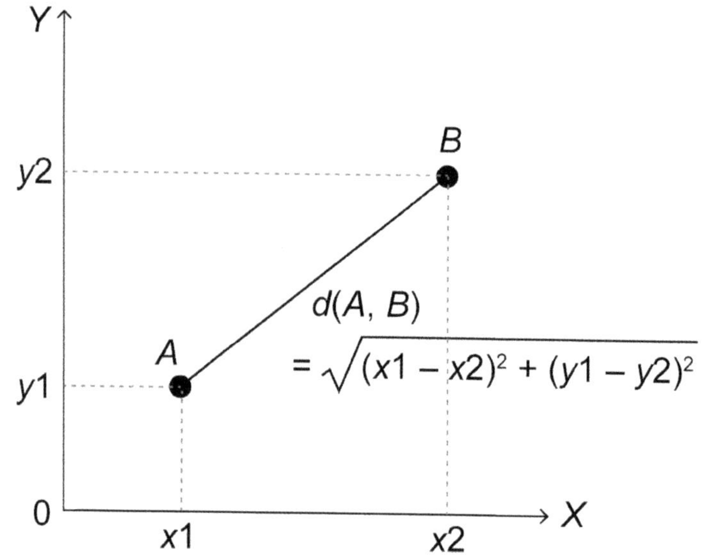

Figure 4.22 **In Euclidean space (an instance of metric space), the distance between two point is given by the square root of the sum of squared differences between the coordinates.**

in reference to them (e.g., "your lodge is *next to the church*" or "you will see my house *when you pass the gas station*"). Burroughs and Sadalla presented participants with pairs of buildings on a university campus or buildings in the city of Phoenix, Arizona. The pairs either differed in the degree of cognitive salience (one was higher than the other, or one was used as a reference point but the other was not) or not (neither of them was cognitively salient). Participants were then given sentence frames of "(location A) is close to (location B)" and "(location A) is essentially next to (location B)," and asked to judge for each pair which building fits into the first position (A) and which fits into the second position (B). In these sentence frames, the second, rear position serves as a reference location. Burroughs and Sadalla found that buildings that participants placed in the second position tended to be high in salience (i.e., used as reference points in the region).

Burroughs and Sadalla (1979) further asked participants to judge distances for the pairs of buildings. On a response sheet, the name of one building was printed at the origin of a semicircular grid, and participants indicated where the other building of the pair was to be located on the grid. When neither of the pair was a reference point, judged distances did not differ depending on which building of the pair was set at the origin. For pairs of buildings that differed in the degree of cognitive salience, judged distances were longer when the more salient building (i.e., a reference point) was set at the origin than when the less salient building (i.e., a non-reference point) was at the origin. Burroughs and Sadalla summarized the findings as: "nonreference points are judged nearer to reference points than are reference points to nonreference points" (p. 420). This shows that cognitive salience affects distance judgments and leads to the asymmetry in cognitive distance.

The discussions so far concern violations of metric properties in people's spatial knowledge. There is also research evidence for violations of Euclidean properties (properties that are to hold in Euclidean space) in people's spatial judgments. Think about three points that are not lying on the same line: they define a triangle; and in a Euclidean space the sum of the three angles equals 180°. In cognitive space, however, the sum of angles judged by people between three locations that form a triangle does not equal 180°. Moar and Bower (1983) asked students in Cambridge, England, to estimate directions for triads of locations at road intersections that formed a triangle (Figure 4.23). For example, participants were instructed to point "from the junction of St. Andrew's Street and Downing Street facing the direction in which Downing Street leaves the junction to the junction of Pembroke Street and Trumpington Street" (p. 109). Participants estimated three directions for a triad, for a total of three triads. Moar and Bower found that the sum of mean estimates of the three angles for each triangle was 208°, 249°, and 220°, all of which were significantly greater than 180°.

Non-Euclidean properties of spatial knowledge have also been revealed in the analysis of estimates of interpoint distances in multidimensional scaling. Golledge and Hubert (1982) examined distance estimates between 48 places in the city of Columbus, Ohio, by its residents, to address the question of how different types of metric spaces approximate the geometry of internally represented environments. That is, Golledge and Hubert were interested in the degree to which different distance metrics (or distance functions) fit people's estimates of distances: they compared MDS configurations based on Euclidean, city-block, and dominance metrics, which define metric distance between points (x_1, y_1) and (x_2, y_2) as

$$d = \left(\left|x_1 - x_2\right|^p + \left|y_1 - y_2\right|^p\right)^{1/p},$$

where $p = 1, 2$, and ∞, respectively (see Figure 4.22). Golledge and Hubert found that distance estimates by 64 percent of the participants were best fit by a MDS configuration with the Euclidean metric, 18 percent with the city-block metric, and 18 percent with the dominance metric. Golledge and Hubert also examined a rank order correlation between the interpoint distances measured on each of the three MDS configurations and correct distances. For 19 percent of the participants, the highest correlation was observed for MDS with the Euclidean metric; for 49 percent, the city-block metric; and for 32 percent, the dominance metric. In light of these results, Golledge and Hubert discussed that "it would seem appropriate to speculate on some of the different forms of geometry that may be suitable for investigating the structure of cognitive representations" (p. 110).

These empirical studies reveal inconsistencies in people's spatial knowledge, with respect to the violations of assumptions of metric space and Euclidean space for internal representations of the environment (i.e., asymmetry, the sum of three angles of a triangle greater than 180°, and a better fit of MDS configurations with non-Euclidean metrics). These results point to the inadequacy of assuming an isotropic space for internal representations, and some researchers proposed other metaphors than cognitive maps that reflect the nonmetric or non-Euclidean nature of spatial knowledge. For example, Tversky (1993) proposed a metaphor of "cognitive collages." Tversky pointed out that spatial knowledge

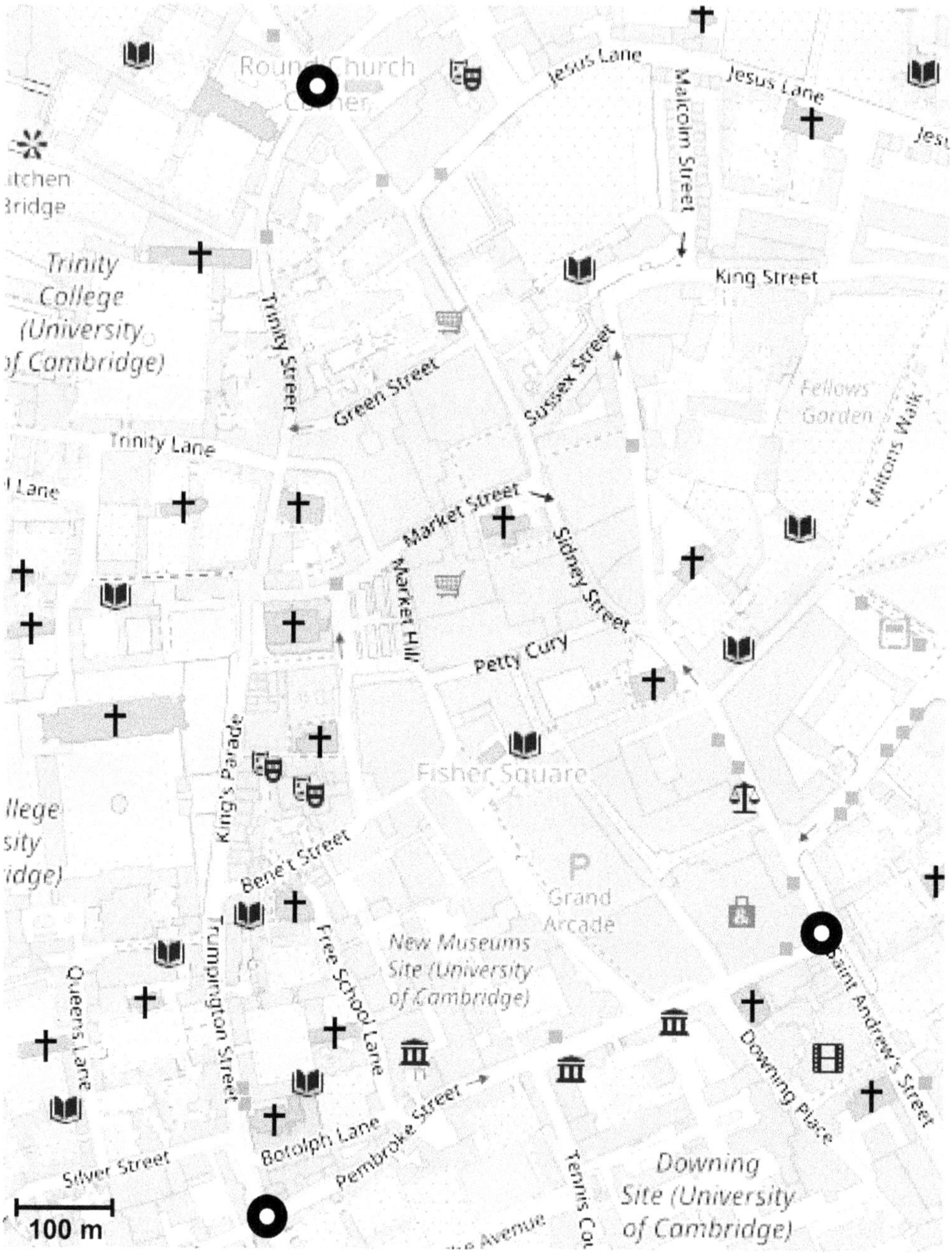

Figure 4.23 **Students in Cambridge, England, estimated directions for triads of locations at road intersections that formed a triangle (denoted by circles). The students' direction estimates were found to violate a Euclidean property. The sum of mean estimates of the three angles was significantly greater than 180°. © OpenStreetMap contributors.**

tends to be systematically distorted and hence it is unlikely that information on internal representations is organized into a single, coherent maplike structure:

> Rather than resembling maps, people's internal representations seem to be more like *collages*. Collages are thematic overlays of multimedia from different points of view. They lack the coherence of maps, but do contain figures, partial information, and differing perspectives.
>
> (p. 15)

Likewise, Hirtle (1998) proposed a metaphor of a "cognitive atlas," to incorporate the idea that spatial knowledge is represented in various formats, including spatial, visual, and propositional, that differ in scale and resolution, and oftentimes contains errors. Hirtle also suggested geographic information systems (GIS) as a possible metaphor for spatial memory, by referring to the varieties of spatial-temporal operations that GIS apply to geographic data, such as overlay, network analysis, distance and direction estimation, zooming in and out, or projection and coordinate transformations (see Golledge and Stimson, 1997, p. 236). To appreciate the metaphor "cognitive GIS," it may be of interest to compare a definition of GIS as "automated systems for the capture, storage, retrieval, analysis, and display of spatial data" (Clarke, 2001, p. 3) with the definition of cognitive mapping that we saw above: "a series of psychological transformations by which an individual acquires, codes, stores, recalls, and decodes information" (Downs and Stea, 1973, p. 9). Basically, humans do what computers that are specifically designed to deal with spatial information do (we will see how accurately people can do that in the next chapter).

Montello (1992) cautioned, on the other hand, that these observations of inconsistencies do not necessarily indicate that spatial knowledge is non-Euclidean or nonmetric. Importantly, Montello argued that a distinction should be made between knowledge representations activated during the processing of information when conducting spatial tasks and knowledge representations that are stored in memory (technically, representations activated in working memory and stored in long-term memory). Montello stated that environmental spatial knowledge should not fit any single metric unless it comes from a unitary representation. That is, if one uses one representation to estimate the distance from A to B and a different representation to estimate the distance from B to A, then the estimates could be asymmetric. In such a case, the person does not experience the asymmetric estimates simultaneously, and it is possible that the two representations are mutually inconsistent but internally metric (p. 144). Montello argued that behavioral data based on multiple representations that are not simultaneously activated in working memory would reasonably exhibit results that violate metric axioms. It does not necessarily imply that "cognitive space is non-metric or non-Euclidean, only that there is no such omnibus 'space' in memory" (p. 146).

These statements clarify that empirical data from spatial tasks (e.g., distance and direction estimates and sketch maps) are products of externalization processes (Liben's, 1981, spatial products), and that human memory entails an interplay of active processing and long-term storage (Figure 4.24). With such a caveat, research has shown the limitation of discussing the structures and processes of spatial knowledge with a unitary concept of cognitive maps. As we will see in the next chapter, studies of the development of spatial knowledge in a new environment provide an important alternative view that acknowledges the existence and extent of individual differences in spatial cognition.

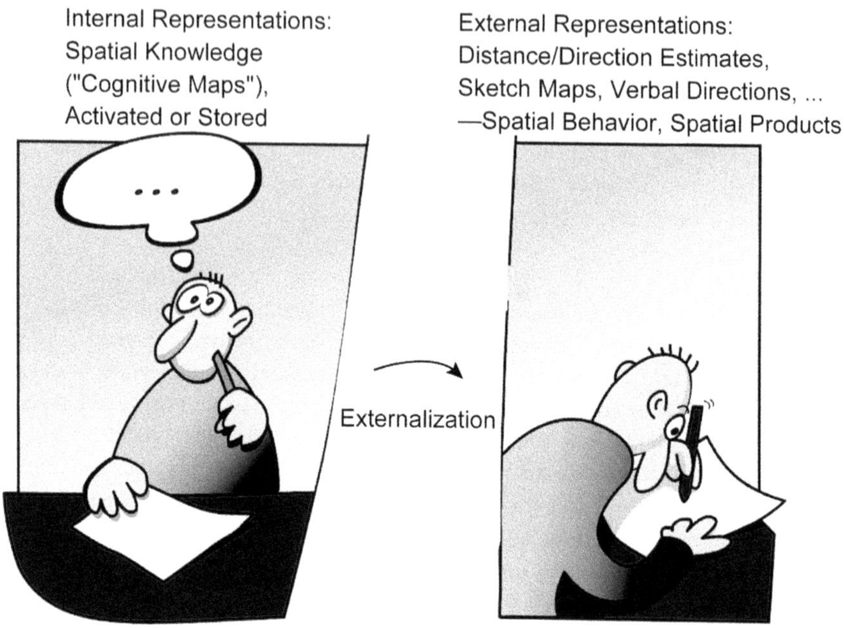

Figure 4.24 **Internal representations and external representations of spatial knowledge. Observed spatial behavior (e.g., distance and direction estimates and sketch maps) is a product of externalization processes, and taps into spatial memory that is stored in long-term memory or activated in working memory. Pictures from https://pixabay.com/**

Categorical and numerical coding of spatial locations

Distortions in spatial memory have been examined through a Bayesian model-based approach that posits stimulus encoding at two levels of detail: as members of a category and fine-grained numeric values (Duffy et al., 2010). For instance, we often remember that we left a key on a kitchen table or parked a car in a parking lot, but not exactly where on the table or in the parking lot. Thus the memory of locations of objects in space typically contains uncertainty. To cope with uncertainty and minimize potential errors in spatial judgments, one estimates spatial locations by strategically combining information about categorical membership (on the table or in the parking lot) and numeric values (6 inches from the right edge of the table or 30 m away from the front gate). The weight given to the categorical and numeric information varies depending on the variability of each information, or the degree of uncertainty: the greater the degree of uncertainty in numeric information, the more heavily categorical information is weighted. Importantly, uncertainty about an exact location leads to the adjustment of a numeric value by a categorical judgment, which in turn causes a systematic and predictable bias in the estimates of spatial locations. Namely, people's spatial memory is biased toward the location of a category prototype (Newcombe, 2018). The prototype may be a central point in a categorical region: for example, people's memory of dots located in a circle is biased toward the centroid of each quadrant of a circle divided by horizontal and vertical lines (Huttenlocher, Hedges, and Duncan, 1991). In another case, the prototype is a category's running mean, as revealed by people's estimates of a series of visually presented line lengths from memory (Duffy et al., 2010).

The categorical adjustment of estimates of spatial locations has been demonstrated for objects in 3-D environments as well. Holden, Newcombe, and Shipley (2013) asked 22 undergraduate students to view and memorize point locations marked by a laser pointer on natural and human-made objects in a campus garden (e.g., a plant pot, a tree stump, or a column). After a short delay (retention intervals of 10–20 seconds), they estimated the memorized locations by indicating them on the corresponding objects by themselves. Holden et al. found that participants' responses tended to be biased toward the center of mass of the 2-D shape surrounding a target location. For example on the surface of a tree stump, participants estimated the locations of previously viewed point locations to be shifted toward the center of the surface: the central point is thus considered a category prototype, toward which their responses are biased. Furthermore, the degree of bias toward the category prototype (i.e., the magnitude of error toward the center of mass) increased linearly as the time length of a retention interval increased. This is another indication of the influence of the degree of uncertainty on the weight given to category information: categorical information was more heavily weighted as the memory of numeric values (or metric locations) became degraded with the passage of time.

In the aforementioned study, categories were defined perceptually on the basis of physical features of objects (see also Friedman, Montello, and Burte, 2012), but categories can also be defined conceptually. Holden et al. (2016) asked experts in structural geology (n = 9), organic chemistry (n = 11), and English literature (n = 9) to conduct a location memory task as in the Holden et al. (2013) study, using images of landscapes that contained geological structures (e.g., a geologic fault or a fold). In some images, expert geologists identified larger categories than did novice students; in other images, expert geologists identified smaller categories than did novice students. An example of an expert category is an outcrop surface of a folded structure (knowledge based), and an example of a novice category is an area on the outcrop surface bounded by visually distinctive lines, such as white lines in the background of dark soils (perceptually defined). Estimates by expert geologists were biased toward the center of geologically defined categories when they were smaller than novice categories; but when geologically defined categories were larger than novice categories, estimates by expert geologists were biased toward the center of novice categories. Estimates by experts in other disciplines (organic chemistry and English literature), on the other hand, were biased toward the center of perceptually defined, novice categories all the time. That is, expertise in structural geology enabled a flexible use of conceptual and perceptual (visually defined) categories, whichever constrains potential errors of locational judgments to a greater degree (a category that has the smaller spatial extent, or offers less uncertainty).

Review box for Chapter 4

1. Discuss the characteristics of mental representations, particularly when they are considered from the perspective of mental rotation.
2. Give a few examples of situations in which your behavior is guided by your existing knowledge, or schemata.
3. Find examples of image schemata (or spatialization) in your daily cognition and behavior.
4. Discuss the hierarchical structure of spatial knowledge. Try the "San Diego–Reno" question with some of your friends and see how they respond.
5. Compare spatial knowledge acquired from maps with knowledge from direct navigation in the environment.
6. Ask some of your friends to do the imagined rotation and translation tasks and see which they find more difficult.
7. Summarize the methods of externalizing people's spatial knowledge. How do you analyze data you obtain from these methods?
8. Ask a couple of people to estimate environmental distances with the method of magnitude estimation. Analyze their estimates using Stevens's power law. Did you find an exponent greater than or less than 1? What does that mean?
9. If you are interested, ask people to estimate distances between pairs of places and analyze their estimated distance matrices with MDS (you can use, for example, the programming language R to do that). MDS will output two-dimensional configurations of the places in the respondents' minds, or visualize maps in their heads.
10. Discuss major research findings about the geometry of spatial knowledge and your thoughts about the metaphor "cognitive maps."
11. Discuss the characteristics of categorical and numerical coding of spatial locations.

Chapter 5

The acquisition of cognitive maps

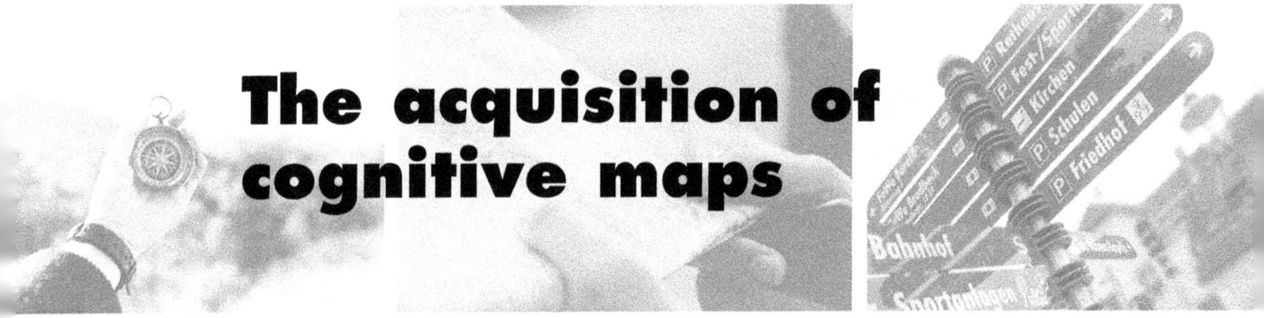

The concept of development

What do you think when you hear the term "development"? Suppose that you make a weekly report to your boss. If the work of your responsibility is coming close to completion, you say, "I'm making progress"; if several subgoals are set toward the completion, you may say, "I'm halfway through it." Or consider a child who is learning to ride a bike. She may first balance on the bike; then ride down a hill supported by her father, apply the brakes, steer left and right, pedal; and finally ride freely without support. Or consider a child who is learning the concept of "animal." He at first calls all furry, four-legged animals dogs; he then differentiates between dogs, cats, and raccoons; or even between breeds of dogs, Beagles, Collies, Labradors, and Shibas. Or think about your ancestors; not your grandfathers or grandmothers, but ancestors of the human species who appeared on the earth two to three million years ago. They have since evolved through natural selection and adaptation, walking on two legs, learning to use tools and hunt, acquiring facilities with language, and discovering new lands; humans and animals that were adaptive and able to live and reproduce in their habitats survived. Or think about a person who enters a room that appears the same as it used to be, but, inside, the wallpaper has been changed and the furniture rearranged. He may first think he has entered the wrong room, but gradually he recognizes things; for example, pictures on the wall or a familiar view from the window. He then gets oriented (or reoriented) in the room with respect to these familiar objects and views, and figures out which pieces of furniture have been moved.

These all exemplify a process of becoming better, acquiring a new skill, or heading to completion. The people in the examples had little or no knowledge, skills, or experience to start with and attained them over the course of time. These examples are of different kinds, and are referred to by different names, such as progress, learning, adaptation, or development. They also differ in time scale; some in the order of minutes (in the case of an altered room), days (riding a bike), months (working on a project), years (concept formation), or millions of years (human evolution). Such multiplicity of examples aside, there is a main theme in the use of the term "development" in the research into human cognition—changes toward the better or completion through qualitative shifts in distinctive stages. It is differentiated from growth or mere accretion. Hart and Moore (1973) contrasted learning and development, discussing that learning "involves *quantitative* changes in the reception and retention of information or subject matter," whereas development "implies *qualitative* changes in the organization of behavior" (p. 250, emphasis added). Werner (1957) argued that the study of development postulates a regulative principle (which he called an orthogenetic principle), which posits "that wherever development occurs it proceeds from a state of relative globality and lack of differentiation to a state of increasing differentiation, articulation, and hierarchic integration" (p. 126). Werner further discussed that the study focuses on understanding the developmental processes that are composed of successive stages; proceeding from a stage in which organisms are dominated by immediate concrete situations to a stage in which they are less context- and stimulus-bound (p. 128). Importantly, the sequential order of the developmental stages is invariant, a subsequent stage encompassing the properties attained in previous ones.

The idea of development as a series of stages of increasing levels of differentiation and organization characterizes Piaget's developmental theory, particularly in the context of the development of spatial concepts in children. Piaget distinguishes two aspects of cognitive functions: *figurative* and *operative*. In figurative cognition, the self and the object are not differentiated, and knowledge of space is closely tied to concrete sensorimotor actions and perceptions. In operative cognition, by contrast, actions become coordinated and internalized, which enables the construction of symbolic internal representations of space and abstract operations on the representations. Hart and Moore (1973) called spaces conceived by these two kinds of cognitive functions *space-of-action* and *space-of-contemplation*, respectively. The basic tenet in Piaget's developmental theory (as seen in Chapter 2) is thus a qualitative transition through the stages of sensorimotor, preoperational, concrete operational, and formal operational periods; the transition heads toward increasing levels of abstraction and complexity, and toward the achievement of internalized representations, operations, and transformations that are logical, systematic, and hypothetical-deductive in nature. As infants grow and mature, their concrete and simple ways of thinking become more abstract and complex. Eventually they are able to think logically and systematically, and engage in hypothetical and probabilistic reasoning to attain a rich understanding of the world.

Ontogenesis and microgenesis

In discussing development, particularly in the context of spatial knowledge acquisition, it is important to distinguish between ontogenetic development and microgenetic development. *Ontogenesis* is the process of knowledge development across the lifespan. *Microgenesis* is the process of knowledge development over a comparatively shorter period of time in a new environment. For instance, developmental changes in the conception of space from infancy to adolescence exemplify an ontogenetic issue that researchers have extensively studied (Hart and Moore, 1973; Piaget and Inhelder, 1948/1967). Moving to a new place and learning about the neighborhood over time constitutes an example of microgenetic knowledge development (Montello, 1998; Siegel and White, 1975).

There are other types of development, such as phylogenesis and ethnogenesis. The different domains of development have often been discussed in analogy to one another, as suggested by the phrase "ontogeny recapitulates phylogeny." Likewise, spatial microgenesis

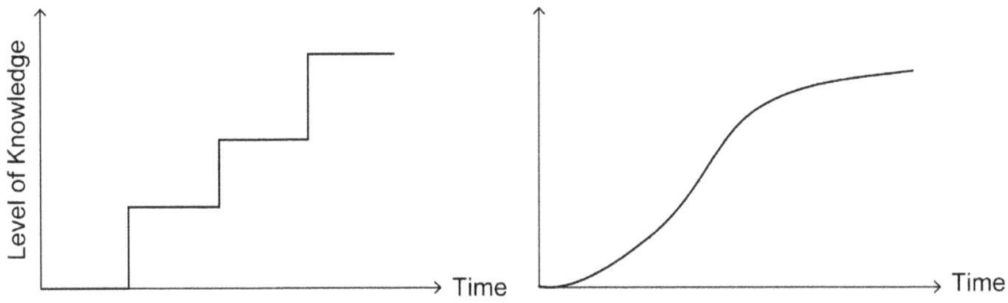

Figure 5.1 **Stage and continuous theories of development. A stage theory posits that knowledge develops through distinct stages of qualitatively different properties and increasing levels of sophistication (*left*). A continuous theory argues that development is more continuous, with the detail and accuracy of knowledge improving gradually and quantitatively (*right*). Reprinted with permission of Edward Elgar Publishing from Ishikawa, 2018a, p. 123.**

has been discussed in recapitulation or parallelism to ontogenesis. Importantly, and influentially, Siegel and White (1975) discussed spatial microgenesis as a transition from landmark to route to survey knowledge, by drawing a parallel with the ontogenesis of spatial concepts (more below).

Stage and continuous theories

In explaining the process of development in any domain, there are two main types of theories: stage theories and continuous theories. A *stage theory* conceives of development as proceeding through qualitatively different stages, with each stage being characterized by a uniform theme. Subsequent stages encompass the properties that characterize the preceding stages, and the transition from one stage to another is relatively abrupt (as illustrated by Figure 5.1, *left*). As seen above, Piaget's developmental theory of cognitive structures and functions is one of the most famous stage theories of cognitive development (Mandler, 1983; Piaget, 1983).

According to a *continuous theory*, in contrast, development is conceptualized as proceeding gradually and continuously (Figure 5.1, *right*). That is, although a stage theory posits that knowledge develops through distinct stages of qualitatively different properties and increasing levels of sophistication, a continuous theory argues that development is more continuous, with the detail and accuracy of knowledge improving gradually and quantitatively. An influential example is the family of information processing theories of development (e.g., Siegler, 1978). Theories of spatial microgenesis have taken either a stage approach or a continuous approach.

Ontogenesis of spatial knowledge

With respect to the conception of space in children, the basic developmental trend conceived by Piaget from figurative cognition to operative cognition maps onto the qualitative shifts of topological, through projective, to Euclidean spatial concepts. Specifically, according to Piaget and Inhelder (1948/1967), the child's conception of space proceeds from topological to projective to Euclidean spaces. For example, infants supposedly do not distinguish between rectilinear and curvilinear shapes, or between circles, squares, and triangles (topological), but come to distinguish views from different perspectives (projective), and finally achieve as adolescents metric understanding of figures and shapes in terms of angle and distance (Euclidean). The acquisition of these three classes of spatial concepts is examined on the basis of children's responses to three spatial tasks developed by Piaget: the three mountains task, the water level task, and the model landscape task.

The three mountains task assesses children's ability to coordinate different perspectives, namely to understand different points of view and differentiate between their own and another person's views. In the task, children view scale models of three mountains (Figure 5.2) differing in size, height, color (green, brown, or gray), and objects located on top (a little house, a red cross, or snow). They then see a wooden doll placed at different

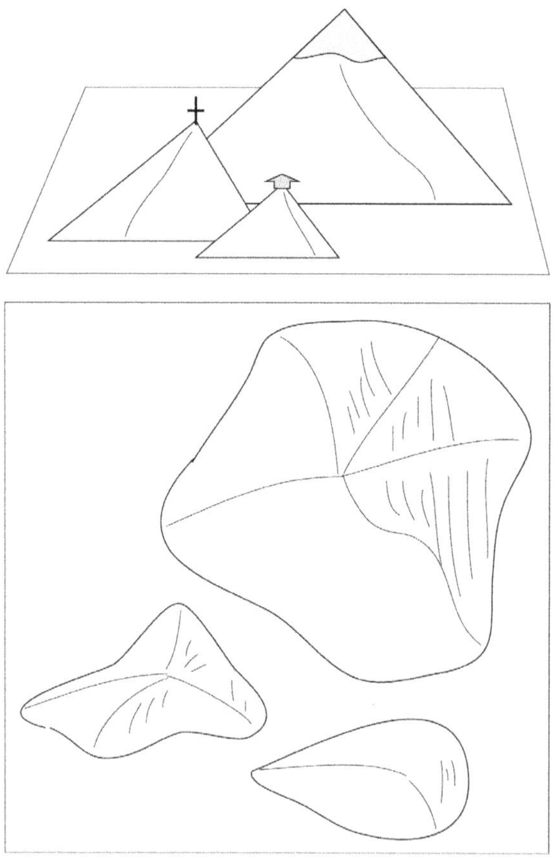

Figure 5.2 **Three mountains that differ in size, height, and color and have different objects located on top (a little house, a red cross, or snow). Children see a wooden doll placed at different locations around the mountains, and are asked to choose among a set of photographs the one that represents the doll's view. Illustration based on Piaget and Inhelder, 1948/1967.**

locations around the mountains, and are asked to choose among a set of photographs the one that represents the doll's view. In a variation of the method, children are presented with a series of photographs and asked to indicate where around the mountains the doll should be placed so that it has the same view shown in the photographs. Responses by children at early developmental stages are *egocentric*: they fail to distinguish between different viewpoints and equate their own view with the doll's view. In the course of development, their responses show "a progressive discrimination and co-ordination of perspectives" and finally achieve "comprehensive co-ordination of viewpoints" (Piaget and Inhelder, 1948/1967, p. 213). The differentiation of perspectives is made possible by the coordination of different viewpoints in a global framework (p. 243), which is the basic prerequisite for the comprehension of projective relations, beyond topological relations that consist of a series of sequential perceptions and step-by-step operations (p. 244). Thus,

> a system of projective relations or perspective viewpoints consists essentially of operations which do not merely assemble perceptual data, but [conceptually] co-ordinate it [perceptual data] in terms of reciprocal relationships . . . and link a particular perspective with the universe of possible perspectives.
>
> (p. 244)

The water level task assesses children's ability to use the coordinate systems of the horizontal and vertical axes, which is an important component of the understanding of Euclidean spatial relations. Children view a series of outline drawings of jars tilted at various angles; they assume that each jar is about a quarter filled with water and draw the position of the water inside (Figure 5.3). In the early stages of development, children's responses are tied to the configurations of the jars (e.g., the surface of the water is drawn parallel to the base of the jar); in the course of development, they become able to respond to a frame of reference fixed outside the jar (i.e., the surface of the water remains horizontal with respect to a coordinate system centered on the environment and aligned with the gravitational up–down direction).

The model landscape task assesses the spatial concepts and operations attained in the final stage of development, namely the abstract idea of an external frame of reference and internalized manipulations of objects and spatial relations. Children view two

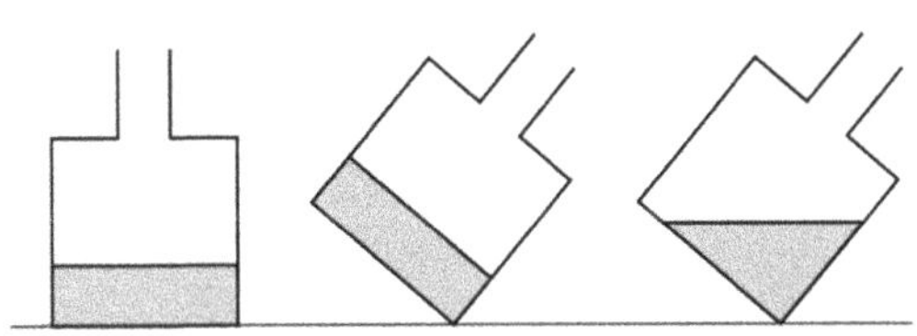

Figure 5.3 **Water level task. Children view a series of outline drawings of jars tilted at various angles. They assume that each jar is about a quarter filled with water and draw the position of the water inside. Children who fail to understand spatial relations in the global horizontal and vertical axes draw a water line parallel to the base of the jar (*middle*) instead of the desk (*right*). Illustration based on Piaget and Inhelder, 1948/1967.**

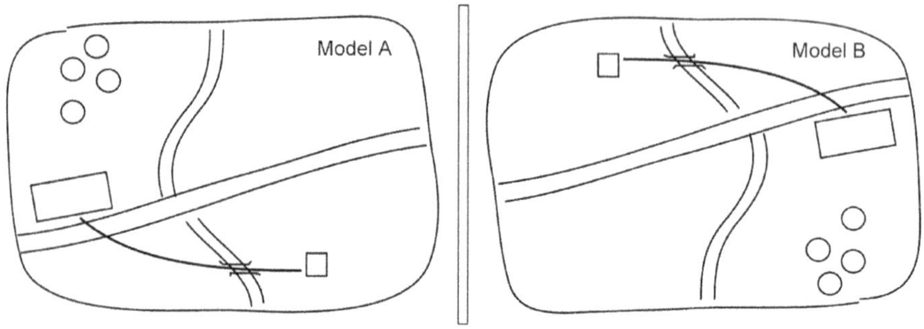

Figure 5.4 **Children view two models of an open country in which objects are arranged to form a unique spatial layout across a partly hilly topography. The two models are identical, but one is rotated 180° with respect to the other; they are separated by a screen so that children cannot see them simultaneously. Children see a small doll placed on Model A, and are asked to place another doll on Model B at the same position as the doll on Model A. Illustration based on Piaget and Inhelder, 1948/1967.**

models of an open country in which houses, roads, a stream, a bridge, and trees are arranged to form a unique spatial layout across a partly hilly topography (Figure 5.4). The two models (Models A and B) are identical, but Model B is rotated 180° with respect to Model A, and separated from Model A by a screen so that children cannot see them simultaneously. A small doll is placed at different positions on Model A successively, and children are asked to place another doll on Model B at the same position as the doll on Model A. Children at an early stage of development place the doll on the basis of what Piaget and Inhelder (1948/1967) called topological relations of proximity and enclosure. If the doll is in a field on Model A, children place a doll in a field on Model B, irrespective of which specific field it is or what objects are found nearby. If the doll is in a stream on Model A, children place a doll along a stream on Model B, anywhere along it. They then begin to take relative positions into account, such as the relations of left–right and before–behind, but still fail to understand the consequence of the 180° rotation of the model. In the course of development, children show progress, through a step-by-step process and on a trial-and-error basis, in the coordination of spatial relations and the determination of the position of a doll in reference to the orthogonal frames of reference in two dimensions (p. 425).

Piaget and Inhelder (1948/1967) discussed that children's responses to these tasks indicate qualitative shifts in the development of spatial concepts in parallel to the stagelike development of intelligence in general:

> This began with elementary sensori-motor and perceptual activity, then evolved into symbolic images, then concrete operations, and eventually abstract or hypothetico-deductive operations. In this respect, the development of layout diagrams . . . illustrates extremely clearly the gradual formation of the whole body of concepts, originally topological, eventually projective and euclidean.
>
> (p. 446)

Microgenesis of spatial knowledge

The foregoing discussions concern *spatial ontogenesis*, that is, the development of spatial concepts in children from infancy to adolescence. An equally important question of spatial cognition concerns the acquisition of spatial knowledge in a new environment, called *spatial microgenesis*. Similar to the parallel drawn between spatial ontogenesis and children's intellectual development, an analogy is made between spatial microgenesis and spatial ontogenesis. An influential framework in this regard is the one proposed by Siegel and White (1975). They stated:

> There are some formal similarities between adult learning and child development deriving from the fact that both children and adults must conform to an essential "Main Sequence" in their organization of spatial representations. . . . Indeed, the literature of developmental psychology is full of instances of developmental parallelism, recapitulations, and regressions.
>
> (p. 31)

On the basis of this basic idea, Siegel and White (1975) described the microgenetic development of spatial knowledge as a "process of going from landmarks, to route-maps, to survey-maps," which accords with the Main Sequence of "going from association to structure, and of deriving simultaneity from successivity" (p. 46). Specifically, their microgenetic framework explains the acquisition of spatial knowledge in a new environment through three distinct stages of landmark, route, and survey knowledge. *Landmark knowledge* concerns discrete objects or scenes that are salient and recognizable when observed. In Siegel and White's terms, landmarks are unique configurations of perceptual events that identify specific geographical locations and constitute strategic foci to and from which a person travels (p. 23). *Route knowledge* consists of sequences of landmarks and associated actions, namely knowledge of how to go from one place to another. Siegel and White described the spatial representation at this stage as sequence knowledge of landmarks connected by routes, saying, "If one knows at the beginning of a 'journey' that one is going to see a particular landmark (or an ordered sequence of landmarks), one *has* a route" (p. 24). Route knowledge is thus sometimes referred to by terms such as "procedural knowledge" (Golledge, 1991) or "topological knowledge" (Allen et al., 1996), terms that emphasize the sequential and nonmetric nature of the knowledge.

Survey (or *configurational*) *knowledge* involves metric and two dimensional understanding of the relationships between separate landmarks and routes. That is, in survey knowledge, the former two types of knowledge, landmarks and routes, are integrated. In such a configurational and metrically scaled representation, landmarks and routes, even those learned in separate travel episodes, are interrelated with each other in a common frame of reference, as if they were seen from a bird's-eye view. Figure 5.5 illustrates a case in which a child gives navigational directions to a driver. The child identifies the bank, theater, school, and park, and guides the driver from the current location to the house through these places, but he does not know the direct spatial relationship between his current location and the house. He has accurate landmark and route knowledge, but not survey knowledge.

Important for Siegel and White's (1975) argument, knowledge in the first two stages, landmark and route knowledge, does not contain metric information—information about distance and direction is not yet present. In route knowledge, as conceived by Siegel and

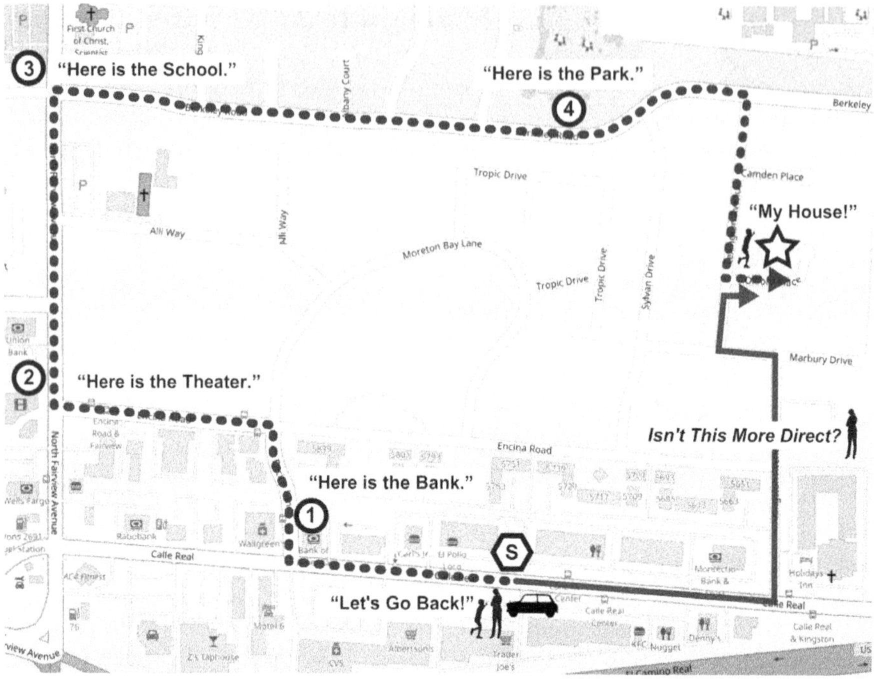

Figure 5.5 **Different types of spatial knowledge. A child successfully guides the driver to his house, in the correct sequence of places along the route. His instructions are unfortunately insufficient, due to the lack of an understanding of the direct spatial relation between his current location and the house. He has accurate knowledge about landmarks and routes, but not configurations. Reprinted with permission of Edward Elgar Publishing from Ishikawa, 2018a, p. 120.**

White, landmarks are organized only sequentially (topologically or projectively) and the space between landmarks is "empty" (p. 29). Siegel and White argued that the empty space may receive metric scaling as experience with the environment accumulates, routes becoming interrelated in a network and represented configurationally (p. 30). In other words, metric knowledge about distances and directions only emerges in the final stage of microgenetic development: survey knowledge.

Development of spatial knowledge: structures and processes

Now we see that knowledge about the environment consists of different types, differing in comprehensiveness and sophistication. How does this knowledge develop? Or how do we acquire knowledge in a novel environment?

Anchor point theory

Golledge (1978) offered a conceptual framework that he called *anchor point theory*, which describes the acquisition of spatial knowledge in an unfamiliar environment as the process of forming and enriching a skeletal representation that is "anchored" by key places. He mentioned:

> The conceptualizations that I wish to put forward here fundamentally accept the idea that learning about an external environment results primarily from interacting with it, and that more complex cognitive representations of physical environments are built up over time. . . . As more ways [to get to and from places] are explored, more places become known; as more places become known, there is a spread effect or spillover in the vicinity of those places.
>
> (p. 78)

Figure 5.6 illustrates the process of "psychological mapping" (Golledge, 1978, p. 78) in spatial knowledge acquisition. When we begin to live in a new environment, the places of home and work, and also places to shop, may first emerge as important places. Similarly, the paths that connect these individual places also have significance for us, as we move between the places along the paths. These places and paths are called *primary nodes* and *links* in our knowledge representation. We become more familiar with the places and paths as the experience of traveling between these places accumulates, and also know about new places and paths around them, for example, new places to shop, favorite places of recreation, and alternative routes to work. With more experience, the areas of familiarity expand as a consequence of our search for retail, recreational, or social-cultural places; and

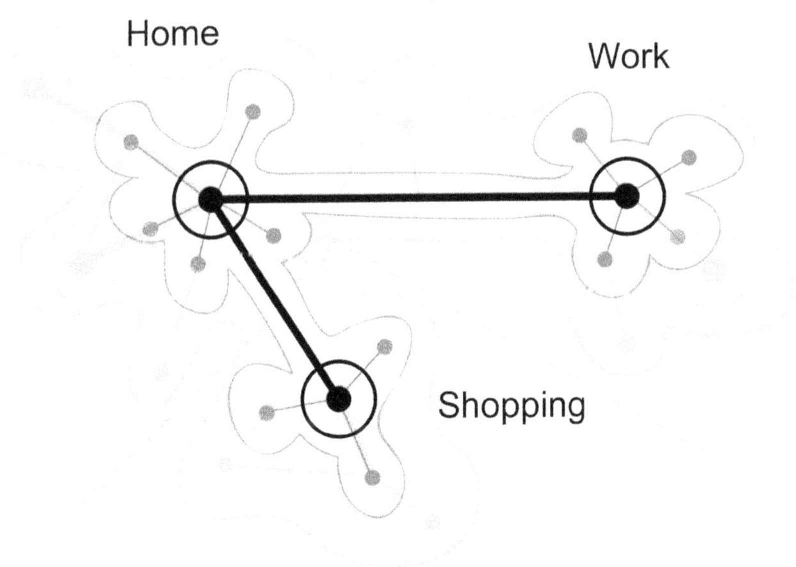

Figure 5.6 **Spatial knowledge development conceived by anchor point theory. Knowledge about places and paths increases with accumulated experience with the environment, forming a hierarchically ordered skeletal structure (filled in over time) according to the level of significance to the person's daily behavioral patterns. Reprinted with permission of Edward Elgar Publishing from Ishikawa, 2018a, p. 121.**

Figure 5.7 **Process of spatial knowledge becoming more accurate with accumulated experience, schematically represented as a distorted grid map. Ideally, people's knowledge becomes less distorted as the length of residence increases (from *left* to *right*). See, however, the discussion of individual differences in spatial knowledge toward the end of this chapter and in Chapter 6. Illustration based on Golledge, Rivizzigno, and Spector, 1976.**

the segmentation of the environment occurs, with the environment being classified into different regions on the basis of their distinctive characteristics (the concept of neighborhood emerges such as "this is my home area" or "the vicinity of my work place").

This framework describes the process in which the knowledge about the places and paths that people frequent and the areas surrounding them increases with accumulated experience and interactions with the environment (Figure 5.7). In the acquired knowledge representation, the places and paths *anchor* the representation as primary, secondary, or tertiary nodes, ordered hierarchically according to the level of significance to the person's daily behavioral patterns.

In relation to the five elements of physical environments discussed by Lynch (1960), any of them can be anchor points. In the learner's mind, they are structured hierarchically, organized as primary nodes and paths, secondary nodes and paths, tertiary nodes and paths, and so on (see Hirtle and Jonides, 1985; Stevens and Coupe, 1978). Golledge's (1978) descriptive theory captures the spatial-temporal dynamics of cognitive mapping and stimulates further discussions about the course of knowledge development, to which we turn below.

Lynch (1960) provided a similar discussion about the developmental changes in the accuracy of a person's spatial knowledge with the accumulation of experience and an increase in familiarity with the environment. Lynch suggested four stages of increasing levels of structural quality of spatial knowledge, with respect to the manner in which different elements of the city are arranged and interrelated in a person's mind (p. 88). Knowledge representations in the first stage lack any structure: various elements are disjointed and freely floating without being interrelated with each other. In the second stage, the individual elements become roughly related by general direction and relative distance, although lacking definitive connections. The representations then, in the third stage, obtain loose and flexible structure. In particular, elements are known as the sequence of events or movements (*routes* in Siegel and White's, 1975, terminology). Paths that are not habitually traveled or directly experienced are still disconnected, though, and their configurations are distorted in the representations. Finally, in the fourth stage, different elements are interconnected firmly and accurately, allowing the person to travel between any places freely, even along new paths. The knowledge representations are now rigidly and densely structured and "take on the characteristics of a total field" (p. 89).

Empirical studies based on sketch maps likewise found an increase in the number of elements and the fine tuning of their locations as a function of the length of residence. Devlin (1976) asked newcomers to the city of Idaho Falls, Idaho, to draw a sketch map of the city on two occasions, (a) two to three weeks after they had arrived in town and (b) about three months after the arrival. Devlin found that initial maps were dominated by the depiction of streets, and that the areas around their residences were drawn in greater detail than other parts of the city. Follow-up maps drawn three months later showed a great increase in the number of streets and landmarks depicted; the areas in the vicinity of their home places did not possess particular prominence in importance any more. Thus participants' sketch maps were filled in over the three-month period, but the relative proportion of streets and landmarks remained similar. Devlin discussed that the sketch maps did not show a big change in the basic structure and the proportion of different elements (streets and landmarks) drawn, but became "more complete, elaborated, extended, and sharpened" (p. 65) with increasing familiarity with the environment.

Beck and Wood (1976) asked American teenagers who came to London to draw sketch maps on the third, fifth, and sixth days of their stay. Sketch maps drawn on the third day primarily showed the regions around their places of residence in more detail and in larger geographic extent than other regions. Sketch maps on the fifth and sixth days depicted the primary regions that appeared on previous maps in further detail and with greater differentiation, and elaborated more connections between places and regions, both old and new ones.

Likewise, Evans, Marrero, and Butler (1981) asked students in two cities (Irvine, California, and Bordeaux, France) to draw sketch maps within the first two weeks of arriving in each new setting and ten months later. Evans et al. observed that the connections of paths between landmarks that served "as initial anchor points" (p. 100) in the students' knowledge representations were elaborated over the ten-month period. They basically found the filling-in and fine-tuning processes that Devlin (1976) and Beck and Wood (1976) observed: "individuals initially comprehend the relative positions of items in space, but fine tune the exact location of items in space with increasing experience" (p. 101).

Stagelike framework of spatial microgenesis

Due to the prevalence of the principle of formal similarities between different kinds of development and the significance of Piaget's theory of intelligence and spatial concepts in developmental psychology, the stagelike framework of spatial microgenesis put forth by Siegel and White (1975) was, and still is, influential in the literature of spatial cognition (Montello, 1998, called it the "dominant framework"). Its influence is revealed, as mentioned above, in the emphasis of the sequential and nonmetric nature of route knowledge, when researchers refer to it as procedural knowledge (Golledge, 1991) or topological knowledge (Allen et al., 1996).

Despite its influence, the Siegel and White (1975) framework has been found to be inconsistent with empirical and anecdotal observations. A particularly important implication of the framework in this regard concerns its treatment of spatial knowledge at early stages of development. It posits no spatiality in adults at the stages of landmark knowledge and route knowledge (as Piaget's theory assumes no metricity in children in early stages of topological space). According to the framework, therefore, people cannot perform tasks that require metric or configurational understanding of the environment after only brief exposure to a new place, for example, taking shortcuts, returning directly back to the starting point, or estimating distances and directions between places, because people do not possess metrically scaled and interrelated knowledge in the initial stages of microgenetic development. If that were the case, people would become lost in an unfamiliar place if they miss a specific turn or find a usual route blocked because of construction. Empirical results point to the contrary, however; people can do these "survey" tasks, at least better than chance would predict (e.g., Klatzky et al., 1990; Landau, Spelke, and Gleitman, 1984; Loomis et al., 1993). As we discuss below, in fact, research shows that individuals differ substantially in respect to how readily they acquire metric knowledge in the environment.

In addition, the dominant framework does not specify the timescale of the developmental course. How long does it take for survey knowledge to emerge? With how much experience in a new environment do people acquire metrically scaled and interrelated knowledge? Do all people reach the final stage of survey knowledge in the course of microgenetic development? These questions remain unanswered in the framework, although the timescale description by Siegel and White (1975) implies that it would be at least days or months, if not years. That said, it should be clarified that the Siegel and White framework insightfully specifies the types of knowledge that people may have about the environment: knowledge of landmarks, knowledge of routes, and knowledge of configurations. Ishikawa and Montello (2006) explicitly acknowledged that "what is questionable about the framework is not so much the content, but the structure and developmental course of spatial knowledge it proposes" (p. 118).

Continuous framework of spatial microgenesis

In response to these problems, Montello (1998) proposed an "alternative framework" that explains spatial microgenesis as a continuous developmental process. It argues by analogy to an information processing theory of cognitive ontogenesis and inspiration from findings in the literature that people acquire metric knowledge from the beginning in a new environment, at better than chance level. The accuracy of metric knowledge develops continuously (or quantitatively) with experience, rather than discretely (or qualitatively) as posited by the Siegel and White (1975) framework.

Specifically, Montello's (1998) continuous framework consists of five tenets:

1. There is no stage at which only pure landmark or route knowledge exists. . . . Metric configurational knowledge begins to be acquired on first exposure to a novel place.
2. As familiarity and exposure to places increases, there is a relatively continuous increase in the quantity, accuracy, and completeness of spatial knowledge.
3. The integration of knowledge about *separately learned places* . . . into more complex, hierarchically organized knowledge structures represents a significant and relatively sophisticated step in the microgenesis of spatial knowledge.

4. Individuals . . . differ in the extent and accuracy of their spatial knowledge . . . [particularly] with respect to the degree of knowledge integration.
5. Linguistic systems . . . provide for the existence of relatively pure topological knowledge . . . *in addition to* metric spatial knowledge, not as a necessary precursor. (p. 164)

Importantly, the alternative framework agrees with the dominant framework that the acquisition of coordinated and comprehensive survey knowledge is an elaborate step in microgenetic development. In fact, it has often been observed that people who are familiar with two separate regions do not understand the spatial relations between them (Lynch, 1960; Rand, 1969). Furthermore, the alternative framework points to individual differences in the extent and accuracy of spatial knowledge that people acquire from direct experience, especially in the ability to integrate separately learned places in a global allocentric frame of reference. Montello (1998) thus argued that the issue of individual differences requires careful investigation on its own, beyond what is revealed by the aggregate analysis of average data.

Empirical research on spatial knowledge integration

The ongoing discussions pointed to the significance that the integration of separately learned places into a configurational representation has in spatial microgenesis. Having recognized it, some researchers empirically examined how and to what extent people can mentally integrate two separate routes encountered in a novel environment (Figure 5.8).

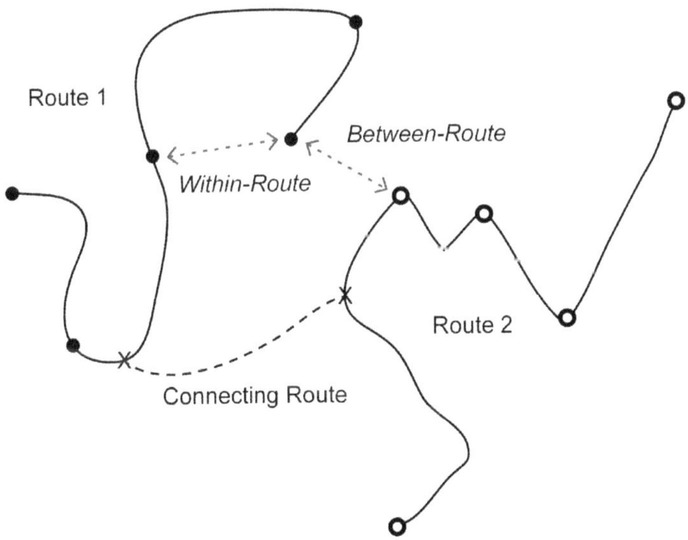

Figure 5.8 **Experimental setting for a study in a route integration paradigm. Participants travel two separate routes that share a common path segment or are connected to each other through a third path, and conduct spatial tasks designed to assess their spatial knowledge of individual routes (within-route knowledge) and of the spatial relations between the two routes (between-route knowledge or integrated routes).**

Basically, these studies exposed participants to two separate routes that shared a common path segment or were connected to each other through a third path, and examined their spatial knowledge of individual routes (within-route knowledge) and of the spatial relations between the two routes (between-route knowledge or integrated routes).

Moar and Carleton (1982) presented to participants a series of slides that simulated a walk along two routes with a common segment, and asked them to estimate directions and straight-line distances between places on the same route and different routes. Moar and Carleton found that within- and between-route estimates did not differ significantly in accuracy (absolute errors), and concluded that the participants mentally integrated the two routes in a network-like representation. Holding and Holding (1988) replicated the results in a similar experimental setting, observing equivalent accuracy and response time for within- and between-route estimation.

Hanley and Levine (1983) examined the integration of two paths on a tabletop scale. In their study, participants wore a blindfold and traced two tabletop paths (three-point paths) separately with their right index fingers, guided by the experimenter. Participants, still blindfolded, were then instructed to imagine the two paths being superimposed on each other in their original orientations, a point on one path being placed over a point on the other path without rotation; they were asked to infer spatial relations between two points on the same path or two points on different paths. Hanley and Levine found that participants' responses were less accurate (although better than chance) and slower between the two paths than within the same path. They concluded that the participants were able to integrate the two separate paths to a certain degree, but that the integrated representation was inaccurate.

Golledge et al. (1993) showed participants slides of a walk along two separate, but partially overlapping, routes in an unfamiliar neighborhood. Participants were asked (a) to arrange photographs of views along the routes in order of appearance, (b) to place the photographs at their correct distances along the routes, and (c) to estimate directions between places on the same route and different routes. Golledge et al. observed that the participants ordered and placed the landmarks along the routes quite accurately; however, they did not estimate directions accurately (although better than chance), doing much worse between the routes than within a route. Golledge et al. concluded that the participants succeeded in mentally integrating the two routes only marginally.

Montello and Pick (1993) examined the integration of two routes that were vertically aligned on two different levels of a campus building. Participants first walked the two routes separately; they were then informed that the two routes were located roughly one above the other, and asked to estimate directions between landmarks on the same route or different routes. Montello and Pick found that between-route pointing was less accurate and slower than within-route pointing. They argued that although the participants integrated the two routes with accuracy better than chance, the representation of the integrated routes was less accurate than the representation of each individual route.

Results of these studies turned out to be inconsistent, some studies demonstrating successful integration (Holding and Holding, 1988; Moar and Carleton, 1982) and other studies showing a failure or inferior performance of integration (Golledge et al., 1993; Hanley and Levine, 1983; Montello and Pick, 1993). Reasons for the inconsistency include differences in (a) the sources of information, (b) spatial scale, (c) the complexity of routes, and (d) the information given to participants to allow for integration.

Furthermore, past studies of microgenetic development mostly employed cross-sectional designs (Herman, Kail, and Siegel, 1979; Stern and Leiser, 1988; Thorndyke and Hayes-Roth, 1982). These studies implicitly equate the length of residence with the amount of exposure to the environment, and thus are lacking in control of the amount of experience and the source of knowledge; that is, how much and in what manner people interacted with the environment. For example, participants may have learned the environment by directly navigating in it or viewing a map, a difference that has been shown to cause important differences in the structures of acquired knowledge representations (Thorndyke and Hayes-Roth, 1982).

Finally, and most importantly, past studies did not examine the microgenetic development of spatial knowledge at the individual level. There are studies of individual differences in spatial cognition with respect to spatial abilities, sense of direction, and neural correlates (Allen et al., 1996; Hartley et al., 2003; Hegarty et al., 2006; see Chapter 6), but few studies have specifically addressed in detail individual differences in the developmental pattern of people's spatial knowledge, in a controlled setting across an extended period of time.

With these issues in mind, Ishikawa and Montello (2006) conducted a longitudinal experiment in which participants traveled in a car along two separate routes in an unfamiliar environment and conducted spatial tasks that assessed their knowledge about the two routes (Figure 5.9). They had never previously visited the study area nor were they shown a map of the area at any time during the experimental period. One route was curved and U-shaped (called the U-route); the other was straighter (the S-route). The two were connected by a short route running between them (the connecting route). On each route, four predetermined landmarks were pointed out and named to participants by the experimenter (the landmarks were not visible from each other). The experiment was conducted once a week for ten consecutive weeks. In the first three sessions, participants traveled each route individually and then estimated directions, distances along the route (route distances), and distances along the straight line between the landmarks (straight-line distances within each route). Importantly, participants were driven blindfolded along a circuitous road (not the connecting route) to get from one test route to the other. At the end of the third session, Ishikawa and Montello also asked participants to estimate directions and straight-line distances between landmarks on the different routes. If participants were in fact disoriented about the spatial relation of the two routes to each other, these between route estimates effectively constituted uninformed guesses. During the fourth session, participants traveled the two routes separately as before, but then they traveled the connecting route for the first time (that is, direct information about the spatial relations between the two routes was given to participants for the first time). From the fourth to the tenth session, participants conducted spatial tasks for the individual routes separately and then tasks between the two routes (straight-line distances and directions from a landmark on one route to a landmark on the other route, requiring route integration). Participants also drew sketch maps of the routes in even-numbered sessions, both before and after integration.

Look at the mean performance first. Figure 5.10 shows mean absolute errors of direction estimates for the ten sessions. People did better for the straight S-route than for the curved U-route and the integrated routes. Performance for the U-route and integrated routes did show a significant trend indicating improvement over the ten weeks. These mean data seem

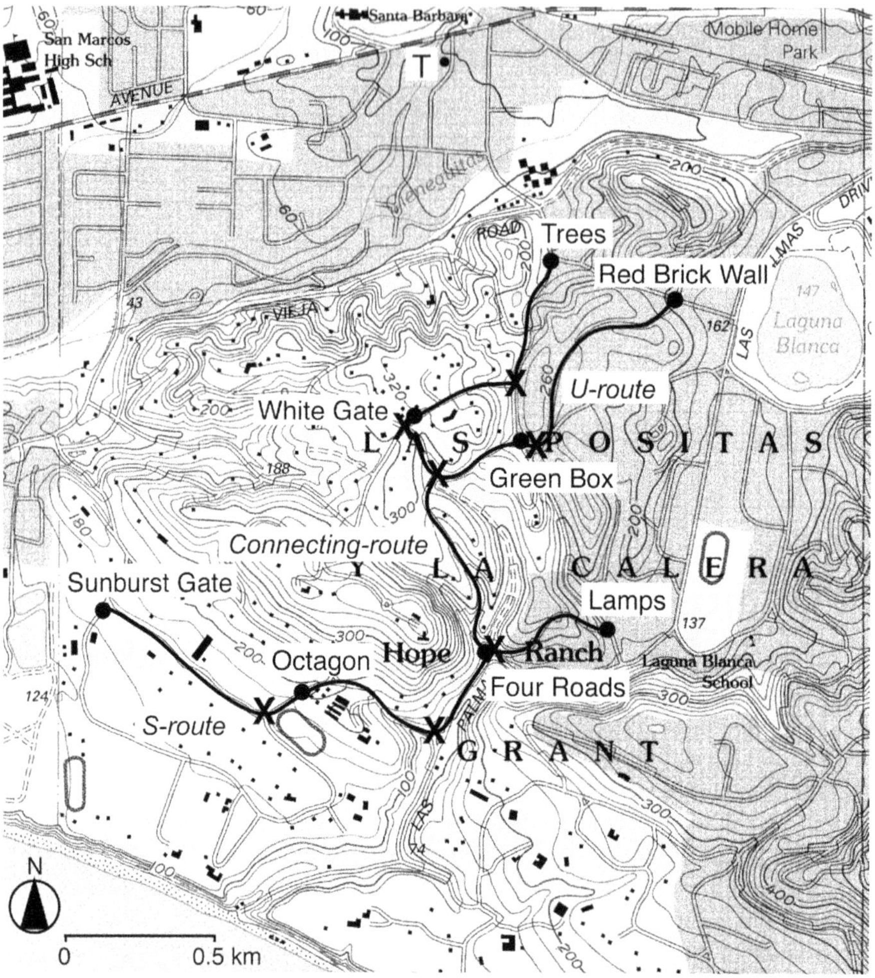

Figure 5.9 **Map of the study area of Ishikawa and Montello (2006). Participants traveled in a car along two separate routes in an unfamiliar environment and conducted spatial tasks that assessed their knowledge about the two routes. One route was curved (called the U-route) and the other was straighter (the S-route). The two were connected by a short route running between them (the connecting route). Reprinted with permission of Elsevier from Ishikawa and Montello, 2006, p. 101.**

to suggest that, although people estimate directions better than chance, they do not acquire very accurate directional knowledge (errors around 40°) or show much improvement over time (decrease in error of only about 10°). Between-route "guessing" on directions after the third session was very poor, only slightly better than chance.

A similar pattern is observed for mean distance estimates (Figure 5.11). People estimated route distances accurately for both routes. They estimated straight-line distances for the S-route very well, with correlations between correct and estimated distances above 0.90. They estimated straight-line distances for the U-route and between the two routes

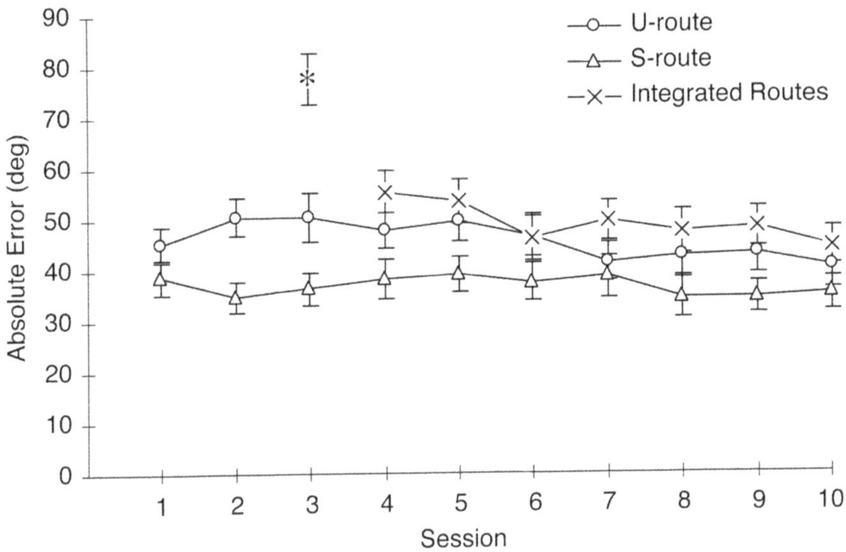

Figure 5.10 **Mean absolute errors of direction estimates for the ten sessions. People did better for the S-route than for the U-route and the integrated routes, but on average, they do not acquire very accurate directional knowledge or show much improvement over time. Reprinted with permission of Elsevier from Ishikawa and Montello, 2006, p. 105.**

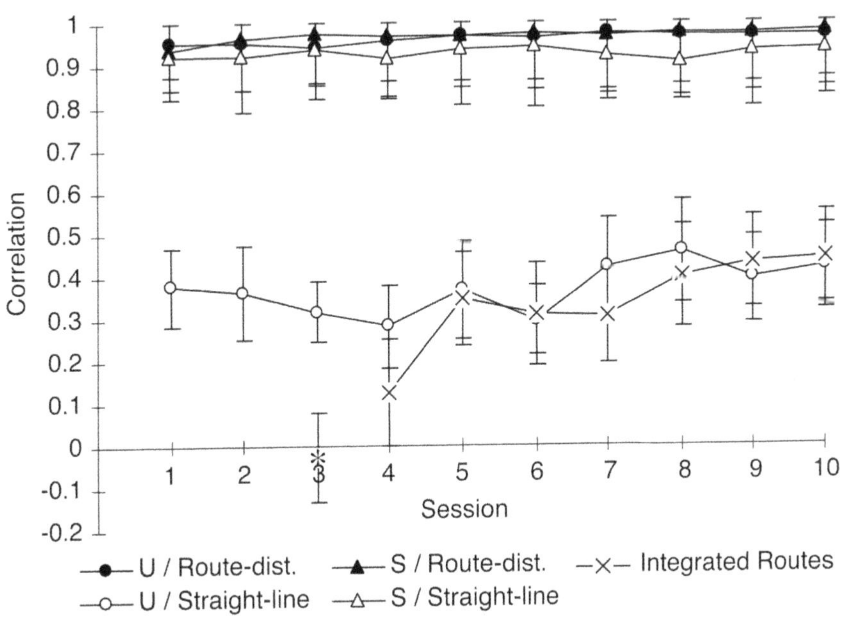

Figure 5.11 **Mean distance correlations for the ten sessions. People estimated route distances for both routes and straight-line distances for the S-route fairly accurately, but estimated straight-line distances for the U-route and between the two routes with much less accuracy. As with direction knowledge, people did not, on average, acquire very accurate straight-line distance knowledge or show much improvement over time. Reprinted with permission of Elsevier from Ishikawa and Montello, 2006, p. 106.**

with much less accuracy, with correlations around 0.40. Again, between-route straight-line distance estimation after the third session and the fourth session was not significantly better than a chance correlation of 0.00. Similar to conclusions from mean direction performance, people do not appear to acquire the accurate survey knowledge that would be required to understand the shape of the curved route and the spatial relations between the two routes. Performance does not improve much over time.

However, if we look at each participant's individual data, a strikingly different picture emerges, revealing rich and large individual differences in the accuracy and developmental pattern of spatial knowledge. Figure 5.12 shows three participants' developmental curves for direction estimates, distance estimates, and sketch maps. Participant P1 showed remarkable performance throughout the experiment and may be called a spatial genius. His (the participant was male) mean pointing error across all sessions averages less than 10° within the S-route and only 17° within the U-route. Even his estimates of directions between routes show only about 20° of error as soon as he travels the connecting route for the first time in the fourth session. Furthermore, his straight-line distance estimates on the curved U-route correlate at 0.73 across all sessions, and his between-route distance estimates correlate at 0.80 after first exposure to the connecting route. His sketch maps are also accurate, bidimensionally correlating with the actual layouts above 0.90 from the very first session.

In contrast, P17 did poorly throughout the experiment, failing to achieve an accurate configurational understanding of the routes even after ten weeks. Her (the participant was female) direction estimates are almost at chance level (absolute errors averaging 70°–90°)

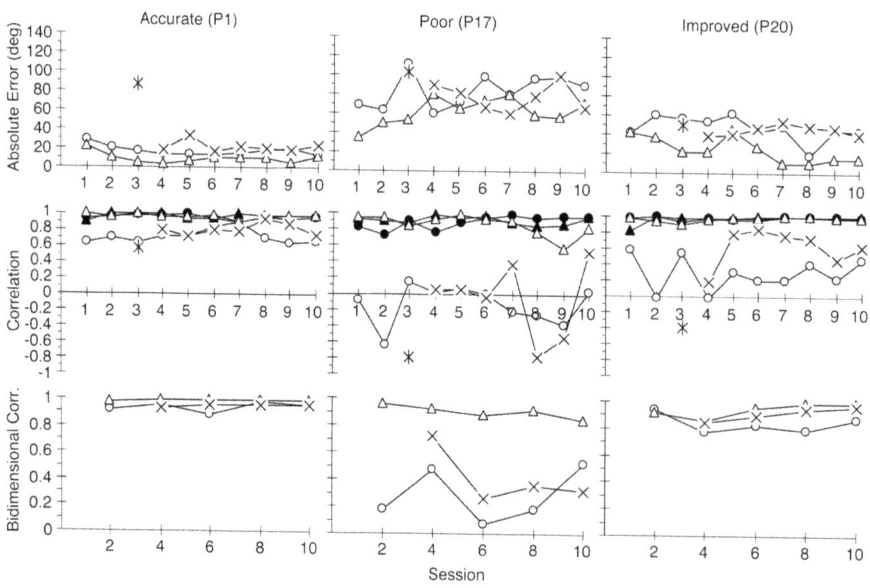

Figure 5.12 **Three participants' developmental curves for direction estimates, distance estimates, and sketch maps. Participant P1 showed remarkable performance throughout the experiment, whereas P17 did poorly, failing to achieve an accurate configurational understanding of the routes even after ten weeks. P20 showed performance that improved significantly over the ten sessions. Reprinted with permission of Elsevier from Ishikawa and Montello, 2006, p. 111.**

and fail to improve over the course of the experiment. Her estimates of straight-line distances for the U-route and the integrated routes are poor, with correlations near 0. Her sketch maps are also inaccurate. Notably, her estimates of route distances for both routes and of straight-line distances for the S-route are quite accurate (correlations above 0.90). Finally, P20 shows performance improving over the ten sessions. His mean pointing error on the S-route improves from 41.2° to 15.2°, his between-route straight-line distance estimates improve from a correlation of 0.17 to one of 0.62, and his sketch maps of the integrated routes improve from a correlation of 0.84 with the actual layout to 0.96. That is, he shows significant and substantial improvement over time.

Participants' sketch maps clearly show the extent of individual differences in the accuracy of configurational understanding (Figure 5.13). P1's sketch map is strikingly

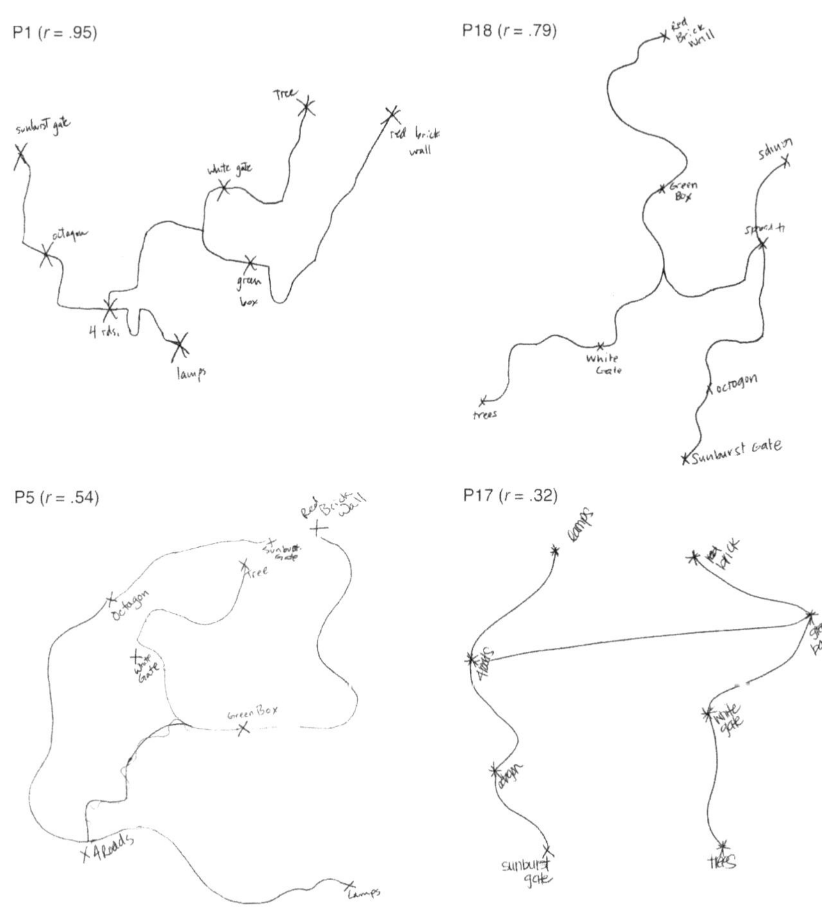

Figure 5.13 **Participants' sketch maps clearly show the extent of individual differences in the accuracy of configurational understanding. P1's sketch map is strikingly accurate (compare it with the correct map in Figure 5.9), whereas P17's sketch map depicts only the fact that there are two routes and they are connected somewhere in between. It should be noted, however, that the identities and sequences of the landmarks are depicted correctly on her map. Reprinted with permission of Elsevier from Ishikawa and Montello, 2006, p. 125.**

accurate, depicting the shapes, even subtle details, of the two routes and their spatial relations. P17's sketch map, in contrast, depicts only the fact that there are two routes and they are connected somewhere in between. It is also evident on her map, however, that she understood the identities and sequences of the landmarks accurately.

Therefore, some people can and do acquire surprisingly accurate metric knowledge, even relatively quickly, including relational knowledge about places between which they have not directly traveled. Other people acquire only a primitive level of knowledge about the environment in terms of metric and configurational accuracy and do not show much improvement over time. Some people are intermediate and show improvement in performance with repeated exposure to the environment. The existence of such large differences between people in the accuracy of metric and configurational understanding of the environment is discussed in detail from the perspective of individual differences in cognitive mapping in the next chapter.

Review box for Chapter 5

1. Describe various ideas (or definitions) of development.
2. Discuss ontogenesis and microgenesis.
3. Discuss stage and continuous theories of development.
4. Describe the development of spatial concepts in children as conceived by the stage theory of spatial ontogenesis.
5. Explain the three types of spatial knowledge about the environment that the stagelike framework of spatial microgenesis distinguished. Give examples of each of the three types of knowledge.
6. Describe the process of spatial knowledge acquisition outlined in the anchor point theory.
7. Compare the stagelike and continuous frameworks of spatial microgenesis, focusing on the ideas of development conceived by the two frameworks.
8. Can you conceive of a world in which no metric information is existent? In such a world, a doughnut and a coffee cup cannot be distinguished.
9. Summarize the results of empirical studies of spatial knowledge integration.
10. Discuss the theoretical and practical implications of the Ishikawa and Montello (2006) study.

Chapter 6

Spatial abilities and individual differences

Spatial ability

We do a variety of things every day—we read books, compare the unit prices of two different brands of the same product, play a musical instrument, socialize with friends and colleagues, shoot a ball through a hoop, or find our way in a new place. You may be good at these activities or may find yourself having difficulty with some of them. You may do better than your classmates on some activities, but you may not do as well on other activities. To consider such variations within a person and between people, one can look at these activities in terms of the ability that they require. One group of researchers noted that people who do well in one domain of activities also do well in other domains; that is, their performances on different tasks are correlated with each other. These researchers posited a model of human ability that consists of a general ability factor that underlies all abilities (general intelligence) and group factors that pertain to specific, individual abilities (Spearman, 1904).

Another group of researchers focused on the existence of multiple abilities, and argued that human ability consists of different components relating to performances in different domains, rather than assuming one general ability (Thurstone, 1934; Eliot and Smith, 1983). For example, Gardner (2006) called these abilities *multiple intelligences*, and discussed individual domains such as linguistic, logical-mathematical, musical, interpersonal, bodily-kinesthetic, and spatial intelligences. Gardner, at the same time, clarified that human activities do not depend on a single intelligence but on a combination of intelligences, describing for example dance as requiring skills in bodily-kinesthetic, musical, interpersonal, and spatial intelligences (p. 22). It is thus important to consider human abilities from the perspective of variations between people and multiplicity within a person. Of the specific components of human abilities, one that has been studied particularly extensively in the literatures of psychology and education is *spatial ability*. It is the ability concerning the perception, retention, and manipulation of visual images (Eliot and Smith, 1983, p. 12; Lohman, 1979, pp. 126–127), as separated from other abilities such as verbal ability. Collectively, it is defined as the ability to "search the visual field, apprehend the forms, shapes, and positions of objects as visually perceived, form mental representations of those forms, shapes, and positions, and manipulate such representations mentally" (Carroll, 1993, p. 304).

Early interest in spatial ability was found in the area of vocational education and personnel selection. Spatial performance tests in the early days thus consisted of the tasks of manual dexterity (e.g., wire bending, tapping, and manipulation of nuts and bolts), object assembly, form board completion, or wooden construction (see Eliot and Smith, 1983, p. 2). Early mental tests, in the context of predicting academic success, mainly concerned verbal ability, but gradually non-language performance tests have been developed. For example, El Koussy (1935) administered a battery of psychometric tests to boys between the ages of 11 and 13 years, and identified a group factor that underlies their scores on spatial tests. El Koussy discussed that the factor represented a unique and specific component of human ability: spatial ability, which pertains to the facility to obtain, use, and manipulate visual spatial imagery. Thurstone (1938) similarly argued for the existence of independent abilities, which he called primary factors or primary mental abilities, and proposed a multifactor model of human ability. Thurstone administered a battery of 56 psychological tests to 240 people, and identified seven major factors that he interpreted and labeled as "visual-spatial," "perceptual," "numerical," "verbal-logical," "word fluency," "memory," and "reasoning–rules." The tests that loaded on the visual-spatial factor included the tasks of mental rotation, cube comparison, and surface development (details below), and Thurstone discussed that the factor represented "facility in spatial and visual imagery" (p. 80).

The identification of the specific factor of spatial ability provided impetus for the development of many kinds of psychometric spatial tests and the administration of the tests to large numbers of people in the context of, for example, military personnel screening or primary and secondary education. Now we look at some major spatial tests that have been developed and used in the literature. One family of tests, called *mental rotation*, assesses the ability to perceive and encode visual shapes and mentally rotate their images into different orientations. An example of two-dimensional mental rotation is the Card Rotations Test (Ekstrom et al., 1976; Figure 6.1). In this test, participants view 20 sets of figures, each set consisting of one criterion figure and eight alternatives. Participants identify for each set of figures whether the alternative figures are the same as the criterion figure (rotated in the picture plane into different orientations) or different (flipped over as well as rotated). Participants are given six minutes to complete the test, and their scores will be the number of correctly identified items minus the number of incorrectly identified items. There is also a three-dimensional version of mental rotation (Shepard and Metzler, 1971; Vandenberg and Kuse, 1978; Figure 6.2), in which participants view sets of line drawings

Figure 6.1 **Example item of the Card Rotations Test. Participants view sets of figures (a criterion figure and eight alternatives) and identify whether the alternative figures are the same as the criterion figure.**

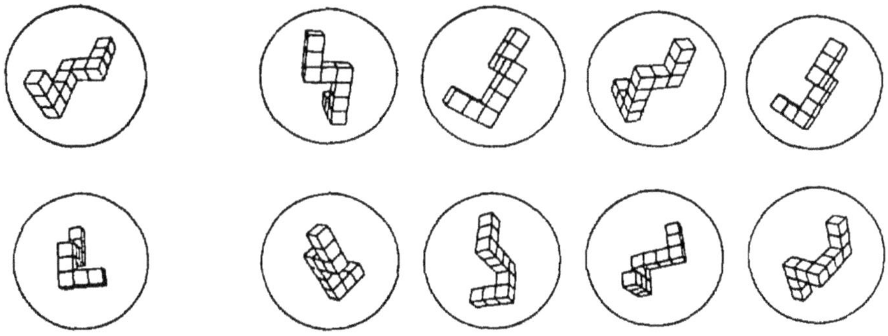

Figure 6.2 **Three-dimensional version of mental rotation. Participants view sets of line drawings of objects, each set consisting of one criterion and four alternative drawings, and answer whether the alternatives are the same as the criterion object. Reprinted from Vandenberg and Kuse, 1978, p. 600.**

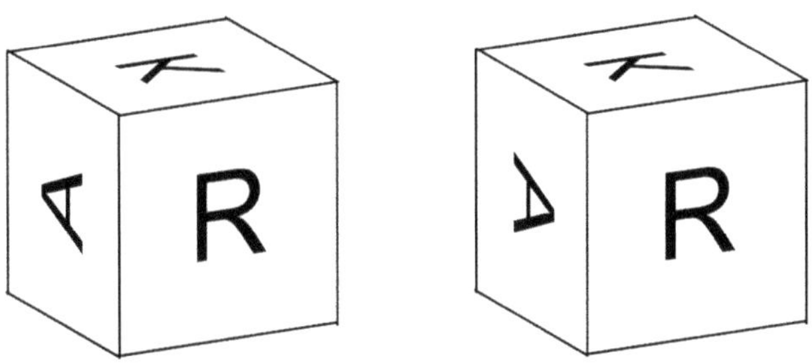

Figure 6.3 **Example item of the Cube Comparisons Test. Participants view pairs of cubes, whose faces have different letters, numbers, or symbols, and answer whether the pair could be drawings of the same cube.**

of three-dimensional objects, each set consisting of one criterion and four alternative drawings. Participants answer whether the alternatives are the same as the criterion object (rotated into a different orientation in the picture plane or depth plane) or different (a mirror image of the criterion object or another different object).

Another example is the Cube Comparisons Test (Ekstrom et al., 1976). In this test, participants view 42 pairs of cubes, whose faces have different letters, numbers, or symbols (no letters, numbers, or symbols appear on more than one face of a given cube; see Figure 6.3). Participants answer whether the pair could be drawings of the same cube, judging from the orientations of the letters, numbers, or symbols written on the visible faces. They are given six minutes to complete the test, and their scores will be the number of correctly identified items minus the number of incorrectly identified items.

There are also spatial tests that assess kinds of mental manipulations other than mental rotation. One major example is the Surface Development Test (Ekstrom et al., 1976). In this test, participants view pairs of drawings; in each pair, a flat piece of paper shown to the left can be folded to form an object drawn to its right (Figure 6.4). Participants answer which edges on the folded object match the edges designated on the flat piece of paper.

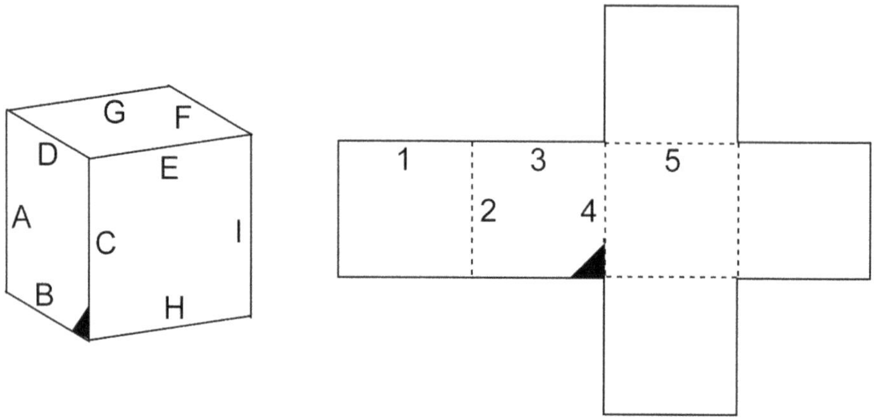

Figure 6.4 **Example item of the Surface Development Test. Participants view pairs of drawings of a flat piece of paper and an object, and answer which edges on the folded object match the edges designated on the flat piece of paper.**

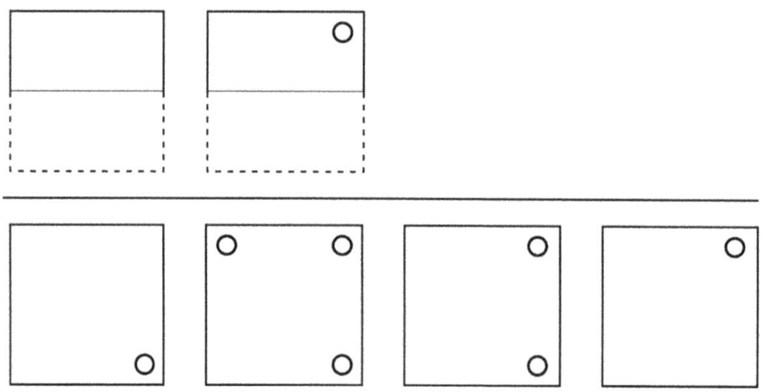

Figure 6.5 **Example item of the Paper Folding Test. Participants view a series of figures that illustrates how a square piece of paper is folded and then punched at a specific location, and choose from five figures the one that shows correctly where the holes will be when the paper is unfolded.**

Another example is the Paper Folding Test, which shows a series of figures that illustrates how a square piece of paper is folded and then punched at a specific location (Figure 6.5). Participants choose from five figures the one that shows correctly where the holes will be when the paper is unfolded. These two tests supposedly involve more complex, multistep mental manipulations than simple, speeded rotation (Linn and Petersen, 1985; McGee, 1979; Pellegrino and Kail, 1982).

Another family of spatial tests assesses the ability to perceive spatial patterns or maintain orientation in relation to other objects in space. The Spatial Orientation Test from the Guilford-Zimmerman Aptitude Survey (Guilford and Zimmerman, 1948) shows pairs of pictures of views from a boat in motion—an original view from the boat paired with a

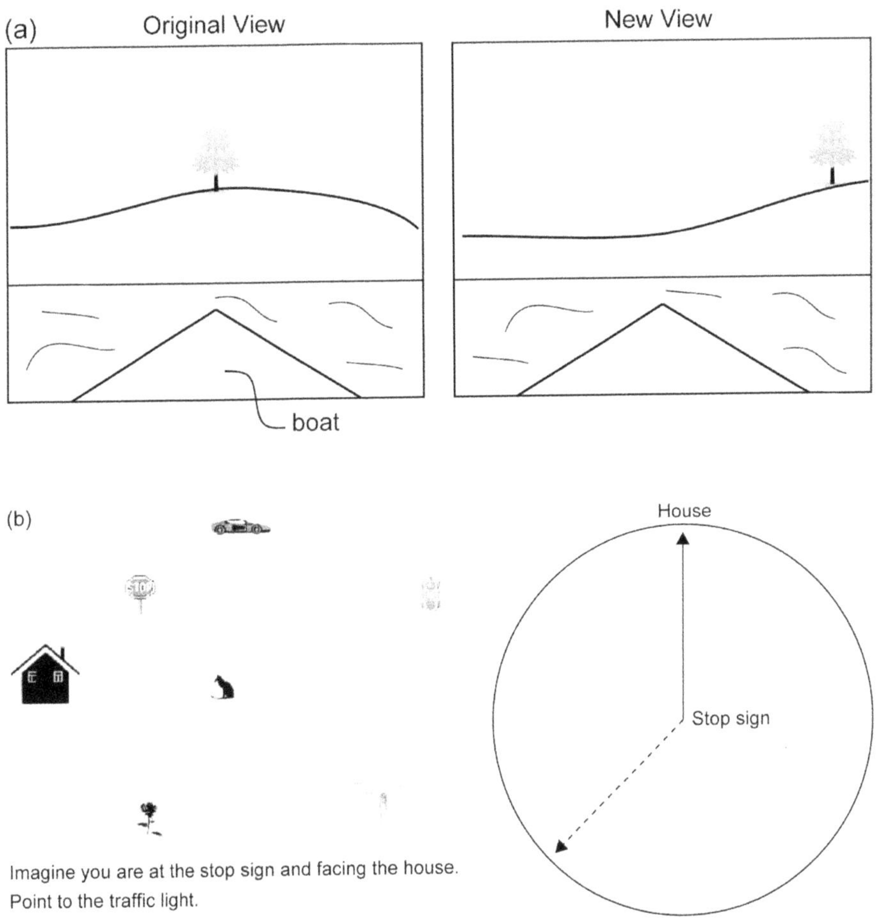

Figure 6.6 **(a) Example items of the Spatial Orientation Test and the Perspective Taking Test. Participants view pairs of pictures of views from a boat in motion, an original view (*left*) and a different view (*right*), and answer how the boat changed its position and direction to get the new view. (b) Example item of the Perspective Taking Test. Participants view an array of objects and are asked to imagine standing at one of the objects and facing another object (*left*). Then they indicate the direction of a third object on a circle printed on a piece of paper (*right*). Reprinted with permission of Elsevier from Hegarty and Waller, 2004, p. 178.**

different view—and asks the participant to figure out how the boat changed its position and direction to get the new view (Figure 6.6a). Similarly, there are tests that assess the ability to imagine a view or stimulus array from a different perspective (Figure 6.6b). A classic example is Piaget's three mountains task (see Chapter 5), in which children view scale models of three mountains and are asked to choose among photographs the one that represents the view of a doll, which is placed at different locations around the mountains. There are variations of perspective taking tests developed in the literature (e.g., Hegarty and Waller, 2004; Kozhevnikov and Hegarty, 2001; Liben and Downs, 1993), and perspective taking has been identified as an ability distinct from mental rotation.

The Hidden Figures Test (Ekstrom et al., 1976) assesses the ability to identify spatial forms embedded in complex spatial patterns. In this test, participants view four sets of simple geometric figures and complex spatial patterns (Figure 6.7). In each set, participants are given five simple figures, and indicate which can be found in each of the complex spatial patterns that are shown below. They are given 12 minutes to complete the test, and their scores will be the number of correctly identified items minus the number of incorrectly identified items. The ability factor that this test loads on is called flexibility of closure (Carroll, 1993, p. 336).

Finally, there is another family of spatial tests, called object and location memory tests (Silverman and Eals, 1992). In the tests, participants view an array of objects first, and then (after the array is removed) view a new array in which additional objects are interspersed or the locations of some of the objects are interchanged (Figure 6.8). They are asked to identify the objects that were added (object memory) or the objects whose locations were interchanged (location memory). Interestingly, women tend to do better on these tests than

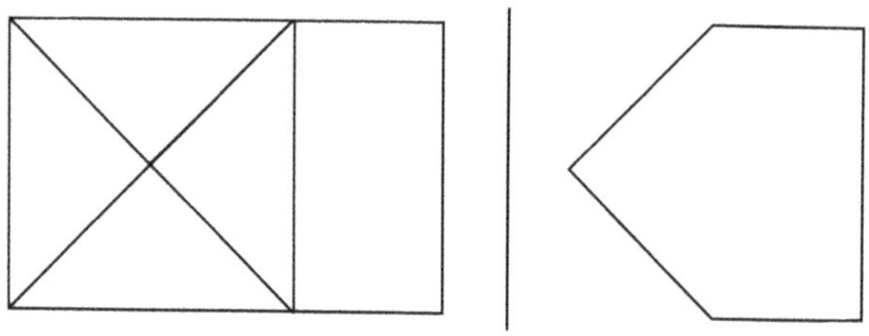

Figure 6.7 **Example item of the Hidden Figures Test. Participants view sets of simple geometric figures and complex spatial patterns, and answer where in the complex patterns the simple figures can be identified.**

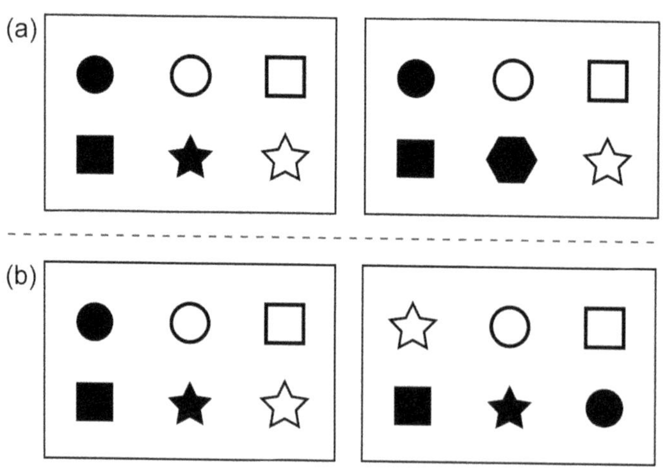

Figure 6.8 **Examples of object and location memory tests. Participants view an array of objects first (*left*), and (after it is removed) a new array (*right*) in which additional objects are interspersed (a) or the locations of some of the objects are interchanged (b). They identify which objects were added (object memory) or interchanged (location memory).**

men, on average. By contrast, men tend to do better on mental rotation than women, on average. The female superiority on object and location memory, in contrast to the male superiority on mental rotation (Linn and Petersen, 1985; McGee, 1979), has particularly attracted the attention of researchers and stimulated discussions from an evolutionary perspective (McBurney et al., 1997; Silverman and Eals, 1992). For example, Silverman and Eals discussed a hunter-gatherer model, which states that there was a male–female division of labor across the evolutionary time between hunting and gathering—men hunt and women forage. They argued that mental rotation and map reading relate to the ability required for successful hunting, in that it enables humans to get oriented in space with a configurational understanding of the environment after a long and winding journey from home, whereas object and location memory relates to the ability required for foraging, in that it enables foragers to locate and find food plants within complex arrays of vegetation. The evolutionary model thus links the supposed labor division in the history of evolution and associated spatial abilities to the male and female superiority on mental rotation and object and location memory, respectively.

Having touched on the issue of male–female differences, it should be cautioned that statements about a comparison at the group level do not equally apply to a comparison at the individual level. For example, saying that men tend to do better on mental rotation than women on average does not mean that every single man does better than every single woman identified. Figure 6.9 shows distributions of scores on the mental rotations test

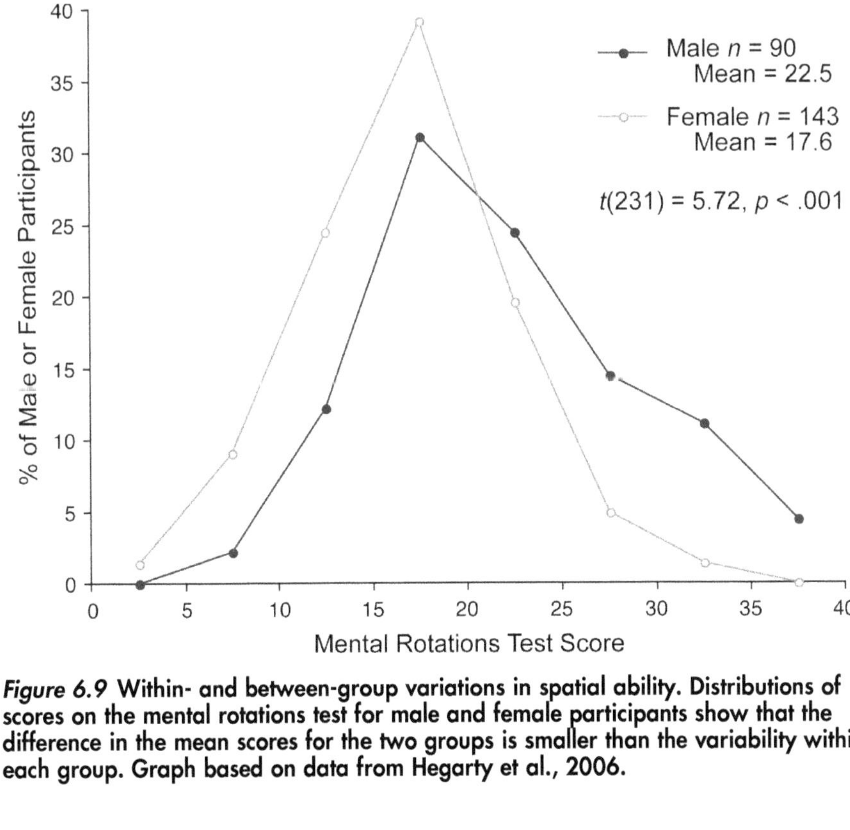

Figure 6.9 **Within- and between-group variations in spatial ability. Distributions of scores on the mental rotations test for male and female participants show that the difference in the mean scores for the two groups is smaller than the variability within each group. Graph based on data from Hegarty et al., 2006.**

(data from Hegarty et al., 2006). The difference in the mean scores for men and women is statistically significant, but the distributions of the two groups' scores overlap considerably, and the variability within each sex is larger than the difference in the means for the two sexes. Therefore, caution should be exercised in interpreting such a statement as "women cannot read maps."

Given the large batteries of psychometric spatial tests, researchers have been interested in the issues of classification of spatial tests and identification of different components of spatial abilities. Through factor analytic studies, researchers have identified three major factors of spatial abilities: spatial visualization, spatial relations, and spatial orientation. *Spatial visualization* refers to the ability to "mentally manipulate, rotate, twist, or invert a pictorially presented stimulus object" (McGee, 1979, p. 893). The factor of spatial visualization is often differentiated from the factor of *spatial relations*, in that the former involves mental manipulations of more complex stimuli and sequential transformations of mental images in a speeded condition (Hegarty and Waller, 2005, p. 129; Lohman, 1979, p. 10). *Spatial orientation* refers to the ability to "comprehen[d] the arrangement of elements within a visual stimulus pattern" (McGee, 1979, p. 893) or to "imagine how a stimulus array will appear from another perspective" (Hegarty and Waller, 2005, p. 129). Pellegrino, Alderton, and Shute (1984) summarized the factor structures observed in factor analytic studies along the dimension of speed–power and the dimension of stimulus complexity (Figure 6.10). In this conceptualization, spatial tests are arranged with respect to whether they are administered in a speeded condition (emphasizing speed) or are relatively unspeeded (requiring both speed and accuracy) and whether the stimuli to be processed

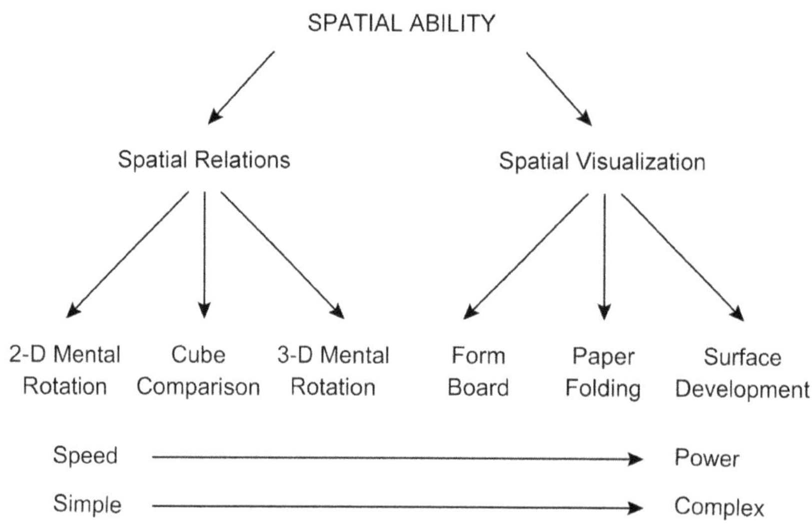

Figure 6.10 **Factor structures of spatial abilities can be interpreted along the dimension of speed–power and the dimension of stimulus complexity. In this figure, you can identify mental rotation and cube comparison toward the speed and simple side, and paper folding and surface development toward the power and complex side, which maps onto the distinction of the factors of spatial relations and spatial visualization, respectively. Reprinted from Pellegrino, Alderton, and Shute, 1984, p. 243.**

are complex and involve many elements. In Figure 6.10, mental rotation and cube comparison are located toward the speed and simple side, and paper folding and surface development toward the power and complex side, which map onto the factors of spatial relations and spatial visualization, respectively.

Some researchers raised concerns about the validity of these factor analytic studies, however. For one thing, factor structures differ depending on which tests are included in the analysis. For another thing, the stability of factor structures hinges on the stability of solution strategies that participants employ. Factor analytic studies assume that people use the same strategy to answer individual items in the same test, but, on the contrary, research has revealed variations in solution strategies between people and within a person on a specific test. For example, people can solve items in the three-dimensional mental rotations test analytically, by counting the number of cubes in different segments or comparing the directions of two end arms of the object (Hegarty, 2010). Furthermore, the same person may use different strategies within the same test depending on the difficulty or complexity of items. Just and Carpenter (1985) discussed that items in the Cube Comparisons Test can be solved through various possible strategies, for example, mentally rotating a cube on the three orthogonal axes (in the picture plane or in depth), mentally rotating the cube on an oblique axis (e.g., an axis connecting the diagonally opposite vertices of the cube), analytically coding the relations between letters on different faces of the cube (e.g., the top of the letter A points to the bottom of the letter E), or imagining an observer moving around the cube (i.e., mental perspective taking). Just and Carpenter observed that people with high spatial ability tended to employ the nonstandard oblique rotation strategy. The nonstandard oblique rotation saves the number of steps in mental rotations of a cube, and hence leads to a decrease in response time. Just and Carpenter further observed that different items in the same test, particularly difficult items, evoked different strategies. Albeit with such caveats about interpretation, the factor analytic studies clarify the characteristics of different components of spatial abilities. As there are multiple human intelligences, there are multiple spatial abilities.

Mental rotation and perspective taking

In the above discussions of different types of spatial abilities, we mentioned the separability of mental rotation and perspective taking; the former entailing object-based spatial transformations and the latter egocentric spatial transformations (Hegarty and Waller, 2004, p. 176). Consider the case of the Cube Comparisons Test: You can either envision the cube rotating around its axis (i.e., mentally rotate it) or imagine looking at the cube from different perspectives (i.e., mentally move around and take different positions around the cube). These two mental operations are logically equivalent, resulting in the same imagined view of the cube to you. In contrast to images of individual manipulable objects, mental representations of spatial arrays may not be equally easily accessible by mental rotation and mental perspective taking. Such a difficulty especially applies to the case of reasoning about spatial relations in a large-scale environment, because you cannot view the whole spatial relations in a horizontal perspective. Research has shown that people conduct these two mental transformations, mental rotation and perspective taking, differently in the cognitive and neurophysiological sense, and find one easier than the other depending on the types of tasks at hand.

For example, Presson (1982) asked 24 undergraduate students to view an array of three or four objects on a 12.7-cm square plywood and, after it being masked by a square cover, to predict how the array would appear after a rotation of either the array or themselves. To respond, participants arranged blocks in a configuration showing how the entire array would look to them if either the array was rotated (array rotation) or they moved around the array (viewer rotation) 0°, 90°, 180°, or 270° clockwise. Presson observed that participants reconstructed the array as seen from new perspectives more accurately and faster in the array rotation condition than in the viewer rotation condition, showing a psychological difference in these logically equivalent transformations.

Hegarty and Waller (2004) asked 61 undergraduate students to conduct a set of mental rotation and perspective taking tasks. Performance on mental rotation was assessed by psychometric tests of spatial visualization, a three-dimensional mental rotations test (Vandenberg and Kuse, 1978), the Card Rotations Test, and the Cube Comparisons Test. Performance on perspective taking was assessed by three tests asking participants to estimate directions from various imagined perspectives. One was the Object Perspective Test developed by Kozhevnikov and Hegarty (2001), in which participants viewed a configuration of seven objects drawn on a piece of paper (e.g., a stop sign, a house, or a traffic light; 12 configurations in total) and were asked to imagine being at the position of one object in the array, facing another object, and to indicate the direction to a third object (e.g., they were instructed, "imagine you are standing at the stop sign facing the house, and point to the house"). The other two perspective taking tests concerned direction estimation for well-known buildings on a university campus and major cities in the United States; the tests asked participants to imagine being at one place, facing another, and point to a third place. A confirmatory factor analysis showed that a model that posited two separate factors for mental rotation and perspective taking explained observed data on participants' test scores better than a model that posited one combined factor, although the two factors correlated 0.81 with each other (Figure 6.11).

We saw in Chapter 4 that response time on mental rotation increases linearly with angular disparity between the orientations of criterion and test drawings (Cooper and Shepard, 1973; Shepard and Metzler, 1971), as an indication of the functional equivalence of mental and physical rotations. Research has shown that this pattern of a linear increase in response times does not necessarily apply to the performance on perspective taking. Zacks et al. (2000) asked 48 undergraduate students to conduct two tasks that were designed to assess performance on object-based transformations (mental rotation) and egocentric perspective transformations (perspective taking). In one task, participants were presented with pairs of line drawings of human bodies in different orientations with one arm extended, and asked to judge whether they were identical or mirror images (called a same–different task). In another task, participants were presented with one line drawing and asked to judge whether it had a left or right arm extended (called a left–right task). Zacks et al. found that response times for the same–different task increased monotonically with the angular disparity between the orientations of paired drawings, whereas response times for the left–right task did not have a significant relationship with the orientations of drawings. Furthermore, response times for same–different judgments were correlated more strongly with scores on the Mental Rotations Test than scores on the Object Perspective Test; by contrast, response times for left–right judgments were correlated more strongly (albeit nonsignificantly) with scores on the Object Perspective Test. Zacks et al. argued

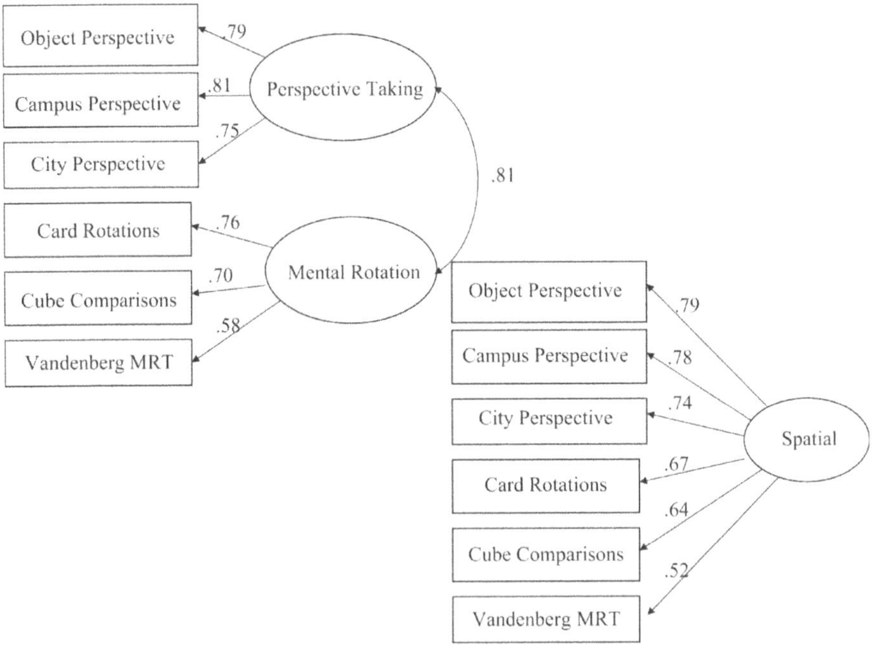

Figure 6.11 **A confirmatory factor analysis of students' scores on a set of mental rotation and perspective taking tasks showed that a model that posited two separate factors of mental rotation and perspective taking fit observed data better than a model that posited one combined factor. Reprinted with permission of Elsevier from Hegarty and Waller, 2004, p. 187.**

that these results indicate that participants conducted the same–different and left–right tasks differently, the former by mentally rotating the images of drawings and the latter by imagining taking the position and perspective of the depicted person: these two mental transformations tap into different psychological processing, revealed by different patterns of relations with stimulus orientation.

Evidence for the dissociation of mental rotation and perspective taking has also been provided in the neuroscience literature. Lambrey et al. (2012) conducted a functional magnetic resonance imaging (fMRI) study to examine the neural bases of array rotation and perspective taking. The researchers presented to 18 adults an array of four objects on a round table in a virtual environment, and asked them to imagine the table being rotated (array rotation) or themselves moving around the table (perspective taking). The current position at the table (which the participants were seated at and faced at the beginning) was marked by a pole and the angle of rotation was indicated by either an arrow or a person (an avatar) located at 112.5° or 157.5° from the viewer around the table. That is, in the array rotation condition, participants imagined that the table would be rotated until the pole on the table came to the location where the arrow or the avatar was. In the perspective taking condition, they imagined that they would move around the table to the location where the arrow or the avatar was (i.e., take the perspective available at the location indicated by the arrow or take the perspective of another person indicated by the avatar). After these imaginary rotations, participants were presented with a new array of four objects, to be available from the new imagined perspective, and asked to identify which one of the four objects had been moved. Participants' responses were found to be faster and more accurate in the perspective taking (or self-rotation) condition than the array rotation condition. In the perspective taking condition, participants responded more slowly when a new perspective was indicated by the avatar than the arrow; that is, it was more effortful to imagine taking another person's perspective than imagining rotational movement to another place. Importantly, in an fMRI analysis, when the perspective taking condition was compared with the array rotation condition, a significantly greater activation was observed in the parieto-occipital sulcus and retrosplenial cortex. By contrast, when the array rotation condition was compared with the perspective taking condition, a significantly greater activation was observed in the intraparietal sulcus. Thus, mental transformations achieved by imagining array rotation and self-rotation constitute psychologically different processes (although resulting in logically the same view) and activate different parts of the brain. Furthermore, a seemingly subtle difference in the instructions to take the perspective available at another *location* and to take another *person's* perspective leads to cognitively different consequences.

Sense of direction

The foregoing sections looked at spatial abilities assessed by psychometric spatial tests. A question of theoretical and practical interest, particularly in the context of spatial cognition, concerns how these spatial abilities are related to environment perception, as Ittelson (1973) defines it. Importantly, psychometric spatial tests deal with perception on a small scale, with figures printed on a piece of paper, whereas cognitive mapping concerns perception at a larger scale, or spatial orientation and wayfinding in the environment (remember the distinction between figural space and vista or environmental space, discussed by Montello, 1993). On the grounds that both concern spatial perception, it is reasonable to assume some correspondence between performances on psychometric spatial tests and environmental spatial tasks. But it remains an open question as to how much and in what way spatial abilities at different scales are related. For example, do people who are good at mental rotation acquire accurate cognitive maps and find their way easily in a new environment? Conversely, do people who have difficulty with spatial orientation and who are prone to disorientation in a new environment have low mental rotation ability? We thus look in this section at the relationship between psychometric spatial ability and environmental spatial learning.

Lorenz and Neisser (1986) addressed the question by asking: What is spatial ability?

> From an ecological point of view, . . . [the definition of spatial ability as the ability to form and transform mental imagery] seems quite limited: what about finding one's way home, giving directions, or taking short cuts? Knowing the layout of the surrounding environment—and your own position in it—contributes importantly to what one of us [Neisser, 1987] has called "a sense of where you are". Despite its adaptive significance, little is known about this group of abilities: neither its factor structure nor its relation to more standard measures has been systematically studied.
>
> (p. 2)

With that aim, Lorenz and Neisser asked 76 students to take a battery of spatial ability tests (e.g., mental rotation, surface development, and cube cutting) and conduct spatial tasks of environmental learning (e.g., pointing to unseen landmarks, giving verbal route directions, and locating landmarks on a map). By factor analyzing participants' performances on the tests and tasks, Lorenz and Neisser showed that psychometric spatial ability and environmental learning loaded on separate factors, and that the latter was further decomposed into factors corresponding to the memories of landmarks, routes, and geographic directions.

Allen et al. (1996) examined the relationship between spatial ability and environmental learning by asking 100 participants to take a variety of visual-spatial tests and conduct environmental learning tasks. Visual-spatial tests included the Surface Development Test, the Cube Comparisons Test, the Hidden Figures Test (see the previous section), and other psychometric tests that assessed the ability to identify partly obscured pictures and the ability to find shortest paths between pairs of points in a grid. There were two more visual-spatial tests, which Allen et al. developed for the study. One task, the maze learning task, assessed sequential spatial memory, by asking participants to learn a pathway through a 6 × 6 matrix of squares printed on paper, and to reproduce the pathway in a new copy of the matrix of squares, either from the start point (maze learning) or from the endpoint (maze reversal). Another task assessed perspective taking in a scale model of a town (1:87 scale) composed of residential and commercial areas, with houses, buildings, factories, a mountain, streets, and street lights. Participants walked around the scale model and observed it from a number of different perspectives; they were then presented with a series of photographs of views of the model and asked to identify the perspectives from which the photographs were taken. They also viewed a series of routes indicated in the model town, and were asked to identify the routes on a map of the town.

Another set of tasks concerned environmental learning or cognitive mapping in the real world. Participants walked a route in a neighborhood area (1,424 m in length), and then retraced the route from the endpoint to the start point. They also estimated directions and distances (along the route and in a straight line), recognized and sequenced photographs of scenes along the route, and located the photographs at their correct locations on a street map.

A factor analysis of participants' performances on these spatial tests and tasks revealed the separation of the factor of psychometric spatial ability and the factor of environmental learning. More specifically, the factor of environmental learning was found to be decomposed into two components: sequential (or route-based) knowledge and configurational knowledge of the traveled route. The former was defined by route retracing, route distance estimation, and scene recognition, sequencing, and positioning; while the latter was defined by direction and straight-line distance estimation. Furthermore, these two factors of environmental learning, sequential environmental learning and configurational environmental learning, were related to psychometric spatial ability, the former being mediated by sequential spatial memory and the latter by perspective taking ability (see Figure 6.12). Allen et al. (1996) replicated the factor structure with another group of 103 participants, except that the path from perspective taking ability to configurational environmental learning turned out to be nonsignificant.

The foregoing discussions show that basic spatial ability assessed by psychometric spatial tests and the ability pertaining to spatial orientation in the environment constitute two

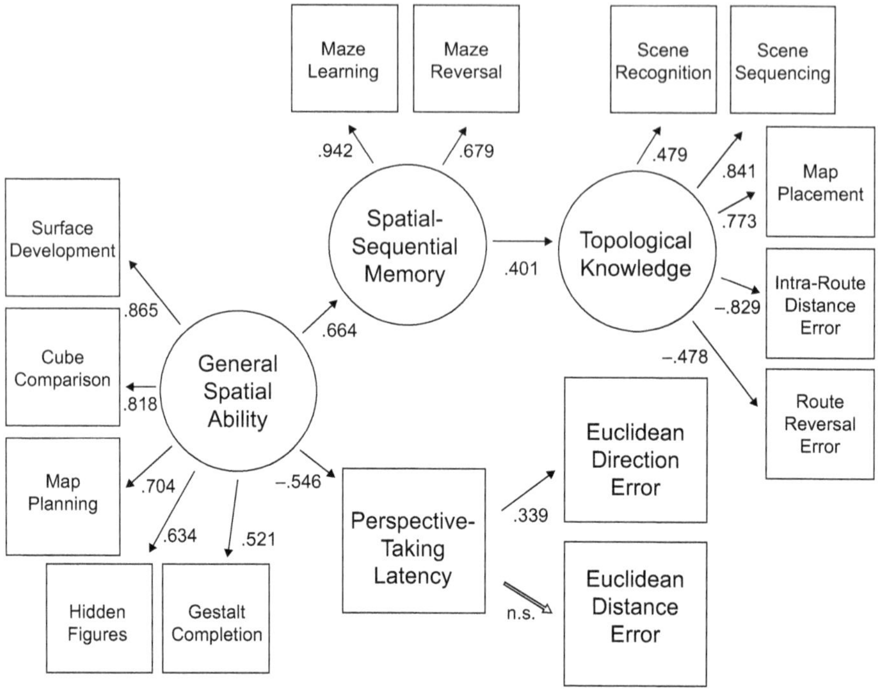

Figure 6.12 **A structural equation modeling analysis of participants' performances on various spatial tasks revealed the separation of the factor of psychometric spatial ability and the factor of environmental learning. The factor of environmental learning was further decomposed into a component of sequential knowledge and a component of configurational knowledge. Reprinted with permission of Elsevier from Allen et al., 1996, p. 344.**

separate factors. A question of concern next is how and whether the two kinds of abilities are correlated. As seen in earlier chapters, some people have difficulty with spatial orientation (e.g., Binet, 1894; DeSilva, 1931; Hudson, 1922; Iaria and Barton, 2010; Peterson, 1916; Trowbridge, 1913; Warren, 1908) and they are described as having a poor "sense of direction." What is sense of direction, how is it measured, and how is it related to cognitive mapping and spatial orientation in the environment? To tackle these questions, researchers conceived the concept of sense of direction as a self-report measure and showed that participants' self-reported scores predict their performances on spatial tasks that require an accurate configurational understanding of the environment. Kozlowski and Bryant (1977) used a simple one-item questionnaire to assess the construct of sense of direction, asking "How good is your sense of direction?" Students' responses to the question on a 7-point scale (1 = *poor*; 7 = *good*) correlated 0.49 ($p < 0.01$) with the accuracy with which they pointed to unseen buildings on campus. When the students were classified into good and poor sense-of-direction groups based on their responses to the question, a mean pointing error for a good sense-of-direction group was 19.3°, while that for a poor sense-of-direction group was 33.2°, the latter group making a significantly larger pointing error. Another group of students was tested in an unfamiliar underground tunnel pathway, with limited visual access to outside views. After walking the underground pathway, participants pointed from one end to the other end of the pathway. Mean pointing errors for a good sense-of-direction group decreased in the course of four trials, but not for a poor sense-of-direction group. Mean errors on the fourth trial were about 25° and 60° for the good and poor sense-of-direction groups, respectively.

The studies mentioned above revealed the separation of performances on psychometric spatial tests and environmental learning tasks, and the predictability of the latter by self-report sense of direction. In light of these findings, Hegarty et al. (2002) developed a standardized scale for assessing people's sense of direction, called the Santa Barbara Sense-of-Direction Scale. The scale consists of 15 statements about navigational abilities, preferences, and experiences:

1. I am very good at giving directions.
2. I have a poor memory for where I left things.
3. I am very good at judging distances.
4. My "sense of direction" is very good.
5. I tend to think of my environment in terms of cardinal directions (N, S, E, W).
6. I very easily get lost in a new city.
7. I enjoy reading maps.
8. I have trouble understanding directions.
9. I am very good at reading maps.
10. I don't remember routes very well while riding as a passenger in a car.
11. I don't enjoy giving directions.
12. It's not important to me to know where I am.
13. I usually let someone else do the navigational planning for long trips.
14. I can usually remember a new route after I have traveled it only once.
15. I don't have a very good "mental map" of my environment.

Participants answer the degree to which they agree with each statement on a 7-point scale, from 1 (*strongly agree*) to 7 (*strongly disagree*). It is scored so that a higher score indicates a better sense of direction (responses to items 1, 3, 4, 5, 7, 9, and 14 are reflected, and the mean of the 15 responses is calculated). Hegarty et al. reported that the scale has a good internal consistency (a coefficient alpha of 0.88) and a high test–retest reliability (a correlation of 0.91 for two administrations separated by 40 days).

Hegarty et al. (2002) examined the relationship of the scores of sense of direction assessed by this scale and the accuracy of spatial orientation assessed by pointing to objects and landmarks from memory at two different scales of space. They asked 24 students to estimate directions between four objects laid out in a laboratory (at the scale of vista space) and directions between four well-known landmarks on a university campus (environmental space). Hegarty et al. hypothesized that sense of direction taps into the ability to stay oriented and understand spatial relations in the environment with locomotion, and hence would relate to the accuracy of configurational knowledge acquired from direct navigation in the environment. Consistent with this hypothesis, sense of direction was correlated significantly with pointing errors on campus (correlations around –0.50, $p < 0.05$), but not with pointing errors in the laboratory (correlations around –0.20, nonsignificant).

Hegarty et al. (2006) examined the accuracy of configurational understanding of environments at different scales in relation to various individual difference measures of visual-spatial ability, sense of direction, and verbal and reasoning ability. Hegarty et al. asked 221 students to learn three different environments by direct and simulated walking along routes in the environments. First, participants viewed a videotape showing a walk through two floors of a local building; along the route, seven landmarks were pointed out to participants. They then viewed the videotape again from the starting point; they stopped at each landmark and estimated directions and straight-line distances to two other landmarks, and finally drew a sketch map of the route. Second, participants walked a single-floor route in a desktop virtual environment, along which four landmarks were pointed out. They then walked the route for the second time, stopping at each landmark, and estimated directions and straight-line distances at each landmark to two other landmarks, and finally drew a sketch map of the route. Third, participants walked through two floors of a campus building and learned eight landmarks along the route. During the second walk on the route, participants stopped at each landmark and estimated directions and straight-line distances to two other landmarks, and drew a sketch map of the route. Thus, in each environment, participants traveled a route and conducted spatial tasks that were designed to assess their configurational understanding of the traveled route.

In addition, participants took visual-spatial ability tests, including the Mental Rotations Test, Embedded Figures Test, Arrow Span Test, and Perspective Taking Test (for the Mental Rotations and Embedded Figures Tests, which are similar to the Card Rotations and Hidden Figures Tests, respectively, see above, Figures 6.1 and 6.7). The Arrow Span Test assesses the capacity of spatial working memory, in which participants were presented with a series of arrows on a computer screen one at a time, in one of eight possible directions, and asked to indicate the directions of the arrows in their order of presentation. Participants viewed 15 sets of arrows in total, the number of arrows in a set ranging from two to six. In the Perspective Taking Test, participants viewed four objects placed at the center of each wall of an 8-m square room. Then, they were asked, out of view of the objects, to imagine standing in front of one of the objects, facing the middle of the room, and to point to another object. There were 12 trials in total. Participants also filled out the Santa Barbara Sense-of-Direction Scale.

Finally participants took verbal and reasoning ability tests, including the Vocabulary Test, Reading Span Test, and Abstract Reasoning Test. In the Vocabulary Test, participants selected from five words the one that is closest in meaning to a target word (24 items in total). In the Reading Span Test, participants read a set of unrelated sentences aloud one at a time and recalled the final word of each sentence at the end of the set (20 sets of sentences in total). In the Abstract Reasoning Test, participants viewed a series of figures sequenced according to a specific rule, and selected from a set of five figures the one that would be the next item in the series (50 items in total).

Participants' responses to these environmental learning tasks and psychometric tests were examined through factor analysis and structural equation modeling. The results showed that measures of configurational learning for the three environments were grouped into two factors: a factor that concerns the ability to learn a spatial layout from visual media (a videotape and a desktop virtual environment) and a factor that concerns the ability to learn a spatial layout from direct experience in the environment (Figure 6.13). The results further showed that psychometric spatial ability was more strongly related to environmental

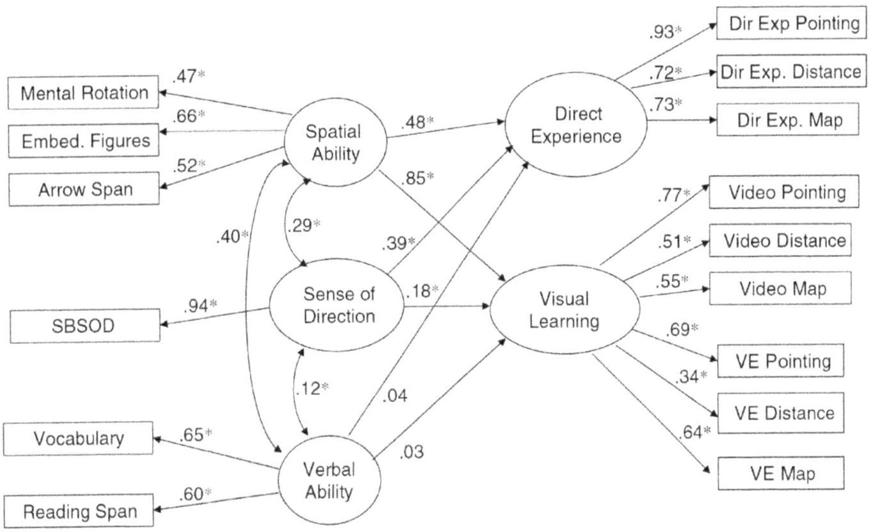

Figure 6.13 **A structural equation modeling analysis of measures of configurational learning of three different environments yielded a factor that concerns the ability to learn spatial layout from visual media (a videotape and a desktop virtual environment) and a factor that concerns the ability to learn spatial layout from direct experience in the environment. Reprinted with permission of Elsevier from Hegarty et al., 2006, p. 166.**

learning from visual media than to environmental learning from direct experience. By contrast, sense of direction was more strongly related to learning from direct experience than to learning from visual media. Verbal and reasoning ability was not significantly related either to environmental learning from visual media or to environmental learning from direct experience (see Figure 6.13). The indirect relationship between environmental learning and psychometric spatial ability mediated by perspective taking ability, which was found significant by Allen et al. (1996), turned out to be nonsignificant in this study.

Spatial ability and spatial thinking

As the early interest in measuring spatial ability was related to the attempt to predict career aptitude and vocational success in mechanical and technical occupations (Smith, 1964), the significance of spatial ability has long been recognized from a practical point of view. In the context of pedagogy, researchers have shown that students' spatial ability correlates significantly with their success in science, technology, engineering, and mathematics (STEM) disciplines, including physics, chemistry, biology, mathematics, engineering, geoscience, anatomy, and surgery (Birchall, 2015; Carter, LaRussa, and Bodner, 1987; Churchill et al., 1942; Downs and Liben, 1991; Hall-Wallace and McAuliffe, 2002; Hegarty et al., 2009; Keehner et al., 2004; Mathewson, 1999; Pallrand and Seeber, 1984; Piburn, 1980; Rochford, 1985; Russell-Gebbett, 1984; Tuckey and Selvaratnam, 1993; Wanzel et al., 2002).

Finally, spatial thinking relates to our everyday spatial behavior, ranging from reading a weather map in a newspaper to finding the way in a newly visited environment. Inasmuch as we live and act in space, our cognition and behavior involve spatial components, as we interact with, learn about, and think about the surrounding environments—space matters to us and spatial thinking matters to our spatial behavior.

The discussion of spatial thinking in the National Research Council's (2006) report centered on three major perspectives: concepts of space, use and understanding of representations, and processes of reasoning. The report described concepts of space as the "building blocks for spatial thinking" (p. 18) and identified spatial concepts that apply across disciplines and ones that are tailored to a particular discipline. Basic spatial concepts include the primitive notions of space, time, identity, location, or magnitude, and higher-order concepts derived from them, such as order, connection, distance, angle, boundaries, shape, or regions. Spatial thinking entails an understanding of these spatial concepts in an interrelated manner, in the sense that the spatial structures, distributions, and relationships of individual objects in space are represented in a given frame of reference. Representations, either in the mind or external to the mind, allow us to comprehend the density, dispersion, and pattern of spatial phenomena, to compare multiple entities in size, distance, orientation, or shape, and to appreciate the direction and speed of movement of entities and phenomena in space. Finally, one can apply operations and transformations to the mental and external representations, and reason about the processes underlying the occurrences and distributions of objects and phenomena observed in space. Spatial thinking is thus, in an elaborate form, a multifaceted endeavor that links between the knowledge of spatial concepts and the use of spatial representations for the purpose of reasoning in analytic, strategic, and context-specific ways. It is often found to be an effortful and difficult process by students, and calls for formal education and instruction (p. 48).

Hegarty (2010) discussed spatial thinking with an emphasis on flexible choice of suitable strategies and adaptive use of spatial representations. When encountering spatial tasks and spatial problems, one can employ different strategies to solve them. One possible strategy is using mental imagery or visualization. For example, consider the tasks of developing a mechanical device, reasoning about the formative process of a geologic structure, and comparing the structures of molecules. In these spatial tasks, mechanics mentally envision the movements of individual gears to consider how the mechanical device works, geologists visualize hidden parts of the geologic structure to understand its formation process, and chemists envision and mentally rotate the molecules in three dimensions to judge whether they are the same or mirror images. Another possibility is using analytic strategies; that is, decomposing a task into subcomponents or applying rule-based reasoning. In the mechanical example above, one can note that any two interlocking gears move in opposite directions. In the chemistry example of comparing two molecules, one can see if two of the bonds to a central carbon atom are identical; if so, one molecule superimposes on its mirror image. Such alternative analytic strategies (or heuristics) augment the effortful process of mental visualization, and are especially helpful for tasks that are complex and for people with low spatial ability. In fact, in each specific domain there are usually analytic strategies that experts in the domain favor.

In spatial thinking, external representations, such as maps, graphs, and diagrams, also play important roles but people tend to have difficulties with them (Ishikawa and Kastens, 2005; Ishikawa et al., 2011). You may have the experience of finding some advanced spatial

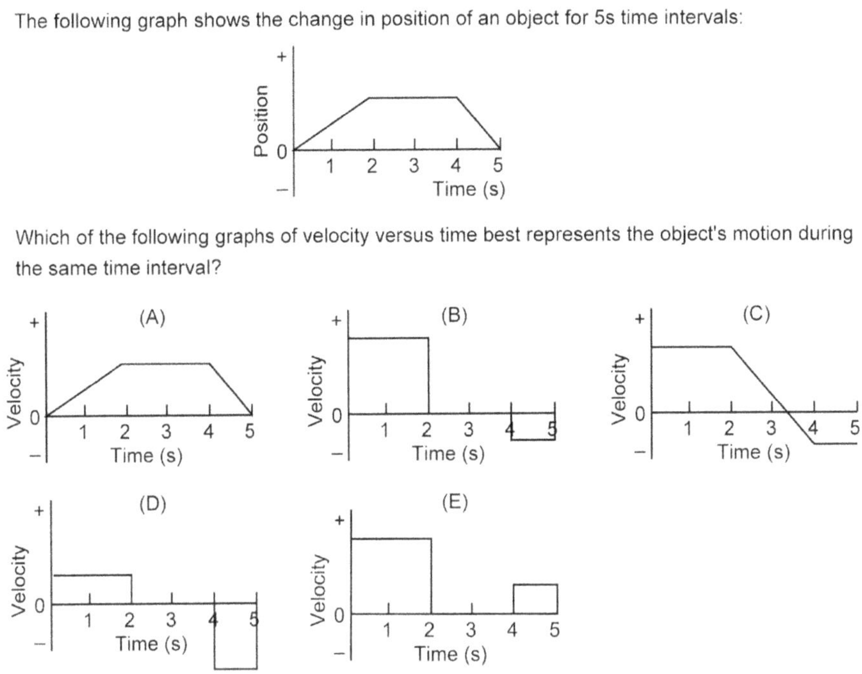

Figure 6.15 **Students with low spatial ability in a physics class had difficulty understanding graphs of the relationship between position, time, and velocity. In this example, the correct answer is D (the object moved with a constant velocity from 0 to 2 s, did not move from 2 to 4 s, and then moved in the opposite direction and twice as fast as in the first interval, before coming to a stop at 5 s). They tended to interpret a graph depicting an object's position as a function of time literally as a picture of an upward slope, a flat surface, and a downward slope (and choose alternative A). Reprinted with permission of John Wiley & Sons, Inc., from Kozhevnikov, Motes, and Hegarty, 2007, p. 554.**

visualizations such as 3-D images or animations difficult to understand. McDermott, Rosenquist, and van Zee (1987) found that students in an introductory physics course had difficulty relating features on a graph to the physical concepts that the graph represented. For example, some students did not understand that a position versus time graph for uniform motion on a level track is sharply inclined; they expected that the shape of the graph should be horizontal, resembling the shape of the track. Kozhevnikov, Motes, and Hegarty (2007) similarly showed that students with low spatial ability had difficulty understanding graphs of physics problems that depicted the relationship between position, time, and velocity; they tended to interpret a graph depicting an object's position as a function of time literally as a picture of an upward slope, a flat surface, and a downward slope (Figure 6.15). People thus differ in the ability to understand and use external spatial representations.

People also differ in the ability to choose spatial representations suitable to the task at hand. Hegarty (2010) showed that people tend to prefer realistic representations to abstract ones (e.g., detailed, three-dimensional, or animated representations vs. simplified, two-dimensional, or static representations). People also choose representations that are more realistic than is necessary, although extraneous realism added often interferes with the user's understanding and reasoning (e.g., Berney and Bétrancourt, 2016; Höffler and Leutner, 2007; Tversky and Morrison, 2002). In a task of comparing the pressure and temperature in different regions on weather maps, people preferred maps with extraneous information (features of terrain and state boundaries) or irrelevant information (e.g., temperature information in the task of identifying pressure). When using such maps, novices responded more slowly and less accurately than when they used simple maps with only task-relevant information. Being able to decide on an appropriate type of representation and level of detail constitutes an important part of being a good spatial thinker.

Now consider spatial thinking as it pertains to our everyday activities. Packing a suitcase, arranging furniture in a room, walking around in a multistoried shopping mall, and making a detour when a usual route is blocked—these are but a few examples of our everyday spatial behaviors. We live in space, perceive the surrounding space, and acquire knowledge and feelings about it. This is why we repeatedly discuss in this book that our behavior is based on what we know and how we feel about the environment. Importantly, however, as we discussed in Chapter 5, people's knowledge about the surrounding environments shows large individual differences: although being in the same environment, some people have an accurate cognitive map and some people have an incomplete and distorted knowledge representation (Ishikawa and Montello, 2006; Newcombe, 2010). This stimulates research into spatial thinking, and particularly cognitive mapping, from an individual differences perspective.

The idea of individual differences in cognitive mapping

To start a discussion of spatial knowledge acquisition from an individual differences perspective, it is worth looking at the fact that people's confidence in spatial orientation and wayfinding differs greatly. Some people enjoy visiting new places and find their way easily anywhere. Others are reluctant to visit new places because they have difficulty knowing where they are and staying oriented in unfamiliar environments. The latter people unfortunately have poor cognitive maps; their internally represented environments are distorted, fragmented, and schematized (Golledge and Spector, 1978; Lynch, 1960; Stevens and Coupe, 1978). In a phrase that is commonly used, they have a poor sense of direction (Kozlowski and Bryant, 1977).

Sense of direction concerns the ability to orient oneself in the environment by comprehending the two-dimensional (or higher) layout of the surrounding space. It is thus an important measure of environmental spatial ability, or cognitive mapping, and correlates moderately (and better than psychometric spatial ability tests do) with people's configurational understanding of the environment (Hegarty et al., 2006; Ishikawa and Montello, 2006). To orient yourself in space and know where you are, you need to have an accurate representation of the space in your head and be able to use the representation flexibly by aligning it with the surrounding space.

Such accuracy and flexibility require that individual places be interrelated with each other in a common frame of reference and integrated into a whole, maplike image. As discussed above, people do differ in their ability to acquire such comprehensive knowledge. In fact, it is commonly observed that people's knowledge of the environment is incomplete. For example, people know the name of a building but do not know where it is located; people have been to a certain theater but cannot describe how to get there; people cannot reach a frequented restaurant if they deviate from a usual route; or people are familiar with two neighborhoods but do not know how far apart they are and in which direction they are located from each other. Difficulty with spatial orientation has long been reported (Binet, 1894; DeSilva, 1931; Hudson, 1922; Iaria and Barton, 2010; Peterson, 1916; Trowbridge, 1913; Warren, 1908). One may ask then, why?

This question points to an important step in spatial learning and orientation, namely the acquisition of survey knowledge (see Chapter 5). In survey knowledge, whether it emerges in the final stage of a qualitative shift or as a result of quantitative detailing and fine tuning, separate landmarks and routes, even those learned during different travel experiences, are integrated and interrelated with each other, with at least approximate distances and directions between them being available. Metric scaling and integration renders survey knowledge a comprehensive and flexible representation, and is realized by elaborate cognitive processing. But how well can people accomplish the difficult task of acquiring integrated and comprehensive survey knowledge?

The results of Ishikawa and Montello (2006), which we discussed in Chapter 5, provide the insight that spatial knowledge acquisition, particularly the progression from route to survey knowledge, is more correctly characterized as an individual difference trait rather than a universal developmental progression. There are large individual differences in the accuracy and developmental pattern of spatial knowledge—individuals differ strikingly in their levels of performance and their tendency to improve in their configurational understanding of the environment. Some people acquire accurate cognitive maps more or less immediately in a new environment and maintain the high level of performance throughout. Others are quite poor at understanding the environment configurationally and never show much improvement. Still others are intermediate in their performance. In the Ishikawa and Montello experiment, half of the 24 participants improved significantly in performance across the ten weeks, but the amount of improvement was not particularly large. Mostly, people were either good or poor survey learners from beginning to end; a few intermediate learners achieved survey performance comparable to that of good learners after accumulating experience in the environment.

Ishikawa and Montello (2006) concluded that neither the dominant nor alternative frameworks for spatial microgenesis fully explain the large and significant individual variations. People's configurational understanding of the environment can be extremely accurate and precise or quite poor. Whereas the discrete framework by Siegel and White (1975) assumes no metric information in the initial stages of spatial microgenesis, which consists of isolated occurrences and sequential chaining of landmarks, people's route distance estimates are quite accurate right away, with correlations greater than 0.90. Thus, the space between landmarks along routes is not empty in early stages. People's direction estimates are also at least better than expected by mere chance. One need not undervalue what Tolman (1948) meant by the term "cognitive maps" (Ishikawa and Montello, 2006, p. 123) when he described spatial knowledge as "something like a field map" (p. 192) and "not altogether strip-like and narrow" (p. 205). Montello's (1998) continuous framework may apply to spatial learning by good survey learners but the extent of individual differences is larger than it assumed. This important point cannot be appreciated by aggregate data analyses, in which individual variations are relegated to the error term.

Finally, although at least some people's early spatial knowledge is not without metric scaling, it is not necessarily very precise. The fact that people's performance on distance and direction estimation is better than chance only denies the assumption that people have no metric knowledge. It tells us that people have at least minimal metric understanding, but not that they acquire knowledge which is precise or "metric" in the mathematical sense of the term (as in following the metric axioms). To examine the level of metric precision of their participants' spatial knowledge, Ishikawa and Montello (2006) conducted Monte Carlo simulations of various heuristic models differing in the sophistication of their metric understanding. Results showed that participants estimated within-route directions by paying attention to the shapes of the routes, and estimated route distances with accuracy equal to or better than the ordinal level. They estimated between-route directions and drew integrated sketch maps by understanding more than the general relationship of left and right, but estimated straight-line distances, both within and between routes, with accuracy worse than that expected by a categorical (short vs. long) judgment heuristic. Thus, people's spatial knowledge possesses quantitative information but is represented in an approximate way (Ishikawa and Montello, 2006, p. 124; see also Chrastil and Warren, 2014; Foo et al., 2005; Frank, 1996; Yeap and Jefferies, 2000). In this regard, Chrastil and Warren, for example, proposed the idea of spatial knowledge as a "cognitive graph" that is composed of a network of topological connections labeled with local metric information. When considering approximate or qualitative metric knowledge, however, it should be remembered that some people have a configurational understanding that is extraordinarily accurate; a conclusion obscured by aggregate analyses.

The individual difference view of spatial knowledge acquisition has since been further pursued and corroborated (Schinazi et al., 2013; Weisberg and Newcombe, 2016; Weisberg et al., 2014). Weisberg and Newcombe and Weisberg et al. replicated the Ishikawa and Montello (2006) setting of two routes and a connecting route (which Weisberg and Newcombe called a route integration paradigm) on a college campus and in a virtual simulation (Figure 6.16). They examined people's configurational understanding of individual and integrated routes through the tasks of within- and between-route pointing and map making. Based on performance on these survey tasks, they classified participants into three groups: (a) integrators, who performed well on both within- and between-route judgments; (b) non-integrators, who did well on within-route but poorly on between-route judgments; and (c) imprecise navigators, who did poorly on both within- and between-route judgments (Figure 6.17). Schinazi et al. guided 16 participants along two separate routes and a connecting route, and examined their performances on direction and distance estimation and map drawing. Schinazi et al. showed that the size of the hippocampus (gray matter volume in the right hippocampus) was positively correlated with the accuracy of direction estimates, particularly pointing between landmarks on two different routes from memory ($r = 0.61$, $p < 0.05$).

A conclusion from the work in the route integration paradigm and from an individual differences perspective is that it is not very meaningful to ask whether people's cognitive maps are maplike. Rather, it is important to acknowledge that people differ in their ability to comprehend the environment in a configurational manner. It is more fruitful to explore how and why people differ so much in the fundamental ability of cognitive mapping (Hartley and Harlow, 2012; Marchette, Bakker, and Shelton, 2011; Ploran et al., 2015; Wolbers and Hegarty, 2010). In particular, Wolbers and Hegarty discussed variations

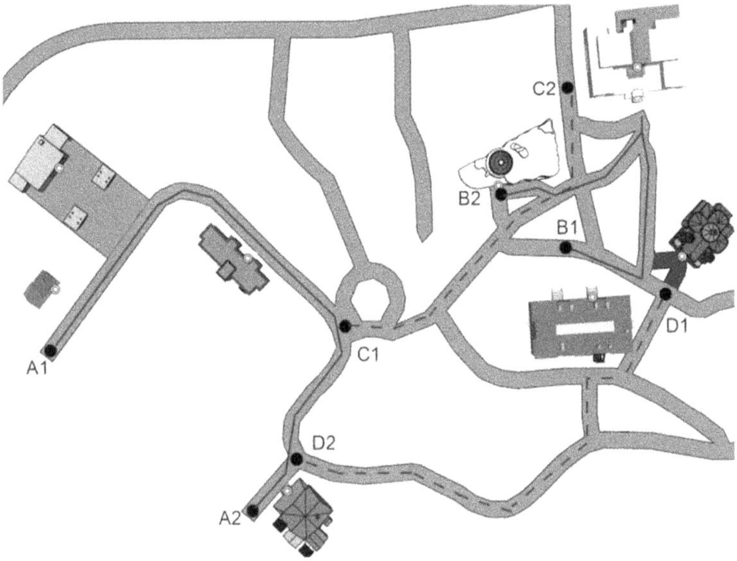

Figure 6.16 **Experimental setting of two routes (A1–A2 and B1–B2) and connecting routes (C1–C2 and D1–D2) in a route integration paradigm, on a college campus and in a virtual simulation. Reprinted from Weisberg and Newcombe, 2016, p. 771.**

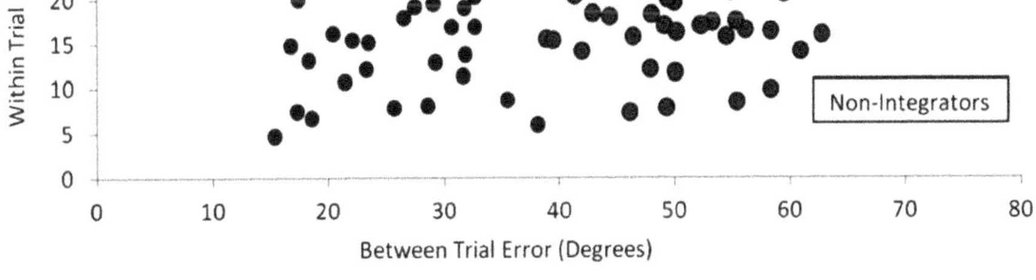

Figure 6.17 **Based on performances on the tasks of within- and between-route pointing and map making, participants were classified into three groups: (a) integrators, who performed well on both within- and between-route judgments; (b) non-integrators, who did well on within-route but poorly on between-route judgments; and (c) imprecise navigators, who did poorly on both within- and between-route judgments. Reprinted from Weisberg and Newcombe, 2016, p. 773.**

in navigational abilities with respect to "the precision with which spatial information is encoded from sensory experiences, the ability to form spatial representations of external environments and the efficacy with which they are used to guide navigational behavior" (p. 138). Specifically, they pointed as possible sources of individual differences to the use of environmental cues (local and distal landmarks, global orientation cues, geometric structure of the environment, and symbolic representations), the processing of self-motion cues (vestibular cues, motor efference copies, proprioceptive feedback, and optic, auditory, or tactile flow), the function of computational mechanisms and processes (e.g., computing distances and directions, imagining shifts in spatial perspective, translating between egocentric and allocentric reference frames, or route planning and decision making), and the accuracy of mental representations (e.g., self-position and orientation, egocentric and allocentric directions and distances, or memories of local views, places, and spatial relations).

Training in the skill of cognitive mapping

Given that spatial knowledge acquisition is more of an individual difference trait than a progression and the extent of individual differences is strikingly large, it is of concern to know whether the trait or ability can be improved through any kind of instruction. The issue of malleability or trainability has long been discussed with respect to basic spatial ability assessed by psychometric spatial tests. Kail and Park (1990) examined the effect of practice on mental rotation for children and adults. Participants in a training condition (eight children and eight adults, with mean ages of 11.1 and 20.1 years, respectively) practiced on a set of 96 mental rotation items three times per session, for a total of 11 sessions. Participants were given feedback on their mean response time and total number of correct responses, and received a monetary incentive for improved performance in comparison to the previous session. They also took a pretest and a posttest, before and after the 11 practice sessions, that consisted of the basic set of 96 mental rotation items. Participants in a control condition (eight children and eight adults, with mean ages of 11.6 and 18.9 years, respectively) took only the pretest and posttest. Kail and Park found that the rate of mental rotation increased substantially for children in the training condition (from 4.52 to 0.83 milliseconds per degree), but not for children in the control condition (means of 2.60 msec/deg). The rate increased slightly for adults in the training condition (from 1.71 to 0.63 msec/deg) and in the control condition (from 2.21 to 1.28 msec/deg). Developmental curves in Figure 6.18 show that mental rotation became substantially faster during the first few hundred trials for children, and became slightly faster over the course of trials for adults; the rates of mental rotation for the two groups were different at first (children were slower than adults) but became comparable with repeated practice.

Uttal et al. (2013) conducted a meta-analysis of 206 studies (for a total of 1,038 effect sizes) investigating the effect of training on spatial skills. The studies included in the meta-analysis used various spatial tests, including the ones that we discussed above, and the methods of training ranged from playing a video or computer game to being enrolled in a semester-long spatially relevant course (e.g., engineering graphics, chess lessons, or geology), to receiving spatial training through practice, strategic instruction, or computerized lessons. The meta-analysis showed that people's spatial skills are improved by a variety of training procedures, with the size of improvement almost one half of a standard

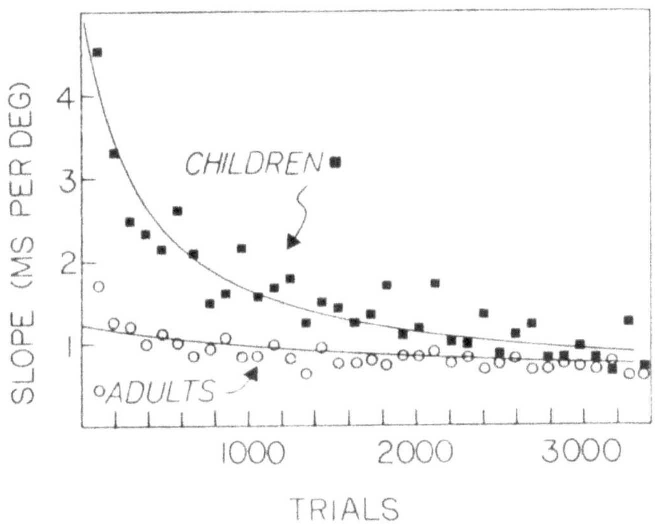

Figure 6.18 **Spatial ability and training. Mental rotation became substantially faster during the first few hundred trials for children, and became slightly faster over the course of trials for adults. With repeated practice, the rates of mental rotation for children and adults became comparable. Reprinted with permission of Elsevier from Kail and Park, 1990, p. 238.**

deviation. It also pointed out that even repeated testing has the effect of improving performance scores for people in control groups.

It does not, however, seem to be so straightforward to improve the skill of cognitive mapping or the ability to learn and understand large-scale spaces configurationally. A hint to this question may be obtained by considering the differences between good and poor survey learners in the encoding process of information about a spatial environment. Wen, Ishikawa, and Sato (2011, 2013) examined the involvement of different components of working memory in spatial knowledge acquisition by people with differing abilities. They showed that people with a good sense of direction used a rational combination of verbal, visual, and spatial processing, while people with a poor sense of direction relied mainly on verbal processing (Figure 6.19). Specifically, people with a good sense of direction encode landmarks and routes verbally and spatially, and integrate them into survey knowledge through all three components of working memory. By contrast, people with a poor sense of direction encode landmarks verbally, without spatial processing, and fail to acquire accurate survey knowledge.

On the basis of this model, Wen, Ishikawa, and Sato (2014) examined the effects of instructing people to learn videotaped routes either verbally or spatially. In a verbalization condition, participants mentioned things in the scenes that they thought were important for learning the route. In a spatial-operation condition, participants moved a model car on a desk according to the traveler's movements and turns, and placed a small object on the desk in relation to the car to indicate the locations of landmarks while viewing a video.

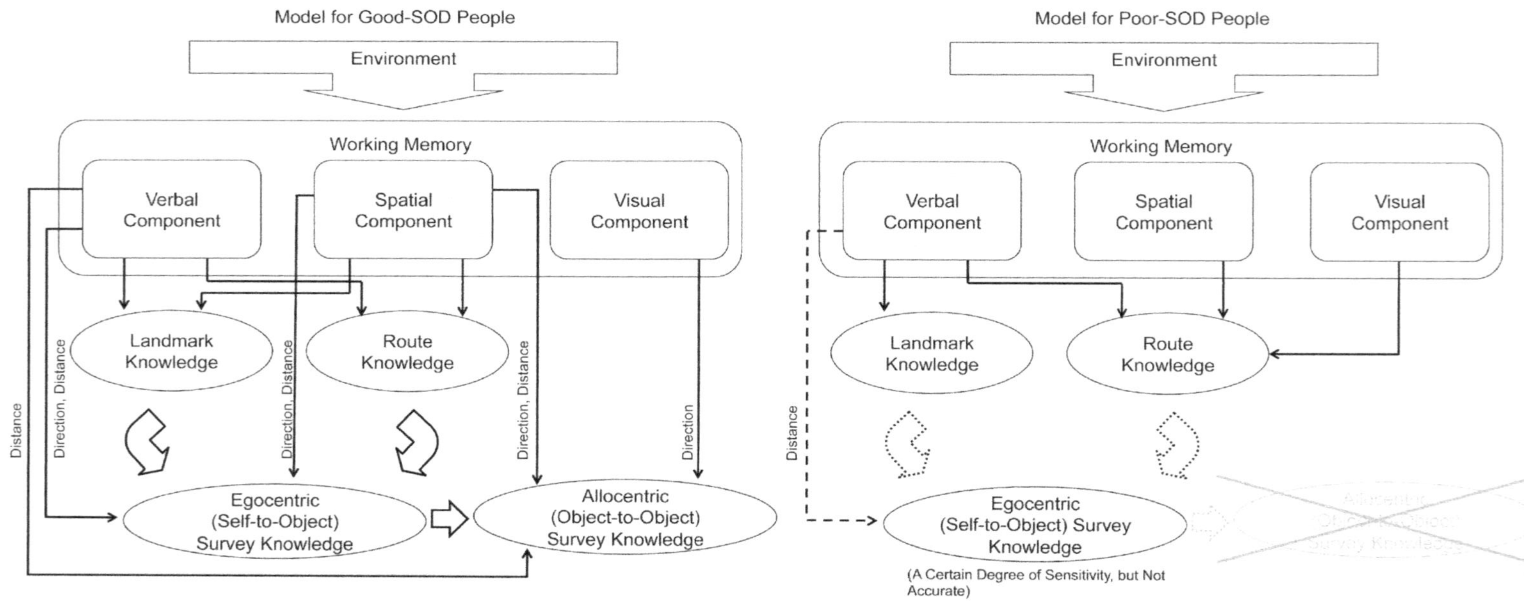

Figure 6.19 **Processing of spatial knowledge by people with differing abilities. People with a good sense of direction use a rational combination of verbal, visual, and spatial processing, and integrate landmarks and routes into survey knowledge; while people with a poor sense of direction rely mainly on verbal processing, and fail to acquire accurate survey knowledge. Reprinted with permission of John Wiley & Sons, Inc., from Wen, Ishikawa, and Sato, 2013, p. 189.**

For landmark knowledge, the learning by people with a good sense of direction was facilitated by both verbalization and spatial operation; learning by people with a poor sense of direction was facilitated by verbalization but hindered by spatial operation. Thus, instructing people to use the process that they usually use works for landmark learning (see Figure 6.19). For survey knowledge, however, learning by the former people deteriorated and learning by the latter people did not change, with either verbal or spatial instruction. These results show that people with good and poor sense of direction differ in the encoding processes of different types of spatial knowledge, but instructing people to use a specific encoding process does not necessarily help. In particular, it is not straightforward to facilitate survey learning by people with a poor sense of direction; they are insensitive to instruction that aims to induce them to use either the process that they are more accustomed to (verbal processing) or the process that good survey learners use (spatial processing). Weisberg and Newcombe (2016) likewise found that manipulating motivation with a monetary incentive for good performance did not enhance participants' survey learning. The ineffectiveness of instruction for survey learning contrasts with the malleability of psychometric spatial ability by practice. To develop effective instructional methods for survey learning is a future research question of theoretical and practical importance.

Review box for Chapter 6

1. List major spatial ability tests, and classify them into groups based on the types of abilities that they assess.
2. Summarize the results of factor analytic studies of psychometric spatial ability tests. Explain the major factors that these studies observed.
3. Discuss the distinction between mental rotation and perspective taking. Conduct the array rotation and viewer rotation tasks by yourself. Which do you find easier?
4. Discuss the relationship between spatial ability assessed by psychometric spatial tests and performance on spatial learning (cognitive mapping) in the environment. What do you think about spatial abilities at different scales?
5. How good is your sense of direction? Measure it with the Santa Barbara Sense-of-Direction Scale. Is your SBSOD score high or low? Does that agree with your intuition or past experience?
6. Summarize research findings about the relationship between spatial ability and STEM learning.
7. Are you high or low on spatial ability, verbal ability, and sense of direction? What do you think is the relationship among these three kinds of abilities?
8. Discuss what spatial thinking is.
9. Summarize the findings of studies of cognitive mapping from an individual differences perspective.
10. On the basis of your experience, discuss whether sense of direction can be improved by training. Why or why not?

Chapter 7

Communicating information about space

The foregoing chapters mainly dealt with internal representations of the environment; that is, what is contained in the knowledge about the environment and how it is represented in the human mind. This chapter focuses on external representations of the environment, namely expressions of the knowledge about the environment in words or as visual representations, particularly in the context of communicating knowledge about surrounding environments to other people and assisting people in navigation and wayfinding. The first section looks at spatial language as a medium of communicating knowledge about space—verbal descriptions or navigational directions—by clarifying the concept of spatial frames of reference as it pertains to the conceptualization of space. The second section discusses the understanding and use of maps, a major medium of geographic knowledge since antiquity, particularly the characteristics and potential difficulty of spatial orientation with maps in the environment. The third section discusses recent technologies of navigation assistance, or satellite navigation systems and applications, and the characteristics of spatial behavior by users of such advanced geospatial tools.

Spatial language

Spatial frames of reference

When you are asked directions to the nearest restroom, you say, "Go down the hall and make a right, then you will see it on your left." When you tell your partner where you are waiting at the airport, you say on the cellphone, "I am in the North Terminal, standing in front of a cafe on the second floor." When you give directions to a taxi driver, you say, "Go down south on Smith Avenue and take the freeway up to Turnpike; the third house after the exit is my place." To describe *where* you are, *where* an object is, and *where* a place is, you need to specify the *location* of yourself, the object, and the place. How do you specify the location, then?—you do so in relation to "something" in space. The something may be a person (e.g., on your left), a concrete object (e.g., in front of a cafe), or an abstract framework imposed onto the environment such as cardinal directions (e.g., down south) or latitude and longitude (e.g., you say "see you at 35° 42′ north and 139° 45′

east tomorrow," if appropriate). In essence, this is a process of describing "where something is with respect to something else" (Levinson, 1996, p. 110). The way in which the *something* and *something else* are related constitutes a system of spatial localization, called a *frame of reference*.

Now look at the scene depicted in Figure 7.1, which consists of a house and a person. How do you describe the location of the person in the scene, or where the person is?

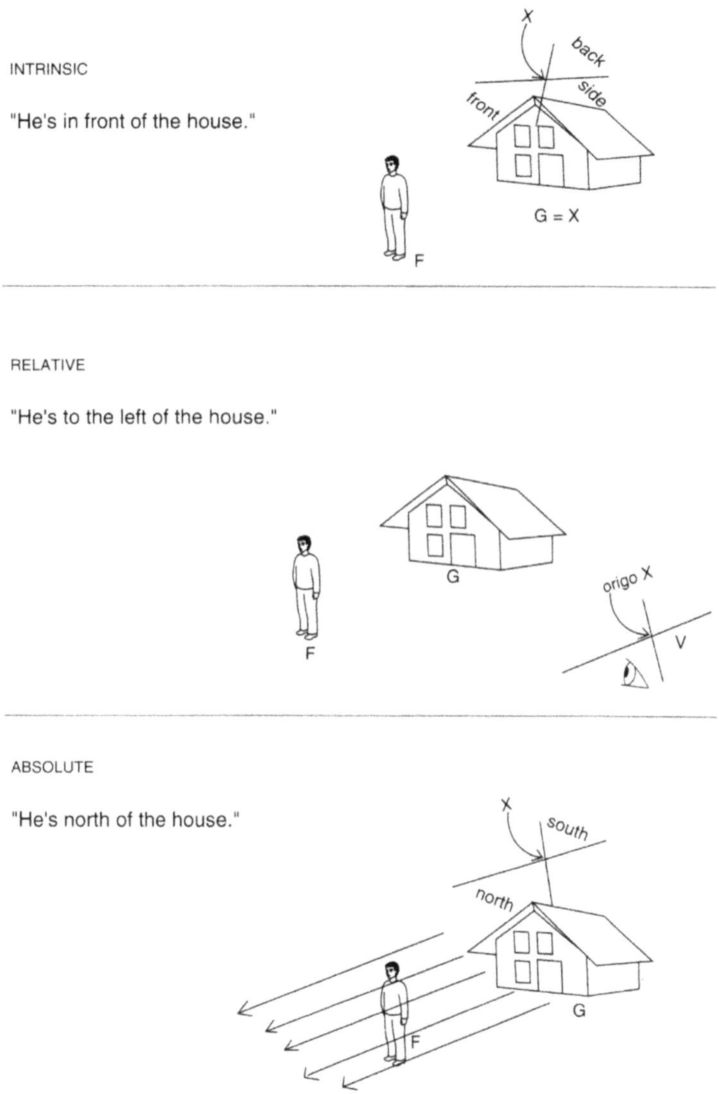

Figure 7.1 **Three kinds of spatial frames of reference: intrinsic, relative, and absolute. The same scene can be described in three different ways, using the expressions "in front of," "to the left of," and "north of." Reprinted with permission of Cambridge University Press from Levinson, 2003, p. 40.**

Levinson (1996) discussed three different ways to do that, namely three kinds of frames of reference.

One coordinate system may be centered on the house, yielding the description "the person is in front of the house" (Figure 7.1, *top*). This description works, or makes sense, because the house has its own intrinsic axis of front–back; the side having the main entrance is usually considered the *front* and the opposite side the *back*. This coordinate system is thus based on an object's intrinsic features and is called an *intrinsic frame of reference*. What gives an object intrinsic properties depends on its shape (e.g., a sphere has no intrinsic front or sides), canonical orientation (e.g., the influence of gravity on the distinction of up–down), or characteristic motion (e.g., right and left doors of a train depend on which direction it is going in).

Or you can describe spatial relations in a coordinate system whose origin is centered on the viewer: "the person is to the left of the house" (Figure 7.1, *middle*). In this system, the space in the horizontal plane is divided into four sections of *front*, *back*, *left* and *right*, tied to the viewer's body or perspective (heading or the direction of gaze), and the spatial relation of the person and the house is described relative to the viewer; thus it is called a *relative frame of reference*.

A third expression is also possible, which uses fixed directions as a coordinate system, frequently bearings fixed in the horizontal plane (e.g., cardinal directions). Suppose that the points of the compass are laid out in the environment as in Figure 7.1 (*bottom*). Then you can describe the scene as "the person is north of the house." The designators *north*, *south*, *east*, and *west* are independent of the viewer's location, and hence the coordinate system is called an *absolute frame of reference*. The fixed bearings may be coordinated with environmental gradients or topography, such as the directions of uphill and downhill. Fixations can also be made to external landmarks, such as "in the direction of the Eiffel Tower," "toward the North Star," or "on the mountain/ocean side of the town."

Spatial relations described in the three frames of reference have their own characteristics. For example, depending on which frame of reference is employed, descriptions may or may not change according to the movement of the viewer. Suppose that the viewer in Figure 7.1 now moves to the other side of the house. The description in the relative frame of reference changes to "the person is to the *right* of the house." The descriptions in the intrinsic and absolute frames of reference, in contrast, stay the same ("front of the house" and "north of the house" do not change depending on the viewer's location).

Consider another case, although somewhat unrealistic, in which the house in Figure 7.1 is replaced by a big ball. The person is still to the left of the ball and north of the ball, but can you say "the person is in front of the ball"? You can, but it becomes ambiguous now because the ball does not have an intrinsic front side; the "front" of the ball now depends on the viewer's position. If the viewer's location, under this condition, is rotated 180° counterclockwise, the scene is then described as "the person is behind the ball"; the description of the spatial relation is now *relative*, not intrinsic.

There are various other classifications of spatial frames of reference. One well-known example is the distinction of egocentric and allocentric frames of reference. In the egocentric frame of reference, the origin of the coordinate system is placed on the subjective body of the organism; in the allocentric frame of reference, the origin is centered on somewhere other than the self. In the context of cognitive mapping, Levinson (1996) elaborated on the egocentric–allocentric distinction by further differentiating between egocentric strip maps (as Tolman, 1948, used the term in contrast to cognitive maps), allocentric landmark-based maps, and allocentric maps with fixed bearings (p. 129). Another distinction that is often made is between viewer-centered and object-centered frames of reference. The distinction was made by Marr (1982) when he described vision as a computational process proceeding from a two-dimensional retinal image, through representations of the changes and distributions of light intensities (the primal sketch) and the orientations of visible surfaces (the 2½-D sketch) in a viewer-centered frame of reference, to a three-dimensional representation of the shape and spatial arrangement of an object in an object-centered frame of reference (the 3-D model representation). Levinson discussed that mental rotation of an experimental stimulus into a different orientation (e.g., mentally rotating an alphanumeric character; see Chapter 6) taps into the ability to represent an image in viewer-centered and object-centered frames of reference. Levinson also discussed the classification of viewer-centered, object-centered, and environment-centered frames of reference, which maps onto the distinction of the relative, intrinsic, and absolute frames of reference (p. 132). These classifications are not mutually exclusive, and they are used in different contexts and for different purposes. For example, Tenbrink (2018, pp. 214–215) discussed that an egocentric frame of reference may be used in a relative sense ("from my point of view, the ball is to the left of the cupboard") or an intrinsic sense ("on my left"). Similarly, an allocentric frame of reference may also be used in a relative sense ("viewed from the door, the ball is to the left of the cupboard") or an intrinsic sense ("on your left").

When you describe space in language, either by directly perceiving it or from memory, you need to "linearize" a two-dimensional spatial layout in a temporal sequence or flow of utterances. Linde and Labov (1975) examined the characteristics of linear encoding of thought into language, by asking New York City residents to describe the layout of their apartments. An analysis of descriptions for 72 layouts revealed that the residents predominantly described the layout in the form of an "imaginary tour," traversing for the listener a network of paths and rooms in the apartment. A "map" type of description detailing the shapes and spatial relations of rooms was, in contrast, a rare exception, constituting only 3 percent of the observed layouts. In the tour descriptions, Linde and Labov identified a series of regularities, in that the tour starts at the front door of the apartment, enters each room in a multi-room branch (if it is a one-room branch, it does not enter it), and moves back to the fork point where other branches originate (instead of turning around and retracing the rooms in the branch). In this way, the spatial information of the apartment is organized in linear form, with rooms mentioned in systematic order. The choice of the outside entrance to the apartment as the starting point allows the speaker to use relative directions of left and right. In a similar vein, studies of the psychology of language examined route descriptions and survey descriptions (Taylor and Tversky, 1992; see below), which respectively describe a spatial environment in a reference system whose origin is moving with the traveler (e.g., left and right turns) or fixed in the environment outside the traveler (e.g., north-south-east-west).

Language and spatial cognition

Spatial frames of reference, particularly the use of the relative and absolute frames of reference, affects people's cognition of space: speakers of different languages differ in linguistic and nonlinguistic coding of spatial relations. Studies that shed light on the differences are a series of experiments conducted by Levinson (1996) with two groups of people. One group of participants were Tenejapans, who speak Tzeltal, a dialect of a Mayan language, in the community of Tenejapa in southern Mexico. In Tzeltal, spatial relations are described in the absolute frame of reference, essentially a system of fixed bearings of cardinal directions. It is aligned with the territory's general topography of sloping toward the north-northwest; thus *downhill* means (approximately) north, and *uphill* means south. Interestingly, the language lacks relative descriptions of spatial relations such as "to the left," "to the right," or "in front." Tenejapans describe spatial relations in absolute terms at any spatial scale, for example, "the boy is on the north (downhill) side of the tree" or "pick out the uphill one of the two cups on the table" (p. 111). Another group of participants in the experiments were Dutch people, whose language predominantly uses relative expressions of front, back, left, and right in spatial descriptions, as English does.

In the basic design of the experiments, the experimenter asks the two groups of people questions about spatial relations between a set of objects and examines whether they respond in the relative or absolute frame of reference. Participants first view an array of objects on a table, such as toy animals (Figure 7.2); in this figure, an array is depicted as an arrow pointing to a specific direction. The arrow points to the right in the relative frame of reference, and it points to the north in the absolute frame of reference. After viewing the array on the table, participants rotate through 180° and view two arrows on the second table: one arrow is pointing to the right (or to the south) and the other arrow to the left (or to the north). Participants are then asked to choose the arrow that is the same as the one that they had seen on the first table. The analysis focuses on which arrow the two

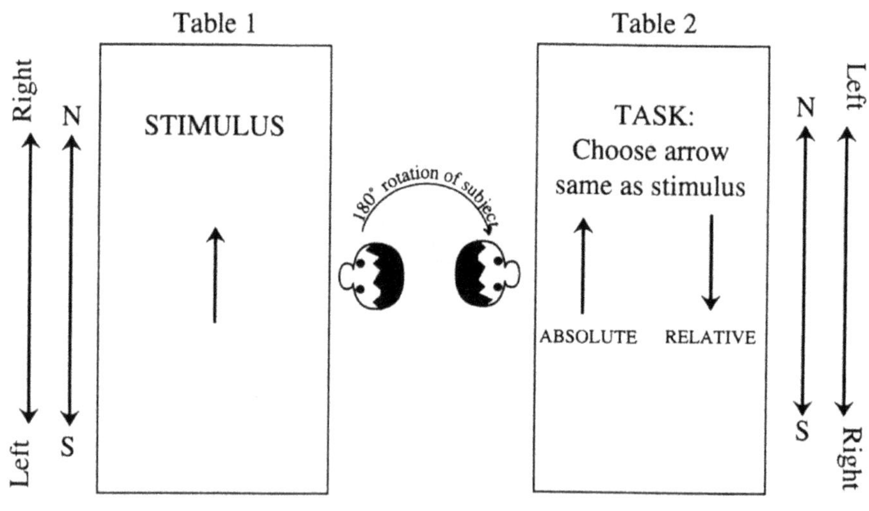

Figure 7.2 **Participants first view an array of objects on a table that are ordered in a specific way (depicted by an arrow in this figure, *left*). After viewing the array on the table, participants rotate through 180° and view two arrows on the second table (*right*). Participants then choose the arrow that is "the same" as the one that they saw on the first table. Reprinted with permission of Cambridge University Press from Levinson, 2003, p. 132.**

groups of people would choose in response to the instructions to choose the "same" arrow as before, namely an arrow that is "correct" in the relative or absolute sense.

Levinson (1996) observed a clear pattern of responses. The Dutch group consistently chose the *relative* arrow pointing to the right on the new table, while the Tenejapan group predominantly chose the *absolute* arrow pointing to the north (or to the left), which accords with the spatial relations encoded in the relative and absolute frames of reference, respectively. Levinson observed similar patterns of results in a variety of tasks including spatial recall, recognition, and inference, and summarized that the results indicate a strong correlation between the dominant frames of reference used in linguistic and nonlinguistic tasks. Levinson stated:

> There is a cross-modal tendency for the same frame of reference to be employed in language tasks, recall and recognition memory tasks, inference tasks, imagistic reasoning tasks, and even unconscious gesture. This suggests that the underlying representation systems that drive all these capacities and modalities have adopted the same frame of reference.
>
> (p. 109)

One should note that the use of the absolute frame of reference requires continuous updating of one's orientation with respect to the directions of fixed bearings. Identification of spatial relations in absolute terms is made possible by the knowledge of the direction of, say, north. Then, the fact that people who speak absolute languages employ absolute spatial encoding at any spatial scale makes it reasonable to assume that they possess good dead-reckoning ability, so that they can orient themselves relative to a fixed bearing even after a long journey from a home base. To examine the possibility, Levinson (2003) asked speakers of absolute and relative languages to point to a variety of locations after traveling tens of kilometers (up to 100 km) from home. They did not know beforehand that they would be asked directions, so the learning was incidental, and the task captured the characteristics of their spatial updating in a naturalistic setting. As shown in Figure 7.3, the accuracy of

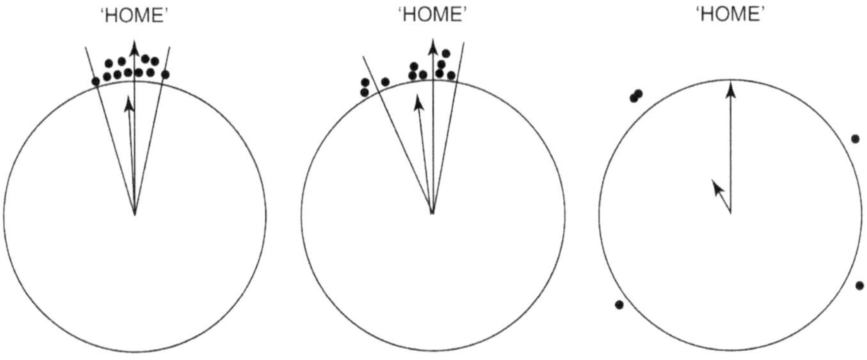

Figure 7.3 **Speakers of absolute and relative languages were asked to point to a variety of locations after traveling tens of kilometers from home. The distributions of estimated directions show that people who speak absolute languages pointed to various locations quite accurately (left and middle panels), whereas people who speak relative languages did it basically randomly (right panel). Reprinted with permission of Cambridge University Press from Levinson, 2003, pp. 233, 235, and 239.**

pointing to various locations was quite high for people who speak absolute languages, but responses by people who speak relative languages (in this case, Dutch people) were basically random. These results suggest that the consistent use of the absolute frame of reference by people speaking absolute languages in linguistic and nonlinguistic coding is made possible by (or at least correlated with) their ability to continuously update their orientations in space with respect to fixed bearings centered on the environment.

Views on linguistic relativism

The findings in the previous section are interpreted as an indication of the influence of language on human thinking and reasoning, a view generally called *linguistic relativism.* The observations are correlational in nature, however, and some researchers discussed the relationship in reverse; that is, being in different environments leads people to employ different strategies or terminologies in reasoning. To further examine the relative response patterns observed by Levinson (1996) for Dutch people (who speak a relative language), Li and Gleitman (2002) examined how speakers of English (a similarly relative language) respond to questions about spatial relations between a set of objects arranged on a table. Li and Gleitman used the same experimental setting and procedure as in Levinson's study, except that they deliberately made various environmental features available that provided landmark or bearing information for the respondents. In one experiment, Li and Gleitman asked 40 English-speaking undergraduates to view an array of three toy animals (e.g., a lion, a dog, and a rabbit) arranged on a table in a straight line (aligned with the left–right and north–south axes); then they turned around 180° to face a new table and were asked to set up the three animals on the table *the same as before.* Half the participants did it in a featureless laboratory room with a window, either with the blinds pulled down or with the blinds pulled up so that participants could have a familiar view of the campus (a university library) if they looked toward the window. The other half of the participants did it outdoors in a grassy area on the campus surrounded by buildings, with the tables and the array of toy animals arranged in the same way as in the indoor condition (i.e., aligned with the left–right and north–south axes). Results showed that participants in the indoor blinds-down condition predominantly chose to arrange the three toys in a *relatively* correct manner (i.e., heading to the left or right), as the Dutch participants in the Levinson's (1996) study did. On the other hand, participants in the indoor blinds-up and the outdoor conditions showed a bimodal pattern in their relative and absolute responses: 40 percent of the participants constantly chose to arrange the toys in a *relatively* correct manner and 40 percent constantly chose to arrange them in an *absolutely* correct manner (i.e., heading to the north or south).

In another experiment, Li and Gleitman (2002) examined the effect of a more conspicuously placed landmark on participants' spatial coding. Another 40 English-speaking undergraduates were asked to conduct the spatial tasks mentioned above in the laboratory room with the blinds pulled up. For half the participants, a symmetric toy object was placed on the right (or south) edge of the first table, and the same toy object was placed on the right (or north) edge of the second table that they viewed after turning 180°. For the other half of participants, the toy object was placed on the right (or south) edge of the first table and on the left (or south) edge of the second table (see Figure 7.4). That is, in the former condition, the object was always to the right of the participant (a relative landmark

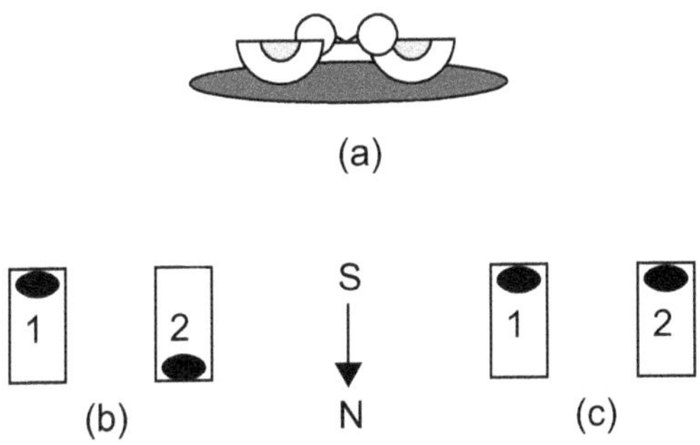

Figure 7.4 **Examination of whether a conspicuously placed landmark affects participants' spatial coding. For half the participants, a symmetric toy object (a) was placed on the right (or south) edge of the first table, and the same toy object was placed on the right (or north) edge of the second table that they viewed after turning 180° (a relative landmark condition, b). For the other half of participants, the toy object was placed on the right (or south) edge of the first table and on the left (or south) edge of the second table (an absolute landmark condition, c). Reprinted with permission of Elsevier from Li and Gleitman, 2002, p. 280.**

condition), and in the latter condition the object was always to the south of the participant (an absolute landmark condition). Li and Gleitman found that participants in the relative landmark condition predominantly arranged the toys in a *relatively* correct manner (as the Dutch participants in the Levinson, 1996, study did), and participants in the absolute landmark condition predominantly arranged the toys in an *absolutely* correct manner (as the Tenejapan participants did). The results from these two experiments demonstrated that landmark information in the environment beyond the tabletop (in the blinds-up and outdoor conditions) and within the reference frame of the tabletop (in the relative and absolute landmark conditions) determined how participants encoded spatial relations: distal landmarks and *absolutely* stable proximal landmarks induced participants to make responses in the absolute frame of reference, while *relatively* stable proximal landmarks induced participants to make responses in the relative frame of reference. Li and Gleitman intentionally left the instructions given to participants in the experiments ambiguous (i.e., to make the arrangement of toys the *same* as before)—people chose from two possible strategies, namely using the relative or absolute frame of reference, by guessing which one the speaker (in this case, the experimenter) had in mind. Importantly, the choice of reference frames is derived from spatial circumstances and environments that people live in; thus people develop terminology that fits the circumstances.

There is another line of research which suggests that people can construct equally accurate mental representations of space from either route descriptions or survey descriptions. In verbally describing an environmental scene and spatial relations in the scene, one can employ two different perspectives. Route descriptions provide tour instructions that are linearly organized from the traveler's horizontal perspective, using egocentric terms such as front, back, left, and right. Survey descriptions present a hierarchically organized overview of the surrounds from a vertical perspective, using allocentric terms such as north, south, east, and west. Taylor and Tversky (1992) presented to participants these two types of descriptions of four environments that differed in geographic scale and the number of hierarchical levels: a county, a town, a zoo, and a conventional center. For example, a route description of a town says, "After turning left onto Maple St. from Mountain Rd., you see, on your right, the Town Park"; a survey description says, "Along the eastern edge of the Town Park runs Mountain Rd. . . . Along the southern border of the park runs Maple St." (p. 268; Figure 7.5). Participants read one route description and one survey description for the larger areas (the county and the town), and one route

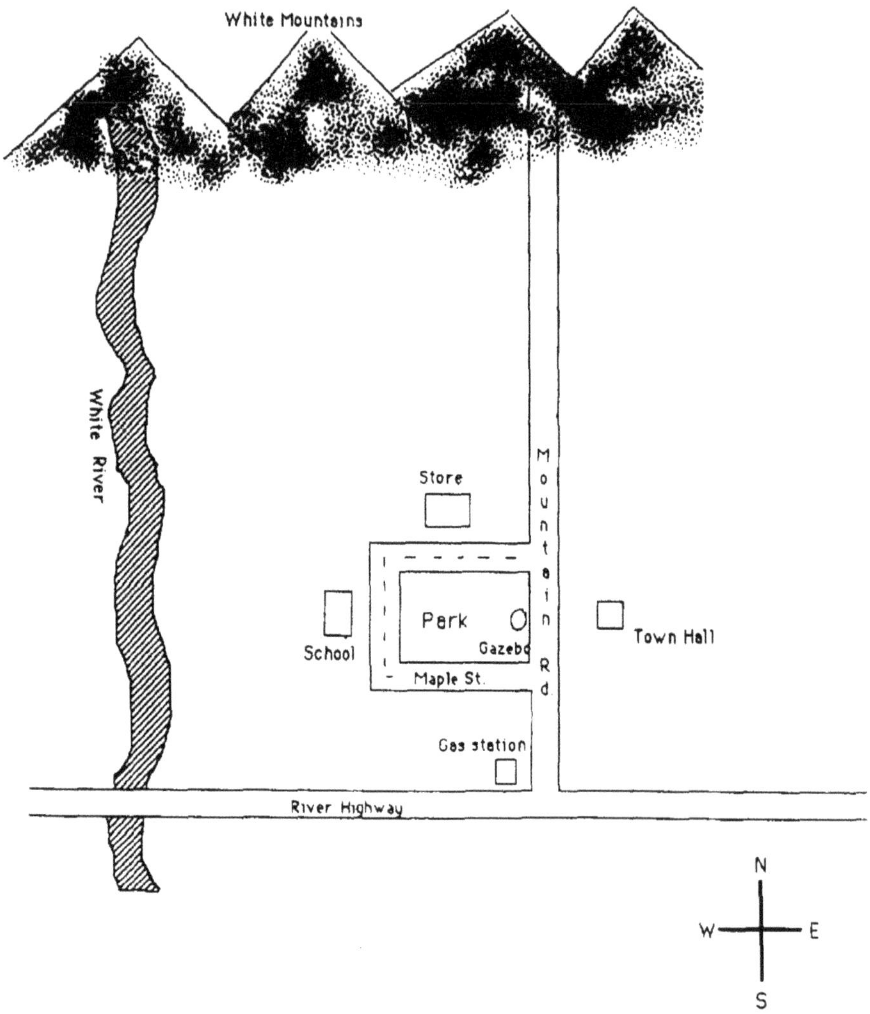

Figure 7.5 Map of a town used in the verbal description study. Spatial relations of places in this town were described either in the traveler's perspective ("After turning left onto Maple St. from Mountain Rd., you see, on your right, the Town Park") or in a bird's-eye view ("Along the eastern edge of the Town Park runs Mountain Rd. . . . Along the southern border of the park runs Maple St."). Reprinted with permission of Elsevier from Taylor and Tversky, 1992, p. 266.

description and one survey description for the smaller areas (the zoo and the conventional center), in random order. Thus, each participant read a description from one perspective only (either route or survey) for a specific environment.

After reading each description, participants judged whether a set of locative and nonlocative statements about the described environment is true. Half the locative statements were made from a route perspective (relating an item in the environment to the traveler's view) and the other half were from a survey perspective (relating an item in the environment to another item in the environment). The statements were either verbatim or inferable from the original description that they had read. Judgments about inference statements from the original perspective assessed the ability to infer spatial relations from information given in the descriptions. Judgments about verbatim statements from a perspective different from the original also assessed the ability to infer spatial relations, because the participant did not learn the environment from that perspective.

An analysis of mean response time and accuracy showed that participants responded faster and more accurately to verbatim statements from the original perspective than to statements from a different perspective. Participants who had read a route (or a survey) description of an environment verified locative statements from a route (or a survey) perspective faster and more accurately than locative statements from a survey (or a route) perspective. On the judgments for inference statements, however, they did equivalently irrespective of whether the statements were made from a route or survey perspective. These results led Taylor and Tversky (1992) to discuss that their participants constructed mental models of the environments in the verbal descriptions; the mental models somehow represented the perspectives of the original descriptions, but also contained information about spatial relations of places in the described environments in such an abstract way to make inferences from either perspective equally possible.

Understanding and use of maps

Another medium of representing space and communicating knowledge about space is the map. It has a long tradition, and history tells us that people had used spatial representations before they acquired language (see Chapter 1), attesting the fundamental human desire to know about the world and to expand the knowledge by discovering new land and exploring terrae incognitae (see Figure 7.6). Maps are a powerful tool in that they allow the user to view the surrounding environment in its entirety, which spans a large geographical extent, with spatial layouts and relations of places represented at a reduced scale, on a piece of paper or on a tablet screen nowadays. You may appreciate the benefit of having such a vertical plan view available when you remember the characteristics of environment perception (Ittelson, 1973): the environment is larger than and surrounds the human body, and thus requires a person to move about in it and integrate individual travel experiences to acquire knowledge about it, which turns out to be difficult (see Chapters 5 and 6). Maps allow people to experience environmental and geographical spaces as figural space, in Montello's (1993) terminology.

Despite the popularity and strength of maps, the understanding and use of them is not as straightforward as, and is often more difficult than, many people assume. One can see

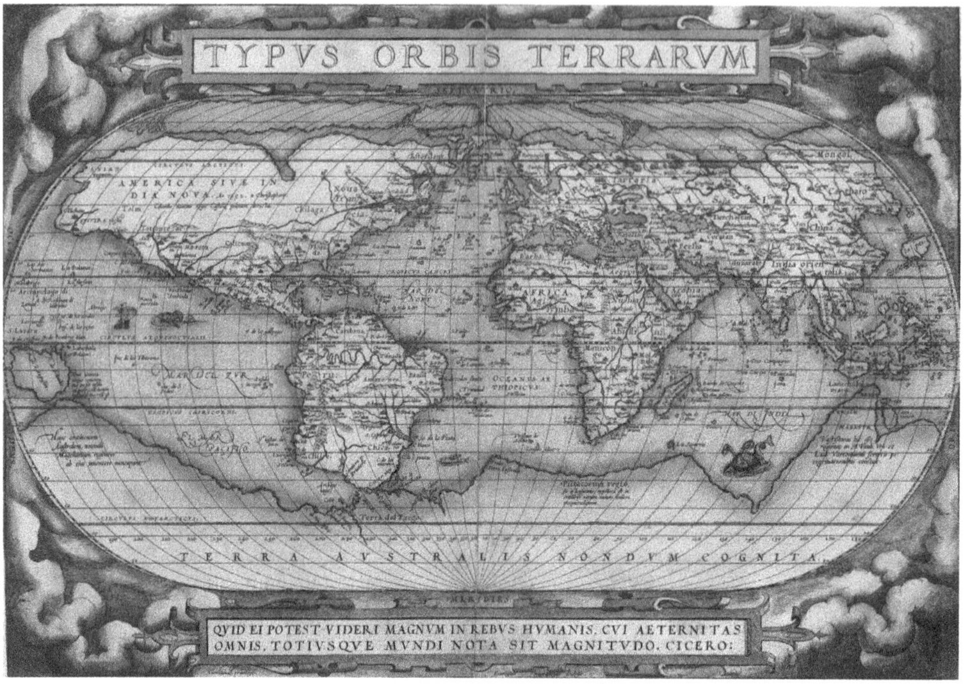

Figure 7.6 **Ortelius's World Map, published in 1570. It shows the knowledge of the world that Western European people had in the sixteenth century, during the periods of the Renaissance, the Age of Discovery, and the Scientific Revolution. Map in the public domain.**

the reason for this difficulty when considering map use from the perspective of interaction between the map, the user, and the environment (Liben and Downs, 1993; see Figure 7.7). Spatial orientation entails understanding your position and orientation in the environment, where you are and which direction you are facing (the person–space relation). To do that you consult a map, but a map does not automatically tell where you are if simply held in the hand. You need to align the map with the environment and locate yourself on the map: you understand the relationship between the map and the represented space (the space–map relation) and the relationship between you and the map (the person–map relation). The alignment of the map and the environment is sometimes achieved physically by rotating the map so that the directions on it (e.g., the upward direction) match the directions in the environment (e.g., the forward direction). Or the alignment is achieved cognitively by mentally rotating the map. The necessity of alignment poses difficulty for people who are not good at mental rotation (Liben and Downs, 1993). This is why you-are-here maps that are misaligned with the environment are difficult to use. You may have the experience of being confused, or seeing other people confused, in a large shopping mall by a floor map that is installed in a misaligned orientation, for example upside down (Figure 7.8).

The understanding and use of maps and the difficulty associated with the understanding of the person–space–map relations have been studied in the context of child development,

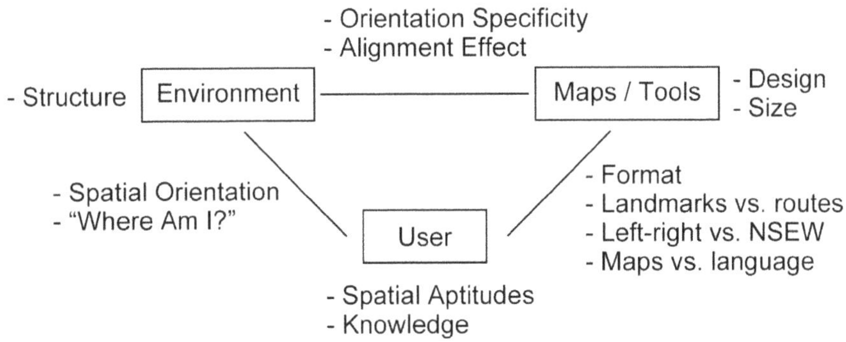

Figure 7.7 **Interaction between the map, the user, and the environment. The characteristics of each component need to be taken into account when the use of a map is discussed. Reprinted with permission of Springer Nature from Ishikawa, 2016a, p. 120.**

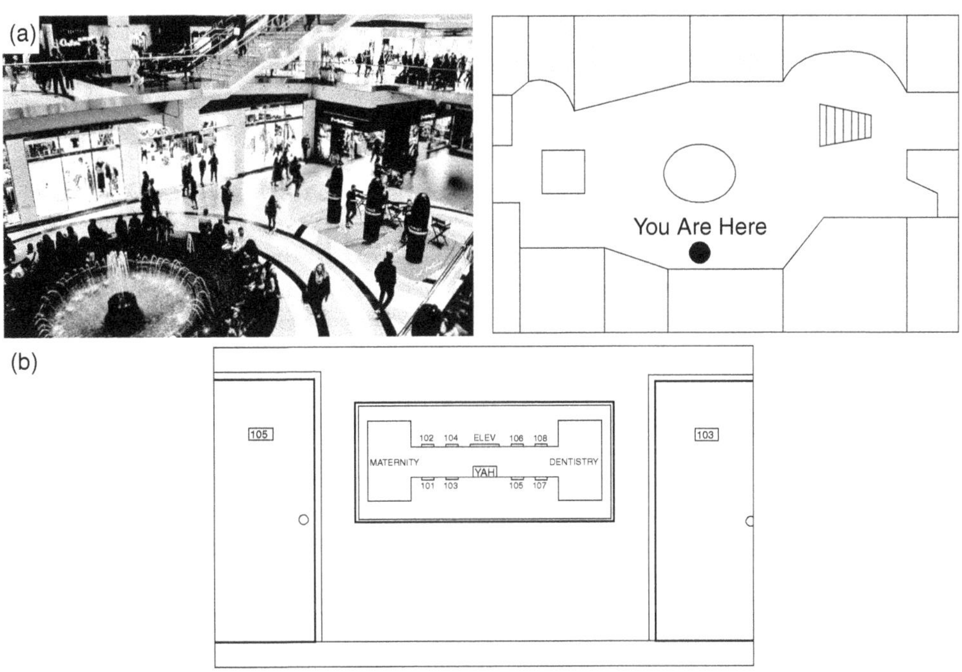

Figure 7.8 **You may encounter difficulty understanding where you are in a large shopping mall, even with a floor map (a). Spatial orientation is especially difficult when the map you see is misaligned with the surrounding space, for example if it has been put upside down (b). Reprinted from Levine, 1982, p. 234. Photograph from https://pixabay.com/**

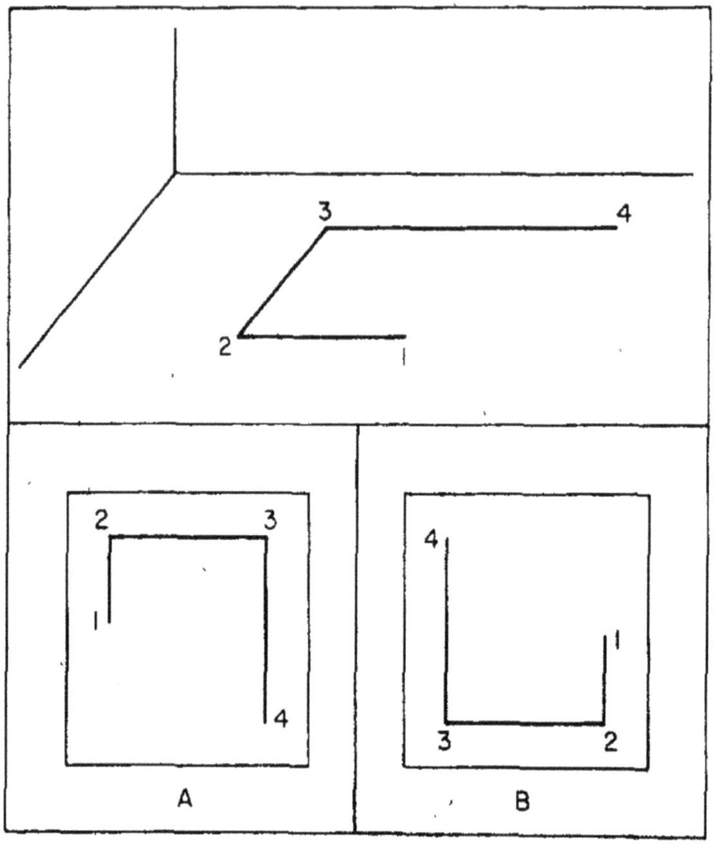

Figure 7.10 **Four-point path laid out on the floor of a laboratory (*top*). Participants viewed a map of the path that was shown in one of two orientations (map A, *bottom left*, and map B, *bottom right*). Reprinted from Levine, Jankovic, and Palij, 1982, p. 169.**

drawn on a 50-cm square card, in either an aligned orientation or a contra-aligned orientation. After studying the map, participants walked the path once from point 1 to point 4. Then they wore a blindfold and were asked to walk between pairs of points on the path directly. For example, in one case, participants stood at point 4 facing point 3, and were instructed to head toward point 1. In another case, participants stood at point 3, facing point 4, and were instructed to head toward point 1. For participants who studied an aligned map, going from point 4 to point 1 would be an aligned task and going from point 3 to point 4 would be a contra-aligned task; and vice versa for participants who studied a contra-aligned map (Figure 7.11). They walked eight paths and conducted the direct-walk task for a total of eight times, half in the aligned condition and the other half in the contra-aligned condition. An analysis of absolute angle errors showed that in the aligned condition, only two of the total of 96 responses yielded errors greater than 90°, but in the contra-aligned condition, by contrast, 29 responses (30 percent) did. Levine et al. discussed that an aligned orientation of a map, or the correspondence of the upward direction on a map

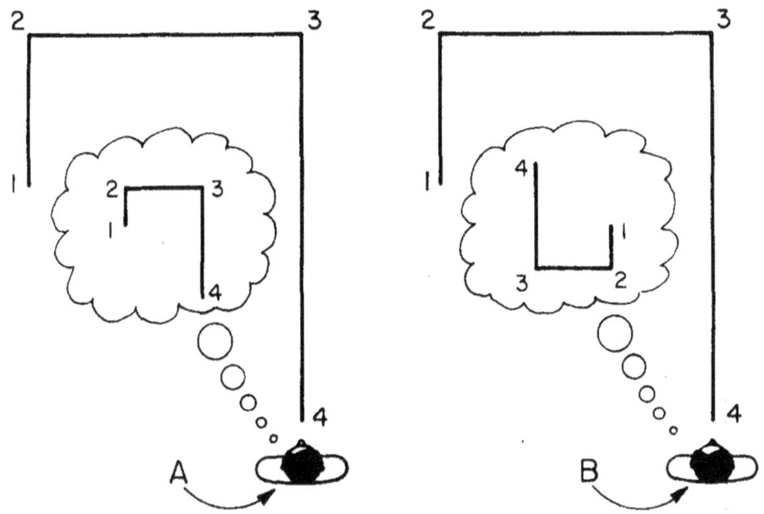

Figure 7.11 **Suppose that participants were standing at point 4, facing point 3, in the laboratory. For participants who had studied map A (or map B) in Figure 7.10, going from point 4 to point 1 would be an aligned task (or a contra-aligned task) and going from point 3 to point 4 would be a contra-aligned task (or an aligned task). Reprinted from Levine, Jankovic, and Palij, 1982, p. 170.**

and the forward direction in the represented environment, facilitates the map user's spatial orientation, and that it is a major property of a you-are-here map (called the forward-up equivalence), which should be considered in its placement.

The orientation of a map thus affects the ease with which the map is used, as it entails a comprehension of the person–space–map relations through mental rotation (Liben and Downs, 1993; see also the argument by Gulliver, 1908). When a map is presented in a physically fixed orientation (e.g., a you-are-here map installed in the environment), you need to rotate your knowledge or internal representations of the environment to attain the forward-up equivalence. It causes difficulty for some people, particularly those with low spatial ability. Liben and Downs examined the understanding of maps by a total of 168 kindergarteners, first graders, and second graders, and found that children with low perspective taking ability (assessed by Piaget's three mountains and model landscape tasks; see Chapter 5) did worse on the arrow-sticker task (see above). The perspective taking ability correlated with angular errors around –0.25 in the aligned condition and –0.35 in the contra-aligned condition ($p < 0.001$).

Difficulty with map use and spatial orientation is also observed among adults. Liben, Kastens, and Stevenson (2002) took college students, during their first semester, to various locations on campus, and asked them to place stickers on a campus map to indicate where they were. As seen in Figure 7.12, the students' performance was not perfect, some stickers misplaced at locations completely deviated from the correct locations. Liben et al. presented to another group of college students a series of photographs of scenes taken on a campus (Figure 7.13, *left*), and for each photograph asked them to place arrow stickers on a campus map to indicate the photographer's location and orientation. The stickers were again found to be placed at scattered locations and in various orientations (Figure 7.13,

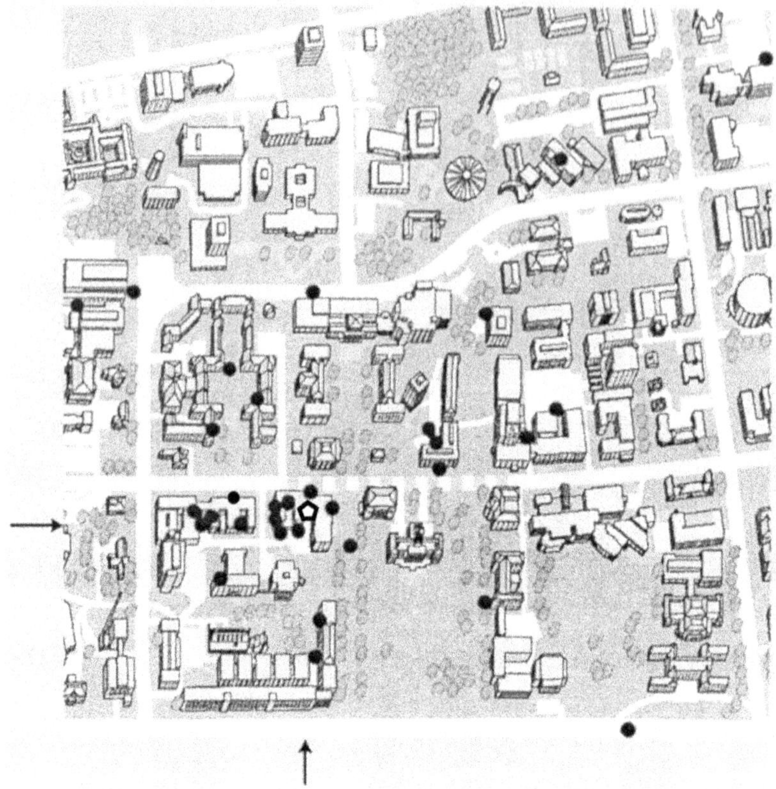

Figure 7.12 College students were taken to various locations on their campus and asked to place arrow stickers on a map to show their locations and facing directions. Their responses show that spatial orientation with a map is not easy for everyone: some stickers are misplaced at locations completely deviated from the correct location (indicated by the two arrows, an open symbol). Reprinted with permission of Elsevier from Liben, Kastens, and Stevenson, 2002, p. 277.

Figure 7.13 Another group of college students viewed photographs of scenes on a university campus, and placed arrow stickers on a map to show the photographer's locations and orientations. Again, you can see that the performance was not perfect and stickers were placed at scattered locations and in various orientations. Reprinted with permission of Elsevier from Liben, Kastens, and Stevenson, 2002, pp. 300 and 302.

right). It is therefore not a good idea to assume, in the provision of navigation assistance, that every user finds a map easy to understand. To give a person a map and say, "We will meet there tomorrow" is not always sufficient. Carefully designed navigational information and wayfinding assistance, preferably adapted to the user's ability and preference, may be needed (see Chapter 10).

Navigation assistance and tools

Findings in the literature of spatial cognition offer important implications for the practice of navigation assistance. In particular, the large individual differences in the accuracy of cognitive maps and the difficulties with spatial orientation and map use have a significant consequence—given such differences and difficulties, it is not viable to provide the same navigational information to everyone. How can we effectively assist poor wayfinders in knowing where they are, and how can we provide navigation assistance in a way that is adjusted to the user and the situation?

The previous section discussed that, although maps are a powerful tool showing spatial relations in plan view and have traditionally been used in wayfinding and navigation, they do not necessarily help everyone to become oriented in space and to find their way in a new environment. Spatial orientation through the coordination of the relationship between the map, the self, and the environment is difficult for people with low spatial ability (Liben and Downs, 1993).

With the advances in information and communication technologies, and especially the development of geospatial data infrastructures and technologies, various location-based services are in popular use now, including digital maps on mobile devices and satellite navigation tools for drivers and pedestrians (Gartner, Cartwright, and Peterson, 2007; Kiefer et al., 2018; Küpper, 2005; Mannings, 2008; Meng, Zipf, and Winter, 2008; Retscher and Kealy, 2006; Werner, 2014). The use of locational information in a digital format has thus become a fundamental part of our daily lives. Information relating to navigation and directions is among the top Internet searches, and 10–20 percent of various Internet searches concern places and geographic locations (Amin et al., 2009; Riegelsberger, 2013). Marketing research (Liu, 2018; Statista, 2019) shows that mobile map applications have high percentages of market reach (90 percent worldwide) and user access (65 percent of monthly users in the United States). A recent survey by a map company showed that during navigation 62 percent of people use maps on a smartphone and 38 percent use in-car navigation systems, while 28 percent use paper maps (Zenrin, 2018).

Location-based service applications, particularly advanced navigation tools, are expected to assist in wayfinding and navigation, as they show the user's position and orientation in real time on an automatically rotating map and direct the user to a destination along the shortest path through turn-by-turn instructions. The user has everything they need for successful navigation available on a mobile device, one may say. Empirical research suggests, however, that satellite navigation tools fall short of our expectations and do not necessarily serve as an effective "navigator" or a "facilitator" of our spatial awareness. People travel longer distances and make more stops, or acquire poorer knowledge of traveled routes and surrounding scenes, when they use navigation tools than when they use maps or are guided along the routes by another person (e.g., Ishikawa et al., 2008; Münzer et al., 2006; Willis et al., 2009).

Willis et al. (2009) compared the accuracy of spatial knowledge acquired by two groups of participants who learned the same environment with two different tools. One group studied a map of the environment in a separate room; the participants were instructed to learn the locations of eight places marked on the map and the layout of paths. The other group directly walked between the eight places in the environment using a mobile navigation tool. The mobile tool displayed the user's location on a map in a north-up orientation. After learning the environment either from the map in a separate room or from direct navigation with the mobile tool, the two groups of participants were taken to each of the eight places and asked to estimate distances and directions to other places. The accuracy of these estimates, Willis et al. reported, did not differ significantly between the two groups.

Münzer et al. (2006) compared spatial knowledge acquired by users of a map and a mobile device. Participants in the map condition learned a series of fragment maps, which showed portions of a route in a zoo in sequence. Participants were given the fragment maps sequentially one by one along the route and memorized it; after it was removed, they were asked to walk to the end of the partial route memorized. They repeated the process of map studying and partial route walking for the entire route. Participants in the mobile condition were provided with route information on a mobile device in one of three forms, either (a) a series of partial route maps with photographs of views at intersections and arrows indicating the directions of travel; (b) a series of partial route maps with photographs of intersections and auditory verbal directions; or (c) photographs of intersections and auditory verbal directions. The experimenter sent these pieces of information to participants' mobile devices sequentially as they passed predetermined intersections along the route. Participants' memory of the route was assessed by the judgments of route directions (which way to go at each intersection) and the placement of photographs of intersections on a road map: participants in the map condition did better than participants in the mobile condition.

Two explanations have mainly been offered for the ineffectiveness of navigation tools. One concerns a lack of conscious decision making on the part of the user. As long as navigation tools direct users on which route to take and where to turn, users simply follow the provided directions; they do not deliberately plan a route or actively control navigational decisions, leading to degraded spatial learning (e.g., Bakdash, Linkenauger, and Proffitt, 2008; Burnett and Lee, 2005; Farrell et al., 2003; Gaunet et al., 2001; Parush, Ahuvia, and Erev, 2007; Péruch and Wilson, 2004).

A second explanation concerns the distraction of the user's attention from surrounding environments to the tools and directions (e.g., Dillemuth, 2009; Gardony et al., 2013; Gartner and Hiller, 2010; Leshed et al., 2008; Willis et al., 2009). The distraction may relate to the difficulty of coordinating the provided information with the surrounding space, especially when the route information is given in a piecemeal fashion on a small screen.

Ishikawa et al. (2008) examined the wayfinding behavior and spatial learning by three groups of participants who learned a new environment in three different manners: one group explored a route with a mobile navigation tool; another group with a paper map; and a third group traveled the route once, guided by another person. The mobile navigation tool, a Global Positioning System (GPS)-enabled cellphone device, showed a map of the surrounding area on a 4 × 5-cm screen, on which the user's position and a route to a destination were dynamically updated according to the movement in space. Due to the limited size of the device, the route was not visible in its entirety on the screen (Figure 7.14a). The paper map, printed on A4-size paper, showed the start and end points, but did not indicate a specific route to take (Figure 7.14b). People in the direct experience condition first walked the route being guided by the experimenter, and then were asked to find their way from the start to the goal by themselves.

(a)

(b)

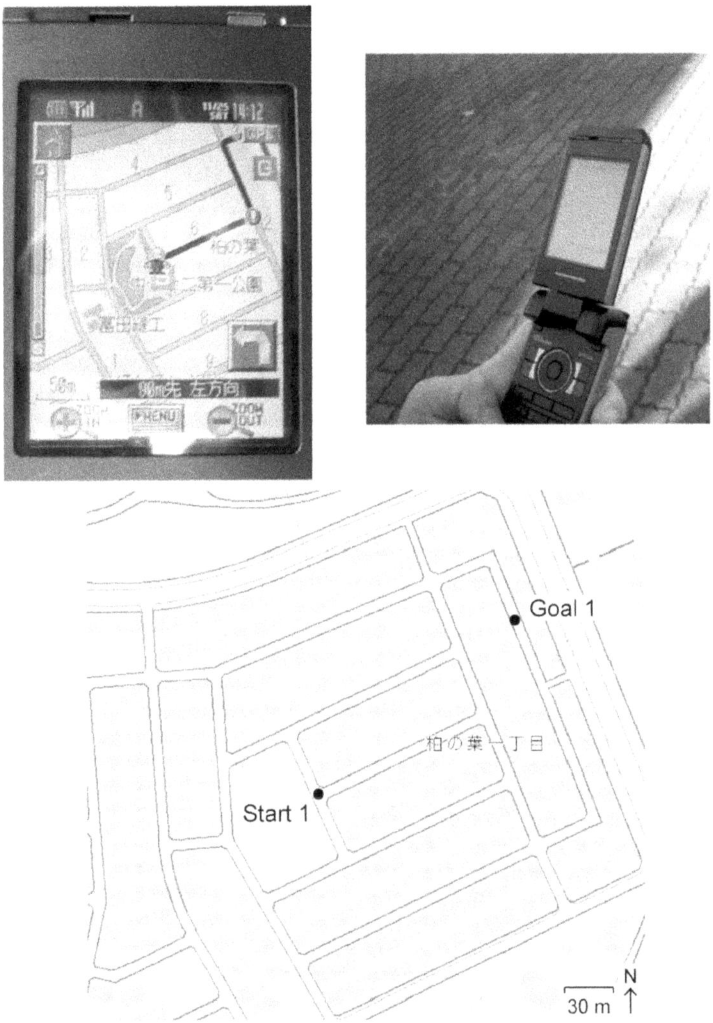

Figure 7.14 **(a) The mobile navigation tool showed a map of the surrounding area on a 4 × 5-cm screen, with the user's current position and a route to a destination dynamically updated according to the movement in space. (b) The paper map, printed on A4-size paper, showed the start and end points, but not a specific route to take. Reprinted with permission of Elsevier from Ishikawa et al., 2008, p. 76. Map based on GSI Maps, http://maps.gsi.go.jp/**

People's performance on wayfinding was assessed by the distance traveled, travel time, and the number of stops (the times that they got lost or disoriented). Their spatial learning was assessed by the accuracy of route memory (the sequence of turns indicated on their sketch maps) and direction estimates from the endpoint to the starting point (configurational understanding of the route). Results showed that mobile users traveled a longer distance and stopped more frequently than did map users and direct experience participants. Mobile users also traveled more slowly and self-assessed the wayfinding task as more difficult than did direct experience participants. Most participants in all three groups did well on finding their way and successfully reached the goal; however, the accuracy of sequential and configurational understanding of traveled routes was lower for mobile users than for direct experience participants. Thus, the navigation tool did direct participants to their destinations, but was less effective in supporting smooth navigation or facilitating accurate spatial learning, when compared with the map and direct experience. In other words, there is still room for improvement for a navigation tool to become a helpful navigator.

Ishikawa and Takahashi (2013) examined the effects of different methods of presenting navigational information on the user's wayfinding behavior and spatial learning. Specifically, Ishikawa and Takahashi looked at the routes that people with different navigational information would take, with a view to examining whether the ineffectiveness of navigation tools found in the literature is related to the lack of the user's decision making: If users of navigation tools passively follow the directions provided, the lack of route planning would be evident in lesser variations in their routes. Furthermore, Ishikawa and Takahashi examined the degree of attention paid to the tool versus the surrounding space, in terms of the time that people spent looking at the tool and the memory of scenes along traveled routes.

In the study, three tools were used to provide navigational information in three different manners. One tool showed a route to a destination, as many commercial navigation applications do. This tool (called a route tool hereafter) displayed on a 3.5-inch (8.9-cm) screen a map of the area within a 150-m radius of the user, with the user's position and the route updating with any movement (Figure 7.15a). Another tool showed the general direction toward the destination, but not a specific route (as a person points by extending his or her finger to the destination, saying, "It's over there"). The tool (called a direction tool) displayed a map of the surrounding area and on the map depicted the user's position and an arrow indicating the direction toward the destination (Figure 7.15b). On both tools, the map on the screen was shown in a north-up orientation. These two tools were compared with a paper map, which was printed on A4-size paper with the locations of the start and the destination marked, but not a specific route (Figure 7.15c). Participants walked between three pairs of places using one of these three tools, being instructed to find their way to the destination along any route of their own choice.

Participants' wayfinding behavior and spatial learning were examined in terms of (a) travel distance, (b) travel speed, (c) recognition memory for the surrounding scenes, and (d) the time spent looking at the tools. In addition, the variability in the routes that the three groups of participants took was assessed in terms of a measure of cosine similarity, which gives the degree of correspondence between two vectors.

Ishikawa and Takahashi (2013) found that users of the route tool remembered surrounding scenes poorly, spent more time looking at the tool, and tended to follow the route as directed by the tool. Their scene recognition memory was 20 percent lower and time of gaze at the

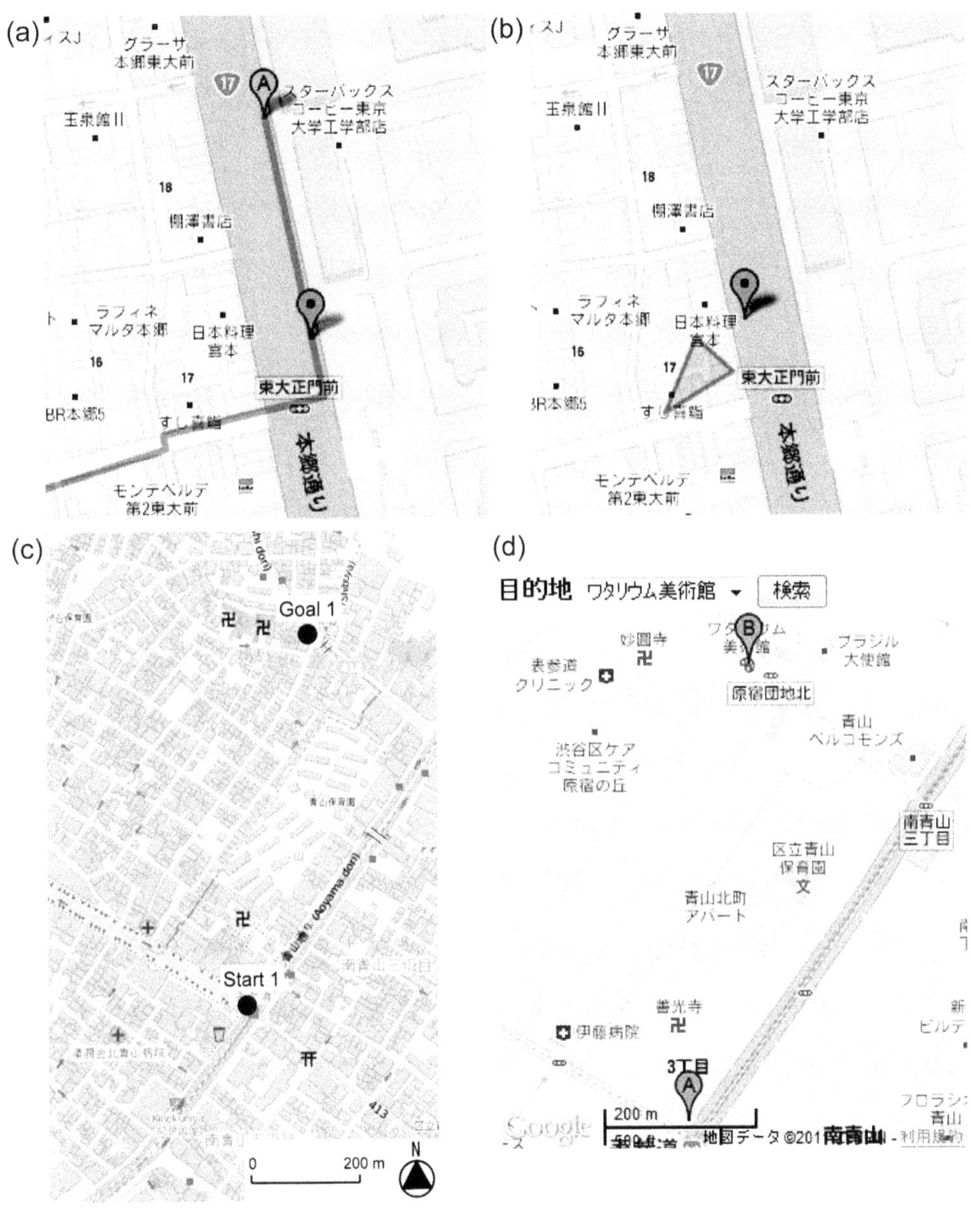

Figure 7.15 **(a) The route tool displayed on a 3.5-inch screen a map of the area within a 150-m radius of the user, with the user's position and the route updating with any movement. (b) The direction tool showed on the screen the direction in which the destination was located, without any route indicated. (c) The paper map, printed on A4-size paper, showed the locations of the start and the destination. (d) On a tool used in a follow-up experiment, a map of the entire route was shown in small size on the device screen. Reprinted from Ishikawa and Takahashi, 2013, p. 20. © OpenStreetMap contributors.**

tool was 30 percent longer than those for users of the paper map. Also, their routes were half as varied. That is, in contrast to map users who attended to the surrounding space rather than the map, users of the route tool paid less attention to the surroundings and simply followed the route instructions. Users of the route and direction tools were equivalent in scene recognition memory and the time spent looking at the tool; but the latter took more varied routes—as varied as the routes taken by map users (Figure 7.16a–c).

In a follow-up experiment, Ishikawa and Takahashi (2013) examined the effect of screen size, with an additional group of participants who viewed a map of the entire

Figure 7.16 **These maps depict the routes that users of the four tools took: (a) the route tool, (b) the direction tool, (c) the paper map, and (d) the device-screen map. The widths of the lines are in proportion to the numbers of participants who took the corresponding paths. The differences in the traveled routes show that external spatial representations, or the manner in which navigational information is given, affects the user's behavior. Reprinted from Ishikawa and Takahashi, 2013, p. 22. Tools developed with Google Maps API. Map data © 2011 Google, ZENRIN.**

route on the small device (Figure 7.15d). The rationale for the condition was that if the distraction of attention from the surrounding space to the tool is due to the small size of a screen, their wayfinding performance and memory of the surrounding scenes would be poorer. In fact, these participants remembered the surrounding scenes poorly, looked at the tool for a longer time, and took less varied routes, compared with the participants who used the paper map printed in larger size above (Figure 7.16d). Thus, the map's strength of allowing the user to attend to the surroundings and remember the scenes is weakened when it is shown in the small size of a mobile screen.

Implications for the design of navigation tools

These empirical findings show that the manner in which navigational information is presented affects the user's wayfinding behavior and spatial learning. In particular, although directing the user on which route to take is a popular method of presenting information on major navigation applications, it renders the act of navigation into simple direction-following; it does not assist the user in remembering the surrounding environments or encourage the user to explore diverse routes freely.

Showing a general direction toward a destination (as the direction tool did) may work in a situation where the aim of navigation is to guide people around different places along various routes, as in tourist navigation or sightseeing. However, when the principal objective is to take the user to the destination, the route tool may be appropriate. Helping the user remember visited places, rather than simply moving between a set of fixed points of interest without any memory of places in between, is desirable for tourist navigation. When the navigator wants the user to leave the visited places with a good memory of surrounding scenes and experience of traveling various routes (some of which are not planned beforehand and come as unexpected discoveries), paper maps may be a good choice. Such experiences of unexpectedness are better avoided in navigation in emergency situations. Thus, designing effective navigation tools requires a careful consideration of the situation and purpose of use.

How can we, then, provide effective navigation assistance to people with a poor sense of direction? The key to this question concerns the possibility that navigational information can be presented in a variety of formats. For example, a navigation tool could direct the user which route to take by showing a map on a screen, highlighting and updating the route according to the movement in space. The route may be shown in its entirety on the screen or portions of the route may be shown in a piecemeal fashion. The map may be in a fixed orientation with the north always up or automatically rotated according to the user's heading in the environment. Navigational directions can also be given in words as turn-by-turn instructions or more globally with cardinal directions. Is there a single best format for every person and in every situation? The answer is probably no, given the large individual differences mentioned above. Some people have a hard time becoming oriented in their surrounding environments by comprehending the spatial layouts in two dimensions and understanding the spatial relations between different places configurationally. Even when provided with navigation assistance, they may still have difficulty. Some people may experience difficulty with maps; they may have trouble understanding where they are on a map or how the map should be oriented in relation to the surrounding space. It is not sufficient that the map has a compass rose—understanding which direction is north in the environment is not an easy task (Levinson, 2003). The ease of using navigation assistance also depends on the structure of the environment: turn-by-turn directions may be suitable for an environment without distinctive landmarks, and cardinal directions may be easier to understand in an environment with a coherent structure (e.g., grid-pattern streets aligned with cardinal directions or an area with the ocean in one direction and the mountains in the opposite direction). The effectiveness of different formats therefore relates to the characteristics of the user and the environment, hence the importance of user adaptation and context awareness in the development of navigation assistance (e.g., Takemiya and Ishikawa, 2011; see Figure 7.7).

Review box for Chapter 7

1. Describe the three frames of reference discussed in this chapter.
2. Ask some of your friends to verbally describe their usual routes to school. Did they describe spatial relations in the relative, absolute, or intrinsic frame of reference? Which reference frame was most popular among the respondents?
3. How would you respond to the arrow choice task in Levinson (1996)? Do you choose a relative arrow or an absolute arrow as the "correct" answer? Why?
4. How would you respond to the toy arrangement task in Li and Gleitman (2002) when a landmark object is placed on the table? What is your take on the question of linguistic relativism?
5. Which do you find easier to understand when you give and receive navigational directions: route descriptions or survey descriptions? Can you construct equivalent mental models from both descriptions?
6. Why do you like or not like reading maps?
7. Discuss the use of a map in the environment from the perspective of interaction between the map, the user, and the environment.
8. Discuss children's map understanding in terms of representational and geometrical correspondences.
9. Discuss the placement of you-are-here maps, focusing on the ease of spatial orientation with aligned and misaligned maps.
10. Find examples of misaligned you-are-here maps in your neighborhood.
11. Summarize the results of empirical studies of wayfinding behavior by users of navigation tools.
12. Tell us your experiences of using satellite navigation tools, both good and bad.
13. Discuss possible ways to design navigation tools that are usable and understandable. What do you mean by "usable" and "understandable"?

Chapter 8

Development of spatial cognition in children

We saw in Chapter 2 Piaget's developmental theory of intelligence in general, and spatial concepts in particular, as one of the major roots in the research on human spatial cognition. A large body of literature therefore exists about the ontogenetic development of spatial cognition, from infancy through childhood to adolescence. Among the issues that have attracted researchers' and practitioners' interest is the coding of spatial relations and the representation of spatial environments in the child's mind.

Chapter 7 discussed spatial frames of reference in relation to spatial language, with a focus on the distinction between the relative and absolute frames of reference. One may represent the surrounding space and objects in the space in relation to one's own body; for example, "I can see the face of my mother when I turn to the right," "the teddy bear is on the bookshelf to my left," or "the keys are put in the box in front of me." Spatial relations in these expressions are conceptualized in a reference system whose origin is set on the self; the system describes self-to-object relations (Sholl, 1996) and is hence called an *egocentric* frame of reference. Alternatively, one may represent spatial relations in a reference system whose origin is located in the environment, outside one's body, called an *allocentric* frame of reference. Allocentric spatial relations are conceptualized in reference to concrete objects, such as salient features in a room ("put a calendar next to the picture on the wall") or landmarks in the city ("the hospital is across from the post office") depicting object-to-object relations (Sholl, 1996); or in reference to an abstract reference system laid over the environment such as a street addressing system (101 Highland Street), cardinal directions (north, south, east, and west), or latitude and longitude coordinates (41° 00′ N and 74° 00′ W).

In the Piagetian framework of spatial ontogenesis, children's conception of space is first tied to sensorimotor actions and hence is inherently egocentric. Space, to children, is a concrete entity, which they perceive and act on from their own viewpoint. Therefore, in the course of ontogenetic development, spatial coding in an egocentric frame of reference should emerge before spatial coding in an allocentric frame of reference becomes possible. If so, in the identification of spatial locations of objects by children, the use of external landmarks would be seen later than the exhibition of egocentric responses. Furthermore, in Piaget's theory, the term "representation" possesses a specific meaning, signifying a step in intellectual development from the stages of concrete and simple actions to the stages of abstract and complex reasoning, or transition from figurative to operative cognition. The ability to imagine perspectives other than one's own and to coordinate views from different perspectives, and accordingly a survey understanding of the environment not tied to traveled routes, as well as the ability to understand symbolic representations of the environment such as maps, should therefore be absent in the early stages of ontogenetic development. Is that true? How and when do these different types of spatial cognition and behavior, some being rudimentary and some advanced, manifest themselves in children?

Egocentric and allocentric spatial coding by children

One of the major methods of studying infants' and children's spatial learning is a variant of a hide-and-seek play. In Acredolo's (1977) study, 3-, 4-, and 5-year-old children learned to find an object hidden under one of two identical cups that were located in a room without any landmarks. Children started from one end of a room and walked straight ahead, and then turned left (or right) to find a cup in which a trinket was hidden (Figure 8.1). Half the children always turned left, and the other half right, to find the trinket. When they learned to do that without errors, they were moved to the opposite end of the room and asked to find the trinket again. By observing which way the children turn, one can tell whether they are tied to the original egocentric response or they correctly reason about

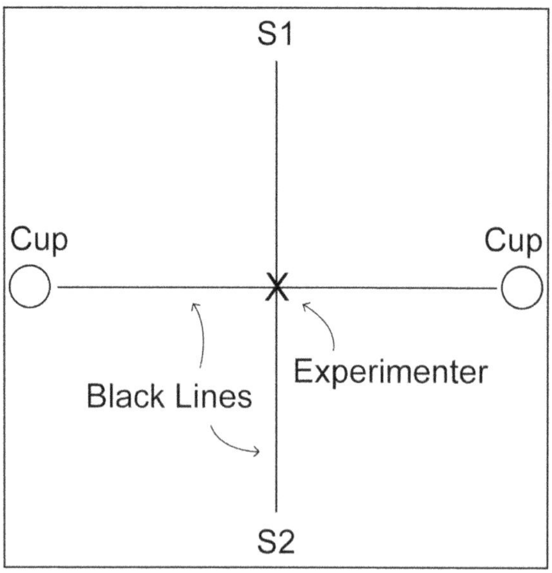

Figure 8.1 **Children 3, 4, and 5 years of age learned to find an object hidden under one of two identical cups that were located in a room without landmarks. They started from one end of a room and walked straight ahead, and then turned left (or right) to find a cup in which a trinket was hidden. Reprinted from Acredolo, 1977, p. 3.**

spatial relations taking the self-rotation into account. That is, if children, starting from the new, opposite side of the room, turn left (or right) as they did in the original learning condition, they respond egocentrically; if they turn right (or left), they correctly update the 180° rotation. Results showed that the 5-year-olds correctly found the trinket from the new starting location, but the 3- and 4-year-olds went to the wrong cup, indicating the dominance of egocentric responses among them. When landmarks were added to the room, for example the locations of the two cups were differentiated by different tablecloths or the starting positions at either end of the room were distinctively marked by a circle and a triangle, however, the tendency for the 3- and 4-year-olds to respond egocentrically disappeared.

Although the above results suggested the ability in preschoolers to use landmarks, Acredolo (1976) noted that their use of landmarks in location identification was fragile and not stable. In the study, 3-, 4-, and 10-year old children entered a rectangular room whose long walls were covered with white curtains and short walls with two distinctly colored curtains (Figure 8.2). Upon entering the room through a door on one of the short walls, children were first led to a table located in the middle of one of the long walls, where they were blindfolded, and then led for a short walk to the middle of a short wall, either the side where the door was or the side opposite to it. In some conditions, during the blindfolded short walk, the experimenter quietly moved the table to the opposite side, in the middle of the other long wall. Children then removed the blindfold and were asked to return to the spot in the room where the walk had begun. Acredolo constructed three conditions with respect to the side on which the walk ended and whether the table was moved: condition 1 (children walked to E1 and the table was moved to T2), condition 2 (children walked to E2 and the table was moved to T2), and condition 3 (children walked

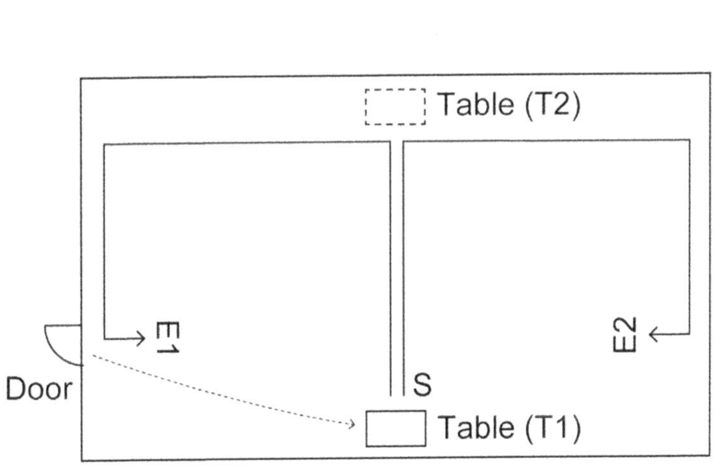

Figure 8.2 **Children 3, 4, and 10 years of age entered a rectangular room, the short walls of which were distinguished by curtains of different colors. They were first led to a table located in the middle of a long wall (and blindfolded there), and then to the middle of a short wall, either the side where the door was or the side opposite to it. In some conditions, the experimenter quietly moved the table to the opposite side during the blindfolded walk. Children removed the blindfold and were asked to return to the spot in the room where the walk had begun. Illustration based on Acredolo, 1976.**

Table 8.1 **Whether the children would succeed or fail when asked to return to the original starting position depending on which frame of reference they use**

Frame of reference	*Condition 1*	*Condition 2*	*Condition 3*
Egocentric response (to one's right)	Succeed	Fail	Fail
Object landmark (the table)	Fail	Fail	Succeed
Global landmark (the room)	Succeed	Succeed	Succeed

to E2 and the table remained at T1). When asked to return to the "original" starting position, the children exhibited different behaviors in the three conditions (Table 8.1) depending on whether they relied on egocentric responses, an object landmark (the table), or a global landmark (the geometry of the room or the spatial relations of the walls). Acredolo found that the children did not show egocentric responses and used landmark information to identify a location in the room, either the table (an object) or the room (a container). The 10-year-olds predominantly used the container frame of reference, leading to the correct original position in all three conditions, while the 3- and 4-year-olds were more likely to use the object frame of reference, erroneously returning to a moved table.

Response learning and landmark use by children

Bremner and Bryant (1977) examined the coding of spatial relations in infants, by asking 9-month infants, seated in a baby chair, to retrieve a toy that was hidden in one of two cups. On a table (30 × 60 cm and 38 cm high), half of which was painted white and the other half black (Figure 8.3), two cups were located, one on the white side and the other on the black side. Infants viewed a toy put into one of the cups; after 3 seconds, with both cups covered by cloths, the infants were asked to search for the toy. They repeated this set of learning and retrieval five times, with the cup in which the toy was hidden being the same for all trials (whether the correct cup was located to the left or right of the infant and on the white or black side of the table was counterbalanced).

After completing the five sets of learning and retrieval, the infants went through one of five spatial transformations (16 infants in each condition). Two groups of infants were taken around to the opposite side of the table; at that position, they viewed a toy put into one of the cups, and after 3 seconds they were asked to search for the toy, with both cups covered by cloths. For one group (Group A) the cup in which the object was hidden remained the same as before, and for the other group (Group B) the two cups were interchanged. That is, for Group A, the correct cup was different from the one during the learning and retrieval phase egocentrically, but the same as before in terms of absolute positions on the table and in the room. For group B, the correct cup was the same as before egocentrically, but different in terms of absolute positions on the table and in the room. Successful identification of the toy would require the infants in Group A (or in Group B) to uncover in the egocentric frame of reference a different (or the same) cup. In reference to the table frame of reference (the white or black half) and the room frame of reference (absolute position in the room), however, infants in Group A (or in Group B) would be required to uncover the same (or a different) cup.

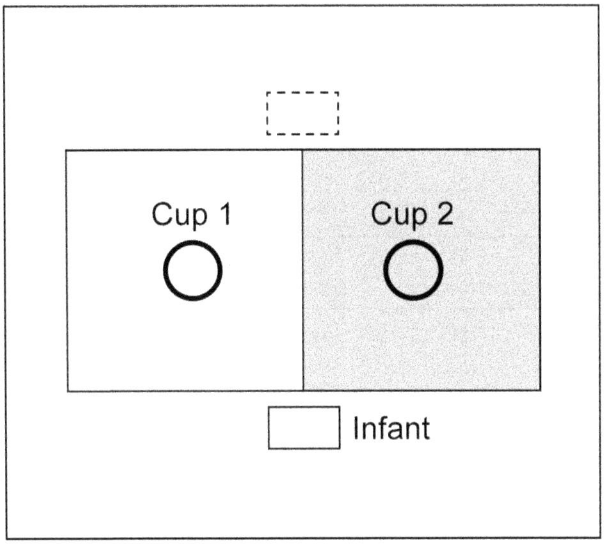

Figure 8.3 **Infants 9 months of age, seated in a baby chair, viewed a toy hidden in one of two cups on a table, and then were asked to retrieve it. Half of the table was painted white and the other half black; one cup was located on the white side and the other cup on the black side.**

For two other groups of infants, after the five sets of learning and retrieval, the table was rotated 180°; they then viewed a toy put into one of the cups, and after 3 seconds they were asked to search for the toy, with both cups covered by cloths. For one group (Group C) the cup in which the object was hidden remained the same as before, and for the other group (Group D) the two cups were interchanged. In these cases, successful identification of the toy would require the infants in Group C (or in Group D) to uncover a different (or the same) cup either egocentrically or in reference to the room frame of reference. In reference to the table frame of reference, infants in Group C (or in Group D) would require to uncover the same (or a different) cup.

Finally, infants in Group E went through no spatial transformations, but the two cups were interchanged. Thus, successful identification of the toy would require the infants in this group to uncover a different cup egocentrically, in reference to the table frame of reference, and in reference to the room frame of reference. Table 8.2 summarizes the necessity of uncovering the same or a different cup for successful identification for the five groups.

Bremner and Bryant (1977) found that infants in Groups B and D successfully found the toy significantly more frequently than did infants in Groups A, C, and E (12 and 13 infants, vs. 5, 5, and 4 infants, respectively). The infants in the latter three groups had difficulty in identifying the correct cup. As shown in Table 8.2, the property that differentiates Groups B and D from Groups A, C, and E is whether to uncover egocentrically the same cup (the former groups) or not (the latter groups). That is, among infants at this age, egocentric responses are dominant; egocentric responses persisted in various conditions even though the infants visually observed (viewed) where the toy was hidden.

***Table 8.2* To uncover the same or a different cup for successful identification of the hidden toy in the five groups depending on which frame of reference is used**

Frame of reference	*Group A*	*Group B*	*Group C*	*Group D*	*Group E*
Egocentric response	Different	Same	Different	Same	Different
Position with respect to the table	Same	Different	Same	Different	Different
Position with respect to the room	Same	Different	Different	Same	Different

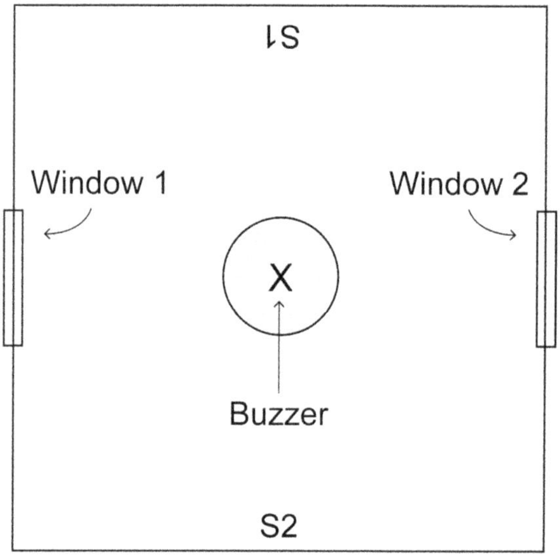

***Figure 8.4* Infants 6, 9, and 11 months of age, seated by one of the walls in a square room enclosed by curtains (S1), viewed the experimenter appearing at one of the two windows on the walls to the infant's left and right. The experimenter appeared at the same window every time a buzzer was heard. The infant was then moved to the opposite end of the room (S2); the experimenter sounded the buzzer and examined which direction the infant would turn to. Reprinted from Acredolo and Evans, 1980, p. 313.**

Acredolo and Evans (1980) examined the use of landmarks by infants aged 6, 9, and 11 months. In a 3.2 × 3.2-m room enclosed by curtains, an infant was seated by one of the walls (position S1 in Figure 8.4) and viewed an experimenter appearing at one of the two windows that were located on the walls to the infant's left and right. The experimenter appeared at the same window every time a buzzer was heard, and interacted with the infant verbally to attract his or her attention. After being trained to expect the occurrence of the event, the infant was moved to the opposite end of the room (position S2 in Figure 8.4); then the experimenter sounded the buzzer and examined which direction the infant would turn to. When there were no landmarks available in the room that could differentiate the two windows, infants at all ages responded egocentrically (80–90 percent of the responses were egocentric): they turned to the window to the left (or to the right) in the testing as they had learned to turn left (or right) during the training trials.

In another condition, the target window at which the experimenter appeared was marked by a yellow star. In this case, the 11-month-old infants showed egocentric responses less frequently (the percentage of egocentric responses dropped to about 40 percent). The 6-month-old infants still responded egocentrically, however, with no decrease in the percentage of egocentric responses. In a third condition, the target window was marked more conspicuously, with a panel of flashing lights around it and stripes on the wall of that side. With such a very salient landmark, the 9- and 11-month-old infants tended not to show egocentric responses, 80 percent of them identifying the correct window. Among the 6-month-old infants, egocentric responses still remained: 67 percent of them looked at both windows either on the same trial or across different trials. These results show that egocentric responses are dominant among infants, but the tendency to use landmarks, when they are salient enough, becomes higher for infants aged 9 and 11 months.

Lew et al. (2004) further examined landmark use by 6-month-old infants in a game of hide-and-seek and peekaboo, a setting similar to the one used by Acredolo and Evans (1980). Infants entered a circular room (2.3 m in diameter) with their parents, and watched the experimenter appear through a slit at the 3 o'clock position after a signal (Figure 8.5); this was repeated three times while the infants were located at the 6 o'clock

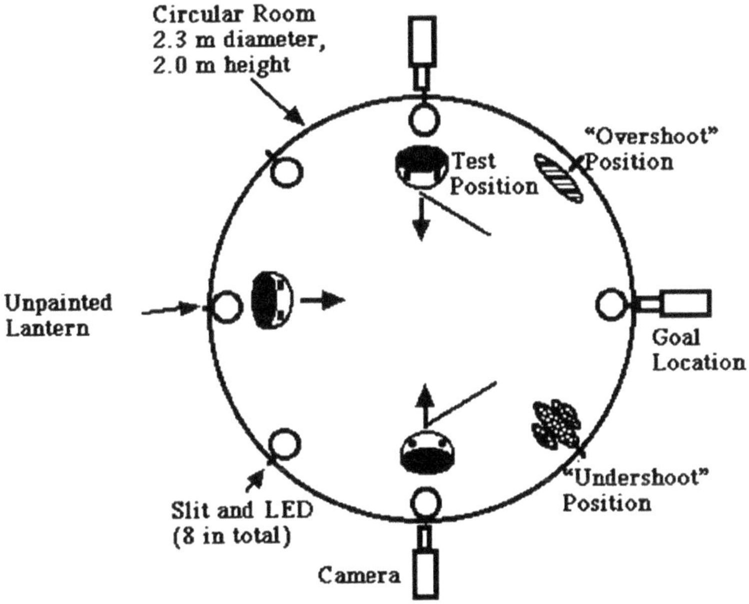

Figure 8.5 **Infants 6 months of age entered a circular room and watched the experimenter appear through a slit at the 3 o'clock position after a signal; this was repeated three times while the infants were located at the 6 o'clock position and three times at the 9 o'clock position. They were then moved to the 12 o'clock position and tested to see which direction they would turn to after hearing the signal. In a direct landmark condition, the target slit was marked by a large black and orange ring. In an indirect landmark condition, the slits on either side of the target slit were marked by the black and orange ring and a large blue lantern. In a control condition, there were no distinctive landmarks available in the room. Reprinted with permission of Elsevier from Lew et al., 2004, p. 84.**

position (looking 45° to the right) and three times at the 9 o'clock position (looking straight ahead). They then moved to the 12 o'clock position and were tested to see which direction they would turn to after hearing the signal (a correct direction would be 45° to the left). Sixteen infants were assigned to each of three conditions. In a direct landmark (or a beacon) condition, the target slit was marked by a large black and orange ring (90 cm in diameter, 60 cm inner diameter). In an indirect landmark condition, the slits on either side of the target slit were marked by the black and orange ring and a large blue lantern (80 cm in length, maximal diameter 43 cm). In a control condition, there were no distinctive landmarks available in the room. Lew et al. found that the infants in the direct and indirect landmark conditions looked at the correct target slit more frequently than did infants in the control condition (55 percent, 43 percent, and 15 percent, respectively). In a further, fourth condition, only one landmark was placed at the slit next to the target slit (at the 4 o'clock position in Figure 8.5); infants in this condition did not do well, doing as poorly as the infants in the control condition above (16 percent). These results show that infants at 6 months of age used both direct and indirect landmarks in spatial orientation and position location when the landmarks were conspicuous enough, but the performance was admittedly fragile, as revealed when only one indirect (non-beacon) landmark was available.

Cognitive maps in children

Thus far in this chapter, we have looked at infants' and children's knowledge about spatial relations in the immediate surroundings, as they view them in a room-sized space. Learning about a space at this scale, or vista space, differs in important ways from learning about a space at a larger scale, environmental space (Montello, 1993). In the latter, a spatial layout is not observable in its entirety from a single point of view, and thus a configurational understanding of spatial relations between places requires mental integration over space and time. We have seen that adults have difficulty constructing configurationally accurate mental representations of the environment, and discussed that frameworks of the microgenesis of spatial knowledge were influenced by the discussions of spatial ontogenesis. What, then, characterizes children's spatial cognition in environmental space or cognitive mapping?

Hazen, Lockman, and Pick (1978) took 3-, 4-, 5-, and 6-year-old children on a walk through a series of rooms constructed in the laboratory. The walls of the rooms were covered by opaque brown paper, and in each room (5.5 × 5.5 × 6 feet) a distinctive toy animal was located at the center, which served as a landmark. The younger children (3-, 4-, and 5-year-olds) walked through four rooms, and the older children (5- and 6-year-olds) walked through six rooms, along either a U-shaped or a zigzag route, as shown in Figure 8.6. The rooms were connected by an opaque curtain door. Children repeated walking through the rooms until they learned the routes and landmark toys correctly; then they were asked to retrace the route, starting from the last room to the first room along the learned route. While retracing the route, they were asked to name the toy to be seen in the next room along the route (a landmark reversal task) and the toy located in another room that was not directly connected by the learned route (a landmark inference task; see Figure 8.6). For either the series of four rooms or six rooms, the children did quite well (almost perfectly) on retracing the route. For the four rooms, the 4- and 5-year-olds did better than the 3-year-olds on the landmark reversal task (around 90 percent vs. less than 80 percent correct), but the three

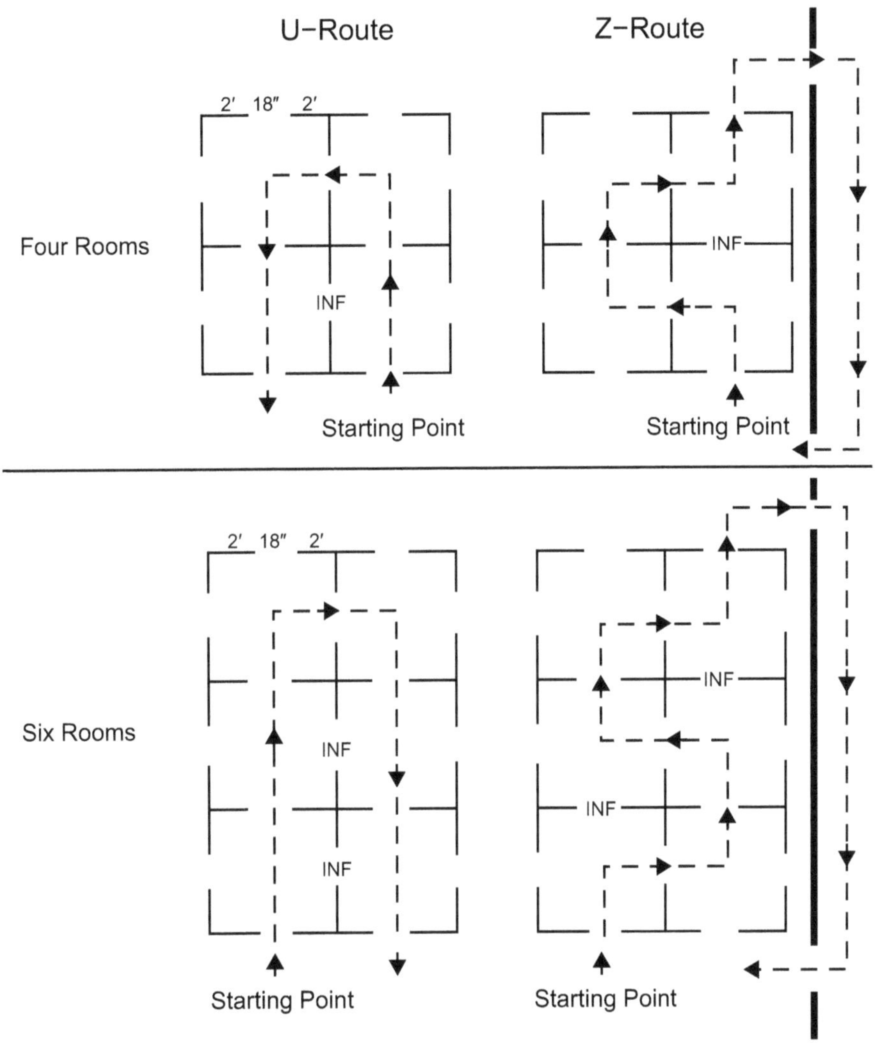

Figure 8.6 **Children 3, 4, 5, and 6 years of age walked through a series of rooms, in which toy animals were located as landmarks. The younger children walked through four rooms and the older children walked through six rooms along either a U-shaped or a zigzag route. Then they retraced the route, from the last room to the first room, and were asked to name the toy to be seen in the next room along the route and the toy located in another room not directly connected by the route. Reprinted with permission of John Wiley & Sons, Inc., from Hazen, Lockman, and Pick, 1978, p. 626.**

groups did equally poorly on the landmark inference task (accuracy in the range of 30–70 percent). For the six rooms, the 5- and 6-year-olds did equally well on the landmark reversal task (80–90 percent correct); they did worse on average on the landmark inference task, although the 6-year-olds did better than the 5-year-olds (80 percent vs. 65 percent correct). For either the landmark reversal or inference task, performance was worse on the zigzag route than the U-shaped route.

Thus, the children in all age groups retraced the traveled routes easily. Children at the age of 4 or older found it relatively easy to reverse the sequence of observed landmarks, but 3-year-old children found it difficult. In contrast to the route reversal and landmark reversal tasks, the children did poorly on the landmark inference task, a task of reasoning about a landmark in a different room that they had walked through but was not directly connected to the current room by the traveled route. That is, unlike a spatial task relating to the sequential experience of traveling a route, a task of spatial inference not tied to the learned sequence poses difficulty for children. Hazen et al. (1978) discussed that children's spatial knowledge is sequential and route-based, not maplike, and fails to allow them to make flexible spatial inferences.

Researchers have also examined children's cognitive mapping from the perspective of the strategies that they use to learn a route in the environment. Lingwood et al. (2015) guided 6-, 8-, and 10-year-old children along a route with six intersections in a desktop virtual environment, and then asked them to retrace the route from the start on their own. In one condition, the route that children traveled had salient landmarks at the intersections; in another condition, the route had no landmarks, so that the children would need to memorize the sequence of turns. The results showed that most of the children at all ages traveled the route successfully by themselves in the landmark condition (90 percent, 95 percent, and 100 percent of the 6-, 8-, and 10-year-olds, respectively); whereas they found the task difficult in the no-landmark condition, particularly younger children having difficulty reaching the goal (10 percent, 40 percent, and 80 percent of the 6-, 8-, and 10-year-olds, respectively). Thus, the 6-year-olds made more navigation errors than did the 8- and 10-year-olds, and found the route learning especially difficult when no landmarks were available along the route. It was also found that the children in the three age groups did better when the landmarks were explicitly pointed out and named by the experimenter during the learning phase. Lingwood et al. discussed that children generally rely on landmarks in wayfinding, and particularly benefit from specific instruction about which objects serve as good landmarks. Around the age of 10 years, children are able to use the strategy of learning a route without landmarks in directional terms, such as turning left, turning right, or going straight ahead.

Cornell, Heth, and Rowat (1992) similarly observed difficulties that young children tended to have with route learning. The researchers asked 60 children aged 6, 60 children aged 12, and 60 adults to walk a route through a university campus and then to retrace it back to the starting point. During the initial walk on the route, participants in one condition received instructions at predetermined locations to look back at the traveled path (or a return path) to remember where they came from, whereas participants in another condition did not receive any instructions. On the second, retracing walk, participants in the look-back condition made correct choices at decision points more frequently than did participants in the uninstructed condition. Children at the younger age, 6-year-olds, tended to make errors at decision points and wandered off the correct return route, and did not benefit from the look-back instructions as 12-year-olds and adults did. By the age of 12, children's route-learning performance becomes equivalent to adults' performance.

Bullens et al. (2010) examined the strategies that children use to navigate in a complex spatial environment, particularly comparing egocentric (response-based)

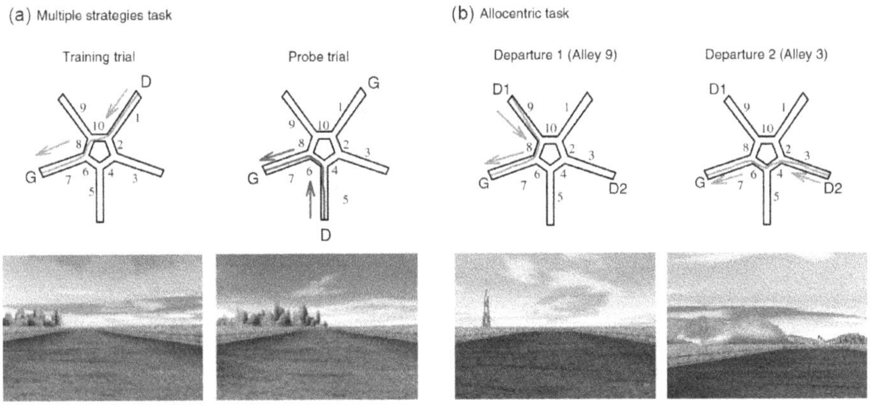

Figure 8.7 **Children 5, 7, and 10 years of age walked through a star-shaped maze in a desktop virtual environment. In a training phase, children started at the end of Alley 1, moving toward the end of Alley 7; this was repeated 15 times. Within the 15 training trials, probe trials were interspersed, in which children started at the end of Alley 5 (without knowing it) and were asked to go to the "goal" as usual. After these 15 trials, children next conducted allocentric tasks, in which they were explicitly told that they would start from either Alley 3 or Alley 9, and were asked to go to the original goal, Alley 7. Reprinted with permission of Elsevier from Bullens et al., 2010, p. 340.**

and allocentric (place-learning) navigation strategies. Bullens et al. asked 5-, 7-, and 10-year-old children to walk through a star-shaped maze constructed in a desktop virtual environment, which consisted of five alleys forming a pentagon at the center and five arms radiating from it (Figure 8.7). As distant environmental clues for orientation, such landmarks as mountains, trees, and antennas were placed between the ends of two adjacent alleys in the virtual environment. In a training phase, children navigated in the maze starting at the end of Alley 1, moving toward the end of Alley 7; they repeated this a total of 15 times. Within the 15 training trials, probe trials were interspersed that were designed to assess the types of strategies that children used in spatial learning and navigation. This time, children started at the end of Alley 5, and were asked to go to the "goal" as usual (the views at the starts of Alleys 1 and 5 looked similar, so the children did not notice that they started at a different alley in the probe trials). If they used a sequential egocentric strategy, they would move toward the end of Alley 1 in the sequence of the same body movements as before; if they used a place-based allocentric strategy, they would move toward the end of Alley 7, as guided by the distant landmarks surrounding it. After these 15 trials, children next conducted allocentric tasks, in which they were explicitly told that they would start from a different location (three times from Alley 3 and three times from Alley 9) and then go to the original goal, Alley 7.

Navigational behaviors observed in the probe trials showed that the children at all ages spontaneously chose a sequential egocentric strategy more often (59 percent, 47 percent, and 43 percent of the 5-, 7-, and 10-year-olds, respectively) than an allocentric strategy (0 percent, 16 percent, and 10 percent, respectively). In the allocentric tasks, about half the children in the three age groups (53 percent, 44 percent, and 44 percent, respectively) successfully reached the goal on the first trial of the tasks. Of these successful children, the percentages of children who had constantly used a sequential egocentric strategy in the probe trials above were 50 percent, 22 percent, and 13 percent for the 5-, 7-, and 10-year-olds, respectively. Thus, children spontaneously tended to use an egocentric strategy in navigation, although they were able to use an allocentric strategy if required, particularly around the ages of 7 and 10 years. The difficulty with allocentric understanding for the younger children was also observed when they were asked to choose the shape of the traveled maze (star-shaped with a pentagon and five radiating arms) from among six alternatives: 61 percent of the 10-year-olds chose the correct layout of the maze, whereas only 13 percent and 16 percent did among the 5- and 7-year-olds, respectively (performance close to chance level). This indicates that an allocentric understanding of space is attained progressively within this age range.

Children's knowledge of distance and direction

Encoding of distance and direction by children

Another important component of spatial cognition is knowledge about distance and direction. For example, you need to know how far and in which direction you have moved from the origin or locate another object in space with respect to your position and orientation. It entails an understanding of spatial relations by metrically scaling the amount of linear separation and the extent of angular disparity in a common frame of reference. We have seen that such a metric understanding of space (conceptions of Euclidean space) is considered, in the Piagetian framework, to emerge in a later stage of ontogenetic development. Is such metric knowledge really absent in children? How do infants and children cope with tasks that require the encoding of distance and direction information?

Landau et al. (1984) examined metric spatial knowledge acquired by a child with congenital blindness, in a series of experiments conducted over a period of 31–54 months of age. The child lost her sight during the first few months of life, and by the time of the study, it was observed that she was able to locate and recognize objects and understand spatial relations and configurations in the surrounding environments. In one experiment, she and her mother entered an unfamiliar room (8 × 10 feet) in which four objects (a chair, a pile of pillows, a table, and a basket containing toys) were located, as shown in Figure 8.8. She learned the four objects by being led by the experimenter from her mother, seated in the chair, to each object back and forth (from her mother to the table and then from the table back to her mother, etc.). She was then positioned at her mother, led to the table, and asked to move between the table, pillows, and basket on her own, finding the routes between them by herself. That is, she was asked to go directly from the table to the toy basket, from the toy basket to the table, and so on. She was tested twice on each of the six routes.

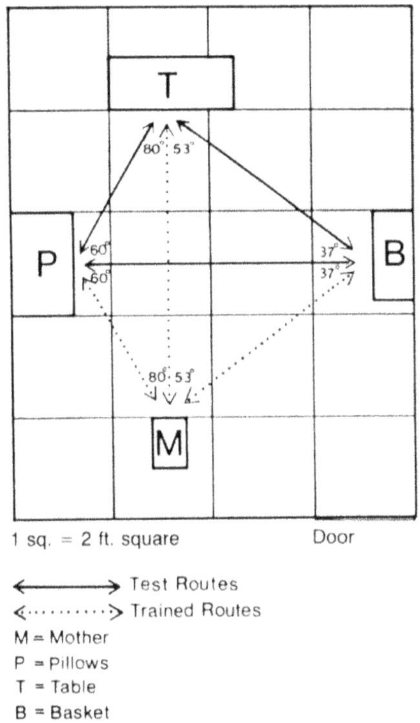

Figure 8.8 **A child with congenital blindness participated in a series of experiments conducted over a period of 31–54 months of age. The child and her mother entered an unfamiliar room in which four objects (a chair, a pile of pillows, a table, and a basket containing toys) were located. She was led by the experimenter from her mother, seated in the chair, to each object back and forth. She was then asked to go directly between the four objects by herself. Reprinted with permission of Elsevier from Landau, Spelke, and Gleitman, 1984, p. 230.**

Figure 8.9 shows the routes that the child took between the four objects (the paths of her movements generated by a series of records of her positions and frontal directions). An analysis of her paths showed (a) that her initial turns were toward the target in 11 of 12 trials; (b) that her final positions fell within 40° of the target (an angle subtended by each of the objects and thus considered as success) in 8 of 12 trials; and (c) that her initial turns generally fell on the shortest straight path to the target. These direct paths seem to reflect her understanding of the distances and directions between the objects, or a configurational layout of the objects.

These results were possible if the child only knew directions but not distances, because she could have stopped once she hit the target object. To examine this possibility, Landau et al. (1984) conducted another experiment in which the target locations were configured the same way as in the previous experiment but there were no physical objects placed at those locations. The child conducted the route finding task the same way as before, and the

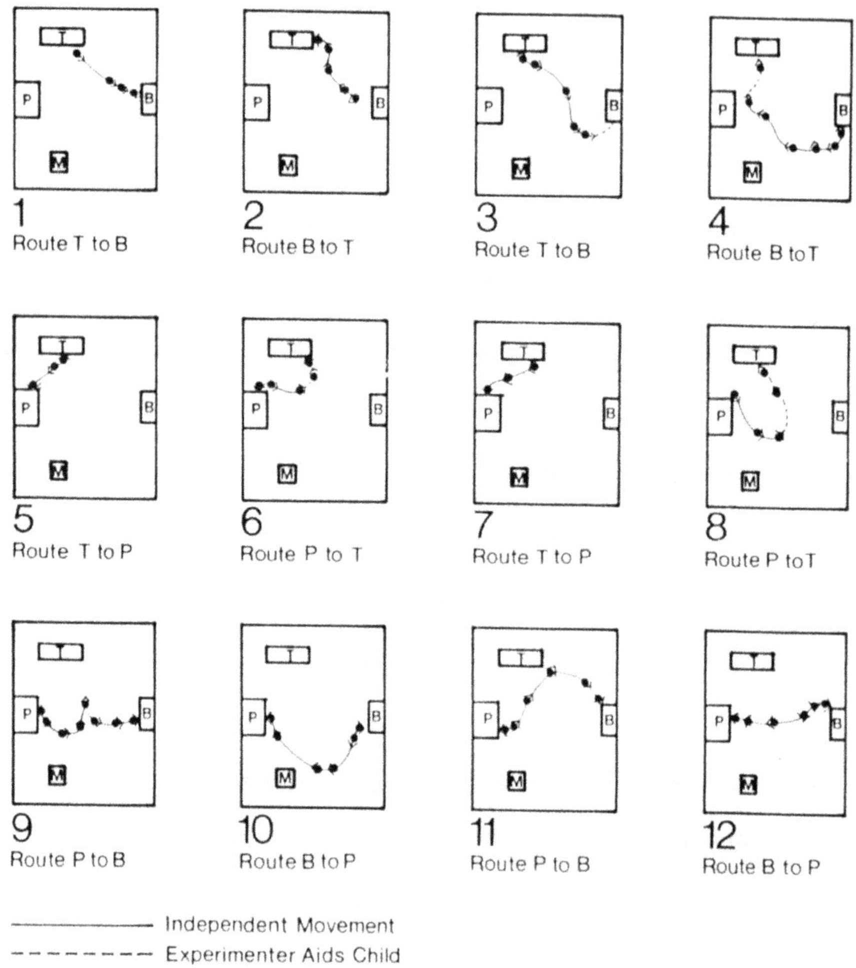

Figure 8.9 **Paths that the child took, indicated in this figure by a series of her positions and frontal directions recorded during her traveling between the objects. Reprinted with permission of Elsevier from Landau, Spelke, and Gleitman, 1984, p. 232.**

locations where she stopped were examined. Her estimates of the target locations were 27 percent off the target in distance, on average, suggesting that her distance knowledge was less accurate than her direction knowledge. In another condition, the experimenter called to the child, "Come find me," at each of the target locations, and the child was asked to go to the place where the voice of the experimenter had come from a few seconds before. The child's initial turns and final positions were similar to those in the previous experiments (initial turns were toward the goal in 9 of 12 trials and final positions were within 40° of the target in 8 of 12 trials). Landau et al. admitted that the child's movements were not along a perfectly straight line, but ascribed this to a lack of direct locomotion rather than a deficit of spatial knowledge.

Landau et al. (1984) also examined possible effects of locational cues from the room, by rotating the entire array of the objects 90° after half the test trials were done and observing changes in the child's paths. Her paths were found to be similar before and after the 90° rotation of the array, in terms of the accuracy of initial turns and final positions. Although there was a slight decrease in her performance, Landau et al. ascribed this to forgetting and distraction during the interval for the rotation. Finally, Landau et al. compared the child's performance with the performance by a group of 3-year-old sighted, blindfolded children, to observe equivalence in their performances in the accuracy of initial turns and final positions.

The results from the series of experiments led Landau et al. (1984) to argue for the existence of distance and direction information, particularly the latter, in the child's knowledge representation. Some researchers criticized Landau et al.'s interpretation of the results, by pointing out that the observed paths deviated from strictly direct straight lines. Also, the child's behavior was examined at a limited spatial scale only. But the results nevertheless point to the child's sensitivity, without past visual experience, to metric information and spatial configurations in the surrounding space, around the age of 3 years.

Similarly, Huttenlocher and Newcombe conducted a series of experiments to examine children's knowledge of distance (scaling of metric distance) and the ability to encode spatial information after self-motion (or dead reckoning). First, Huttenlocher, Newcombe, and Sandberg (1994) examined how children between 16 and 24 months of age encode the locations of objects in terms of distance. The children saw the experimenter burying a toy in a sandbox (41 × 152 cm and 30 cm deep) and, after a short delay, were asked to find the toy in the sandbox. The toy was buried at various locations in the sandbox. An analysis of the locations where the children looked for the toy showed that their estimates were quite accurate, with a distance error of about 5 cm. Furthermore, their estimated locations were systematically related to the correct locations, in the sense that the distances measured from one end of the sandbox for the estimated locations closely matched the distances measured for the correct locations (except that the estimated locations were generally biased toward the center of the sandbox, which Huttenlocher et al. discussed was a consequence of categorical coding of spatial locations, the center being a prototypical location of the spatial category of a rectangle). The same patterns of results were obtained when the sandbox was surrounded by a circular white enclosure, so that the availability of external, local landmarks was excluded. Thus, the children at the age of 16 months showed sensitivity to distance information in a bounded space.

Given these results, Newcombe et al. (1998) conducted another experiment to further examine children's (16–24 months and 28–36 months of age) performance on finding a hidden toy in the sandbox. A first group of children did the task in the same way as in the Huttenlocher et al. (1994) study; a second group of children moved to the opposite side of the sandbox after viewing a toy being hidden in it, and were then asked to find the hidden toy (called the no-movement and movement conditions, respectively). There were two further conditions created in each group: in one condition external landmarks were available (e.g., objects in a laboratory), but in another condition the view of the landmarks was occluded by a curtain. For both the younger (16–24 months) and the older (28–36 months) children, the accuracy of locating the hidden toy was better than expected by chance, but was impaired by the movement to the opposite side. The mean absolute distance error for the movement condition was more than two standard deviations above the mean for the no-movement condition. It was also observed that children older than 21 months did significantly better when external landmarks were available (mean distance errors about 13 cm with landmarks and 23 cm without landmarks), but children younger than that did equivalently with or without landmarks (mean distance errors about 23 cm). On the basis of the results, Newcombe et al. argued that children's ability to use information from self-motion (in this case, mentally updating the movement to the opposite side of the sandbox) emerges by 16 months, at least at the level of accuracy better than chance. Although the performance was better than expected by chance alone, however, the deterioration of performance after the self-movement around the sandbox was large. A fine calibration with self-motion is still difficult for children at this age. On the other hand, the ability to use external landmarks appears to emerge later, around the age of 21 months.

Sensitivity to the geometry of space

Another issue of children's spatial orientation that has been extensively discussed concerns their use of the geometry of an environment. Hermer and Spelke (1994) examined how infants 18–24 months of age searched for an object that they saw being hidden in one of the four corners of a rectangular room (1.8 × 1.2 m). Before starting the search, the infants were inertially rotated and disoriented. All four walls of the room were white in one condition, but in another condition one of the shorter walls was covered by a bright blue curtain (Figure 8.10). In the former case, corners were partially distinguished by the shape or geometry of the room (e.g., a corner having the long wall to the left of the short wall); while, in the latter case, corners were fully distinguished by the geometric shape of the rectangular room combined with the landmark of the blue wall (i.e., the corner having the long wall to the left of the short wall and also being adjacent to the blue wall). Importantly, in the former case the geometry information alone cannot distinguish between diagonally opposite corners, which are rotationally equivalent. Infants repeated the process of viewing an object hidden and searching for the object multiple times, with the hiding locations varied among the four corners, in both the all-white and the blue-wall rooms (see Figure 8.10). Results showed that the infants searched the correct corner and the rotationally equivalent corner equally frequently, and searched these corners more frequently than the other two corners, both in the all-white and blue-wall rooms. That is, they used the geometric shape of the room as an orientation clue, but failed to combine it with the landmark information to identify the correct corner uniquely. By contrast, college students searched the correct and the rotationally equivalent corners equally frequently in the all-white room; but in the blue-wall room they searched the correct corner most of the times. That is, adults successfully used both the geometric and landmark information for accurate spatial orientation.

Hermer-Vazquez, Moffet, and Munkholm (2001) further examined the use of geometry by children 3–4 and 5–6 years of age, in a white rectangular room (1.8 × 1.2 m) with one of the shorter walls covered by a blue curtain. In one condition, children viewed a toy hidden behind the center bottom of the blue curtain, and were asked, after being disoriented by rotation, to retrieve the hidden toy. In this condition, the blue wall served as a direct landmark (called a direct landmark condition). In another condition, children viewed a toy hidden under a cloth in one of the corners, and were asked, after being disoriented, to retrieve the hidden toy (an indirect landmark condition). In the direct

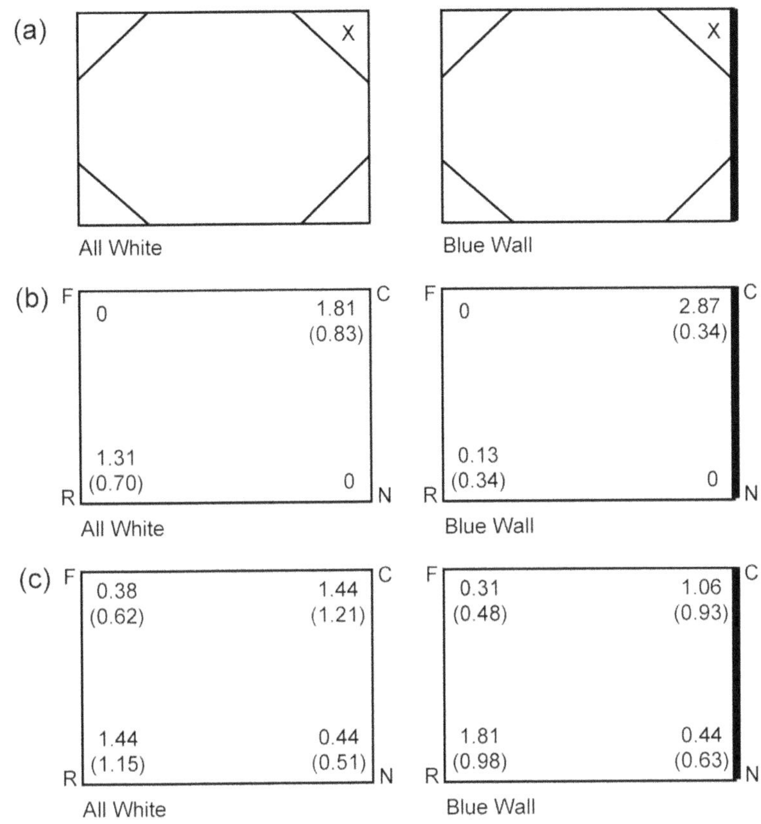

Figure 8.10 **Adults and infants searched for an object hidden in one of the four corners of a rectangular room (b and c, respectively, with mean frequencies and std. dev.). In one condition all walls were white, and in another condition one of the walls was covered by a bright blue cloth (a). In the former case, corners were partially distinguished by the geometry of the room, while in the latter case, corners were fully distinguished by the geometric shape and landmark. Reprinted with permission of Springer Nature from Hermer and Spelke, 1994, p. 57.**

landmark condition, children in both age groups successfully searched the correct location with a significantly and predominantly high frequency. In the indirect landmark condition, the younger children (3–4 years) searched the two geometrically correct corners (the correct and rotationally equivalent corners) equally frequently, while the older children (5–6 years) tended to search the correct corner. That is, the younger children relied on the geometry of the room and failed to use landmark information (proximity to the blue wall) in the indirect landmark condition, as the 18- to 24-month-old infants did in the Hermer and Spelke (1994) study. Moreover, their successful performance in the direct landmark condition does not necessarily imply that they used the blue wall as a landmark: Hermer-Vazquez et al. argued that the 3- to 4-year-old children simply formed an association between the blue wall and the hidden object. By contrast, children around the age of 6 years were able to combine geometric and landmark information in the spatial and configurational sense for accurate spatial orientation.

On the other hand, subsequent research has shown that children can use landmark, as well as geometry, information in spatial orientation, at least in some conditions. Learmonth, Newcombe, and Huttenlocher (2001) asked infants between 17 and 24 months to engage in the hide-and-seek activity used by Hermer and Spelke (1994), in a 3.6 × 2.4-m rectangular room (four times the area of the room used by Hermer and Spelke, with the same ratio of the short to the long walls). When the room was featureless with all the walls covered by white curtains, Learmonth et al. found that the infants predominantly searched the correct and the rotationally equivalent corners with equal frequency (42 percent and 36 percent of the trials, respectively), replicating the findings of Hermer and Spelke. However, when the room had two objects available as immovable landmarks (a bookshelf at the end of one long wall and a room door at the end of the other long wall), the infants searched the correct corner significantly more frequently than the rotationally equivalent corner (60 percent vs. 26 percent of the trials), irrespective of whether the toy was hidden in the corner next to either landmark (directly marked by the bookshelf or the door) or not. Furthermore, when a single landmark (a bookshelf) was placed in the center of one of the short walls, the infants also distinguished the correct corner from the rotationally equivalent corner (62 percent vs. 20 percent). Notably, when one of the short walls was covered by a blue curtain, the infants searched the correct corner significantly more frequently than the rotationally equivalent corner (60 percent vs. 24 percent); namely, they used the blue wall as a landmark to find the toy, which contrasts with the failure to do so observed for the infants in the Hermer and Spelke study.

Learmonth et al. (2001) thus showed that children under the age of 2 years were able to use landmarks to locate objects and orient themselves in a rectangular room. Learmonth et al. discussed that the successful performance observed in their study and the failure in Hermer and Spelke's (1994) study related to the difference in the size of a room—in Hermer and Spelke's original experiment, the room was very small, measuring 1.8 × 1.2 m. To corroborate the effect of room size, Learmonth, Nadel, and Newcombe (2002) examined spatial orientation and object location by older children (3–4 and 5–6 years) in two rectangular rooms differing in size (1.8 × 1.2 and 3.6 × 2.4 m) with one of the shorter walls covered by a blue curtain. The researchers found that in the smaller room, children 3, 4, and 5 years of age searched the correct and rotationally equivalent corners equally frequently (around 40 percent of the trials each); however, in the larger room, they searched the correct corner more frequently than the rotationally equivalent corner (around 60 percent vs. 20 percent). The 6-year-old children predominantly searched the correct corner, either in the small room (71 percent of the trials) or in the large room (84 percent). Thus, the children in the three age groups were able to use the blue wall as a landmark, not confined to the use of the geometry or shape of the room, as an orientation clue in the large room, whereas children under the age of 6 could not combine landmark and geometry information in the small room. The reason for the dependence of children's landmark use on the size of an experimental space awaits a clear explanation based on further examination (Newcombe, 2017); but, as well as providing insights about landmark use by children, these findings offer implications about the effects of scale on spatial cognition. In a large-scale space, configurational information is rarely available in its entirety to the explorer, unlike a bounded space of room size that offers a view of its shape to the observer. This consideration signifies the importance of studying spatial orientation and navigation in the environment in its own right (Jacobs and Menzel, 2014; Wolbers and Wiener, 2014).

Children's understanding of symbolic representations

A symbolic and representational understanding of the world makes a significant step in children's cognitive development, in the sense that children achieve the function of abstract and operative cognition. A question of interest to researchers thus concerns when children are able to understand symbolic representations. Psychologists and geographers have extensively studied children's understanding of various symbolic representations, including maps, aerial photographs, and scale models. Blaut, McCleary, and Blaut (1970) showed an aerial photograph of a small town to 107 first graders, and examined how they understand and use it. Blaut et al. asked the children to tell what they found in the photograph by naming and pointing to it, and to indicate routes between designated pairs of houses by tracing them on the photograph. Of the 107 children, 105 correctly perceived the aerial photograph as representing a view of a landscape as seen from above, and the majority of them identified cars, streets, houses, and trees on it (97 percent, 96 percent, 93 percent, and 83 percent, respectively). Of the 17 children who conducted the route finding task, 16 correctly identified the houses mentioned by the experimenter and traced the roads between them on the photograph. Blaut et al. argued that these findings attested to the ability in 6-year-old children to interpret an aerial photograph, to understand the mapping transformations of scale and projection, and to conduct real-world tasks of map reading and map use.

DeLoache (1987) examined children's understanding of a scale model in the context of a hide-and-seek play. Children 2 to 3 years of age were first familiarized with a room (4.80 × 3.88 × 2.54 m), and also with its scale model (71 × 65 × 33 cm) located in an adjoining room. They saw a miniature toy hidden within the scale model (e.g., a toy dog was hidden behind a small couch in the model); then they were told that an analogous, larger toy had been hidden in the "same" place in the room (a larger stuffed dog was hidden behind a full-sized couch in the room) and asked to retrieve it from the room. The children were grouped into a younger group (30–32 months) and an older group (36–39 months); half the children in each age group saw a toy hidden in the scale model and retrieved it in the room, and the other half saw a toy hidden in the room and retrieved it in the scale model (four trials were conducted in total). Children in the older group successfully retrieved the toy 75 percent of the times on average, while children in the younger group did so only 15 percent of the times. Younger children clearly had difficulty understanding the symbolic and representational relationship between the scale model and the room.

DeLoache (1987) assumed that the difficulty was related to the fact that the scale model was a real, three-dimensional object that children could manipulate; the children may have found it difficult to regard the model as a representation that stood for something else (the room). To examine the possibility, DeLoache used photographs in place of the scale model. Children about the same age as the younger group in the above experiment (30–33 months) were shown a set of four photographs of items of furniture in the room. The experimenter pointed to one of the photographs and said that a toy had been hidden under the item of furniture. Children were then asked to retrieve the toy from the room (four trials were conducted in total). The children also conducted the task as the participants in the above experiment did, using the scale model. Results showed that the children's performance was as low as 30 percent correct when they used the scale model. However, they did well when they used the photographs, with a performance of about 75 percent correct, which was comparable to the performance by the older children in the previous experiment. DeLoache discussed that the 3-year-old children understood the representational correspondence between the scale model and the room, whereas the 2.5-year-old children had a limited understanding of the symbolic relationship, which manifested itself in the photograph task, not in the scale model task.

Blades and Spencer (1994) pointed out that in DeLoache's (1987) experiments, the toy was always hidden at different pieces of furniture (a couch, table, chair, pillow, dresser, and shelf), and hence children could retrieve the toy if they identified individual items of furniture in the model and room, without necessarily understanding the correspondence of spatial relations of the two sets of items. To assess the possibility, Blades and Spencer examined children's performance on toy retrieval in basically the same setting as in the DeLoache study, except that two of four items of furniture were identical (a bed, a wardrobe, and two chairs). Children 5 years of age successfully retrieved a toy, but children 3 years of age failed to do so when a toy was hidden at one of the two identical items. That is, 3-year-old children's symbolic representation was limited to an understanding of the correspondence of object attributes but not spatial relations.

Huttenlocher, Newcombe, and Vasilyeva (1999) applied the setting of a sandbox used by Huttenlocher et al. (1994) to an examination of children's representational understanding of maps, or the scaling of distance between a map and a larger represented space. A group of 3-year-old children and a group of 4-year-old children (n = 25 each) viewed a map on 21.6 × 29.2-cm white paper. The children were told that the experimenter had hidden a disk (2 cm in diameter) in the sandbox and that the map showed where it was. The map depicted the sandbox as a rectangle (20.3 × 5.1 cm) and indicated where the object was hidden with a dot inside the rectangle. Huttenlocher et al. (1999) conducted a set of eight search trials, with the hiding places varied along the length of the sandbox, and accordingly the locations of dots on a map varied along the length of the rectangle. As in the sandbox experiment that we discussed above, children's responses were sensitive to the correct locations measured by distance from one end of the sandbox. All 4-year-olds demonstrated a good performance on the identification of hiding places by interrelating the distance information in the map and in the sandbox, with a mean distance error of 8.9 cm. The 3-year-old children were classified into two groups on the basis of their performance: 15 of them did it correctly with a distance error of 10.4 cm on average, while a second group of 10 children did it poorly with a mean distance error of 37.1 cm. The latter group's responses greatly deviated from the correct locations and did not even preserve correct ordering. They failed to code metric distance and to figure out the representational correspondence between the map and the sandbox.

These results point to the ability in 4-year-old (and some 3-year-old) children to understand representational relations and metric scaling in a small enclosed space and along a single dimension. How can these findings be extended to orientation in a larger space and localization in two dimensions? Vasilyeva and Huttenlocher (2004) administered the map-to-sandbox search task used by Huttenlocher et al. (1999) to groups of 4-year-olds and 5-year-olds (n = 50 each). A map depicted a referent space as a rectangle (12.7 × 17.8 cm) and indicated a target location in the space as a dot inside the rectangle. Two referent spaces of different sizes were used as a two-dimensional spatial layout, 76.2 × 106.8-cm and 243.8 × 341.4-cm light brown rugs. The children viewed a map and were asked to identify the target location on the rug, being told that a toy person had a favorite spot on the rug and the map showed with a dot where on the rug the toy wanted to stand. Children identified a total of ten target locations that were varied in the two dimensions; half the children did it on the smaller rug, and the other half on the larger rug. An analysis of mean absolute distance errors showed that both 4- and 5-year-olds did worse in the larger space (a mean error of 84.1 cm and 54.1 cm, respectively) than in the smaller space (a mean error of 21.3 cm and 13.9 cm, respectively). The 4-year-olds did worse than the 5-year-olds on average, and the difference in performance was greater for the larger space, because the 4-year-olds' responses became particularly inaccurate in the larger space. An analysis of ordinal accuracy of their responses along each axis revealed that about 60 percent and 90 percent of 4- and 5-year-olds, respectively, preserved ordinal relations in the scaling of distance. These results suggest that although 4-year-old children are successful at scaling along a single dimension (Huttenlocher et al., 1999), they have difficulty with two-dimensional scaling. By the age of 5, they generally become able to understand scaling in two dimensions.

Frick, Möhring, and Newcombe (2012) further examined 4-year-old children's sensitivity to geometric properties in the map-to-sandbox search task, using three target locations that were arranged in a straight line, a right triangle, and an isosceles triangle. For the former two arrays, the children searched correct locations at least better than expected by chance. They did not do well for the array of an isosceles triangle, with an overall performance not significantly different from chance (i.e., 33 percent correct). These results provide support for the contention that the understanding of symbolic maps and metric scaling emerges early in development (in preschool children), although it is in a preliminary and limited form with the difficulty in accurately integrating and interrelating distance, angle, and direction information.

Stage theory of child development revisited

We have now looked at the development of children's conceptions of space with respect to spatial coding in different frames of reference: the use of landmarks, the accuracy of cognitive maps, and the understanding of symbolic representations. One may describe the findings as a qualitative transition from the period of a concrete and undifferentiated sensorimotor understanding to the period of abstract and coordinated representational knowledge, with some possible discrepancies in the exact timing of transitions between stages. However, the findings of the ability in children to understand metric distances and directions and symbolic representations, albeit at a rudimentary level, are at odds with the stagelike conceptualization of ontogenetic development with relatively abrupt changes between stages (particularly the description of children's spatial understanding as purely topological in the early stages).

As an alternative view of conceptualizing the course of cognitive development in children, Siegler (1996) proposed a metaphor of *overlapping waves*, as opposed to a metaphor of a *staircase*. Siegler first looked at the ordinary linguistic use of the terms *learning* and *development*, stating that learning refers to cognitive changes over short time periods, changes of limited scope, changes with known precursors, changes whose existence and timing vary among individuals, and changes in early and middle adulthood (and sometimes in childhood), while development refers to cognitive changes over longer periods, changes with broad scope, changes with unknown precursors, changes whose existence and timing are consistent across individuals, and changes in childhood (p. 21). Siegler also pointed out in these descriptions the tendency to draw parallels between "change within an individual lifetime and change over historical or evolutionary time," as seen in "the slogan 'ontogeny recapitulates phylogeny'" (pp. 23–24). He argued that these discussions of cognitive changes at the level of linguistic use and by drawing analogies between different changes do not make a substantial contribution to the understanding of the mechanism underlying such changes.

For the discussion of cognitive development, Siegler (1996) identified four ideas that were central within the field of evolutionary psychology—variation, self-regulation, adaptive change, and inheritance. In particular, he pointed to the potential significance of the idea that organisms evolved gradually from earlier forms through interactions with their environments and adaptations to changing conditions of the environments. In reference to these basic ideas, Siegler focused on the existence of variability in children's thinking, both within and between individuals, in various areas of higher level cognition, including addition, reading, scientific reasoning, and problem solving. Importantly, Siegler argued that to account for the cognitive development and variability in these various skills, the overlapping waves metaphor offers a better alternative to the staircase metaphor. As we have already seen, the staircase metaphor (see Figure 5.1, *left*) applies to the Piagetian stage theory, the essence of which is the suggestion that "children of a given age think about a given task in a single way" (p. 84). Siegler stated that the visual metaphor of stairs illustrates the basic characteristics of the treatments of change in the stage theory of development:

> Children are depicted as thinking in a given way for an extended period of time (a tread on the staircase); then their thinking undergoes a sudden, vertical shift (a riser on the staircase); then they think in a different, higher way for another extended period of time (the next tread); and so on.
>
> (p. 84)

In contrast to this view, the overlapping waves metaphor depicts children as using multiple approaches to thinking over prolonged periods of time, with the relative frequencies of alternative ways of thinking changing gradually with age. The visual metaphor of multiple waves (Figure 8.11) represents different ways (rules or strategies) of thinking. Multiple series of overlapping waves coexist for an extended period of time (multiple different ways of thinking, rather than a single way) and their relative frequencies of use at a specific time differ in the course of development, with possible introduction of new ways of thinking (rather than an abrupt substitution of one way of thinking for another). The overlapping waves metaphor encompasses the staircase metaphor as an extreme case, where the waves have very steep rises and falls.

Siegler (1996) discussed various empirical data on cognitive change that are better explained by the continuous, overlapping waves view of development than by the staircase view. One such example is the development of the skills of addition in children. When

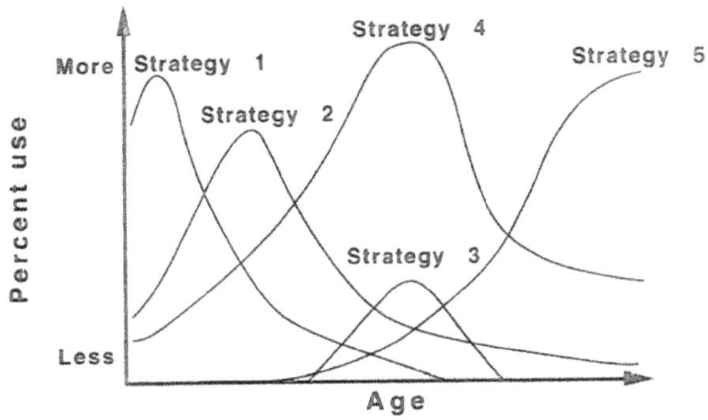

Figure 8.11 **Overlapping waves metaphor for children's cognitive development. The waves in the visual metaphor represent different ways (rules or strategies) of thinking. Multiple series of overlapping waves coexist for an extended period of time and their relative frequencies of use at a specific time differ in the course of development. (For the staircase metaphor, which posits a single way of thinking and an abrupt substitution of one way of thinking for another in the course of development, see Figure 5.1.) Reprinted with permission of Oxford University Press from Siegler, 1996, p. 238.**

presented with an addition problem such as "How much is 2 and 2?", young preschoolers usually guess without employing any strategy (and hence rarely answer correctly). With the experience of solving simple problems, older preschoolers come to answer them on the basis of overt strategies, particularly counting from one. In the first grade, children generally begin to use the *min* strategy, in which they start with the larger addend and count the smaller addend (e.g., counting "5, 6, 7, 8" for "5 + 3"). Advanced children also begin to use *decomposition*, which divides a problem into two simpler ones and adds in succession (e.g., thinking of "18 + 3" as "18 + 2 = 20 and 20 + 1 = 21"). They also begin to simply retrieve arithmetic facts from memory (directly retrieving the answer "8" from memory for "5 + 3"). Empirical data on the relative frequencies of these different strategies for kindergarteners (n = 100), first graders (n = 100), and second graders (n = 101) are shown in Figure 8.12. Siegler identified a rapid increase in the use of retrieval and decomposition from kindergarten to second grade, and in contrast a substantial decrease in the use of counting from one and guessing. The use of the min strategy increases some in this age range, and is much more frequent than that among preschoolers, who consistently start counting from 1 (p. 94). These patterns show that children's thinking entails choosing among different approaches the one that suits a given purpose and situation, in an adaptive and gradual manner tailored to the demand of a particular task.

The continuous, multiple-strategy view of cognitive development contributes to the description of empirical data on children's behavior in the domain of spatial cognition (the continuous framework proposed by Montello, 1998, is an example in the context of spatial microgenesis; see Chapter 5). Newcombe and Huttenlocher (2000) developed an argument along this line, by discussing past research into children's understanding of reference frames, landmarks, and metric distances and directions in terms of four types of spatial or locational coding. One is spatial coding in reference to one's own body, for

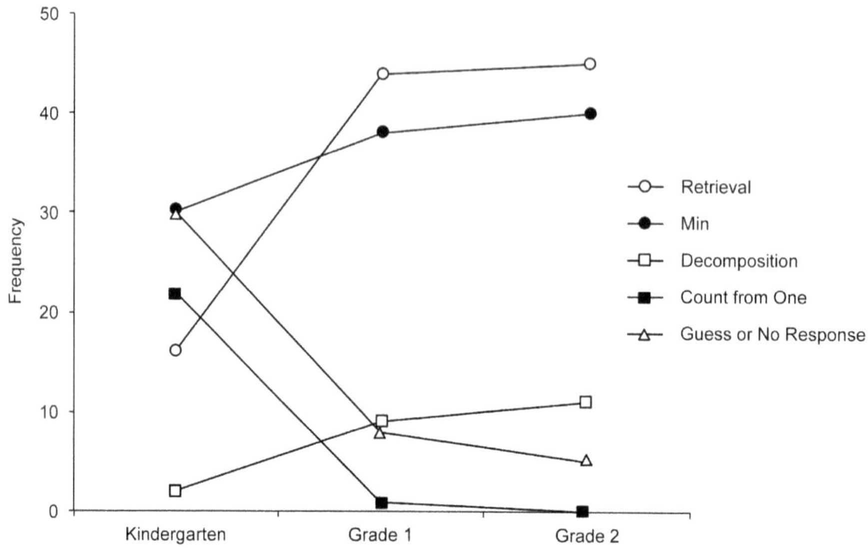

Figure 8.12 **Empirical data on the relative frequencies of different addition strategies used by kindergarteners (*n* = 100), first graders (*n* = 100), and second graders (*n* = 101). Graph based on the data shown in Siegler, 1996, p. 94.**

example, learning to turn left to find a cup in which a trinket is hidden or learning to anticipate the appearance of a person at the window to one's left. It is basically memory of specific motor movements, and Newcombe and Huttenlocher called it *response learning* or egocentric learning. Another self-referenced spatial coding is updating one's position on the basis of body-based information about distance and direction of one's own movement, called *dead reckoning*. Spatial coding is also possible with respect to external landmarks. In one such coding, objects are associated with coincident landmarks, for example, searching a cup on the white or black side of a table, expecting the appearance of a person at the window marked by a yellow star, or finding a toy behind a couch. Another coding is also externally referenced, but now with respect to the distance and direction from a distal landmark, not coincident landmarks. Newcombe and Huttenlocher called the former *cue learning* and the latter *place learning*.

On the basis of this classification of spatial coding, Newcombe and Huttenlocher (2000) discussed that the findings in the literature suggest the existence of the ability to code spatial locations through cue learning, dead reckoning, and place learning, as well as response learning, in children at least by 6 months. In the view proposed by Newcombe and Huttenlocher, development is conceived of as "changes in the importance attached to different types of spatial information as they come into conflict" (p. 49). For example, in the experiment conducted by Acredolo and Evans (1980; see above), when infants moved to the opposite end of the room, spatial information based on response learning (e.g., turning to the window to the left) was in conflict with spatial information based on dead reckoning (self-movement to the opposite side by translation and rotation) and cue learning (turning to the window marked by a yellow star or flashing lights). Infants' choices among these conflicting types of spatial information (or coding systems) depend on the salience of cues, complexity of movement, experience of success or failure, and feedback from the environment. This view of cognitive development is in line with Siegler's (1996) overlapping waves view, and Newcombe and Huttenlocher specifically contrasted it with the qualitative shifts posited by Piaget. At the same time, however, they clarified that early abilities in children are limited and that dominant patterns of coding change through adolescence.

Review box for Chapter 8

1. Discuss the development of egocentric and allocentric spatial coding by children.
2. Discuss children's spatial learning by contrasting response learning and landmark use.
3. What characterizes cognitive mapping by children?
4. Summarize the results of Landau et al. (1984). What do you think of their arguments about the knowledge of distance and direction for the child with blindness in the study?
5. Summarize the results of Huttenlocher and Newcombe's sandbox experiments. What do you think of the children's performance on the coding of metric distance?
6. Summarize the results of Hermer and Spelke's (1994) experiment in a rectangular room, and also the results of follow-up studies. Discuss spatial orientation based on the geometry of a room and on the use of landmarks.
7. Why do you think the study of spatial orientation and navigation in the environment (in a large-scale space, beyond a room-sized space) is important in its own right?
8. Discuss the understanding of symbolic representations by children on the basis of the findings from studies with maps and scale models.
9. Explain the four types of spatial coding (response learning, dead reckoning, cue learning, and place learning) discussed by Newcombe and Huttenlocher (2000).
10. Describe the staircase metaphor and the overlapping waves metaphor for the discussion of cognitive development.

Chapter 9

Feelings about space

In our discussions of people's knowledge about the environment and their interactions with the environment thus far, we have focused on locational information (knowledge about *where*) and the structures and processes of internally represented environments (cognitive maps). Our internal representations of the environment contain another important piece of information: attributive information or affective components of our experiences of the environment.

To consider this, it is worth looking back at the definition of cognitive mapping given by Downs and Stea (1973): "a process composed of a series of psychological transformations by which an individual acquires, codes, stores, recalls, and decodes information about the *relative locations and attributes* of phenomena in his [or her] everyday spatial environment" (Downs and Stea, 1973, p. 9, emphasis added). Downs and Stea thus acknowledge both locational and attributive information in mental representations, distinguishing between descriptive and evaluative attributes with respect to whether they are affectively neutral or charged. Ittelson (1973) also clarified that affective responses to the environment constitute an important component of environment perception, saying that people organize their perceptual responses to the environment around five interrelated levels of analysis: affect, orientation, categorization, systematization, and manipulation. Ittelson discussed that people first make affective responses to the environment (the situation or the ambience) and the emotions thus created direct their subsequent relationships with the environment. These emotions set the motivational tone and determine what people expect, seek, and avoid in their experiences with the environment (p. 16).

This chapter looks at these affective and evaluative aspects of our perception and cognition in the environment (collectively called environmental perception). It thus concerns "hot" cognition that is colored and aroused by feelings, emotions, beliefs, and preferences. Living environments for humans are not merely an abstract space, but places imbued with meanings and values that are either specific to individual residents or shared by groups of residents.

Environmental perception and mental maps

Gould and White (1986) examined people's preferences for places in terms of where they would like to live, and discussed the patterns of preferences observed in their responses through the concept of *mental maps*. They conducted survey analyses in various regions, asking people the question, "If you had a free choice, where would you like to live?" For example, college students in California, Minnesota, Pennsylvania, Alabama, and North Dakota viewed a U.S. state map and ranked the states with respect to personal preferences for living, from the highest to the lowest. Gould and White calculated a preference score for each state on the bases of each student's rank value, weighted by a measure of agreement between different students' assessments, and represented the preference scores across the states as a "mental topographic map." On the map, places of high and low preference were depicted by contours, with hills (or peaks) showing places that people generally like and valleys (or pits) showing places that people generally do not like.

For example, Figure 9.1 shows a preference surface obtained from responses by students in California. It shows a peak in their own state, California, and a high ridge running along the west coast into Oregon and Washington. The surface is trending downward into Nevada and Utah, and then upward over Colorado to form a secondary peak, and then again downward to the states of the Great Plains. A downward trend is observed to the

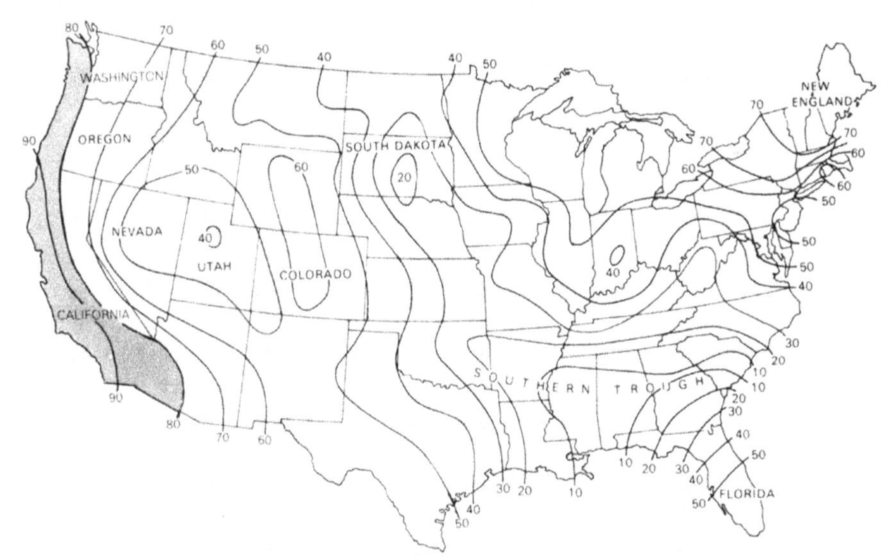

Figure 9.1 **Preference surface obtained from responses by students in California. They were asked, "If you had a free choice, where would you like to live?" On this map, places of high and low preference are depicted by contours (as mental topography). Reprinted with permission of Routledge from Gould and White, 1986, p. 55.**

South, except for Florida, and an upward trend to the states of the Middle West and New England, to form a third peak (Gould and White, 1986, pp. 54–56). Gould and White examined preference surfaces for different countries and regions (p. 52), and observed patterns of common desirability shared by people in a specific country (e.g., a high preference for California) and local desirability reflecting people's feelings about familiar surroundings of their home areas (e.g., a high preference for a birth place).

Building on these findings, Gould and White (1986) discussed how mental maps are influenced by the (invisible) "information space" in which people live. People's preference judgments in part depend on their knowledge about places, and in general the knowledge becomes less detailed (or the degree of ignorance becomes higher) as a function of distance from the location of residence (remember the concept of distance decay that we discussed in Chapter 3). Gould and White stated,

> We know more about the areas close to us, and they tend to become much more important than others about which we know little. Our emotional involvement with other places changes quite markedly with our subjective estimates of how far places are away from us.
>
> (p. 22)

They observed that students in Illinois were less ignorant than students in Pennsylvania about regions close to Illinois and all over the Far West, but vice versa about regions in the East (p. 87). There are exceptions to the general effect of distance of this sort, for example due to barrier effects of natural features (e.g., mountains and long lakes) or political boundaries. Gould and White found that preference surfaces for people in Norway and Sweden had a peak in their home countries, respectively, and showed a smooth decline with distance; when the surface crosses the national border, their preferences dropped down to much lower levels, however (p. 111; see Figure 9.2).

These results show that mental or psychological topographies of people's residential preferences can be conceived, examined, and represented, as in the analysis of physical topographies of countries. The approach can be applied to any country and region, and mental maps thus obtained would provide practical implications for regional planning, tourism, and migration. Gould and White (1986) argued that the study of mental maps has significance for the understanding of people's perception and behavior in space, because human behavior is affected by the portion of the environment that is actually perceived. The spatial environments we daily encounter offer an almost infinite amount of information to us, and so we have to select only information that (we think) is important to us. Gould and White stated, "Our views of the world . . . are formed from a highly filtered set of impressions, and our images are strongly affected by the information we receive through our filters" (p. 28). They emphasized the importance of perception in spatial cognition and behavior and advocated research from the perspective of mental geography:

> An older geography concentrated upon an environmental determinism in the physical realm. . . . Perhaps geographers should focus instead upon the invisible human landscapes which seem to be far more relevant, both for understanding human spatial behaviour, and planning a better spatial environment for people in the future.
>
> (p. 108)

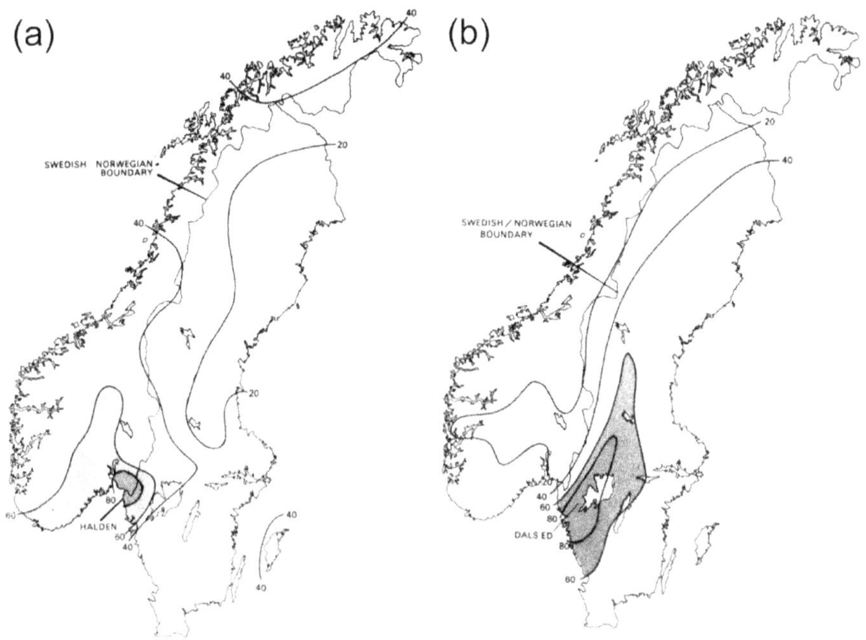

Figure 9.2 **Barrier effects of natural features on preference surfaces for people in Norway and Sweden. Their preferences had a peak in their home countries, respectively, and showed a smooth decline with distance, but dropped down to much lower levels across the national border. Reprinted with permission of Routledge from Gould and White, 1986, pp. 109 and 110.**

People's mental maps have also been studied in relation to the perception of natural hazards. A major impetus for the study of environmental perception in the cognitive-behavioral approaches in geography was provided by the pioneering work by White (1945) on people's adjustments to floods. White argued that a greater reliance on engineering-based solutions to floods such as the construction of levees, floodwalls, and reservoirs might cause further encroachment on floodplains and an increase in annual losses. White's argument stimulated a further study of laypersons' perceptions of natural hazards and attitudes toward floods, earthquakes, and droughts. Burton and Kates (1963) interviewed over 100 people with specialized technical and scientific training, and found that none of them discounted the possibility of a flood occurring in a previously flooded area again. By contrast, 84 of 216 floodplain dwellers who had experienced a flood hazard did not expect to be flooded in the future. There was thus a fundamental difference between technical people and laypersons in their attitudes toward uncertainty. Laypersons have difficulty with a probabilistic or indeterminate view of the world and fail to understand the uncertainty concerning the occurrence of natural hazards properly, hence showing irrational responses such as eliminating the hazard (e.g., "lightning never strikes twice in the same place") or uncertainty (e.g., "floods come every five years").

McPherson and Saarinen (1977) interviewed 162 residents in Tucson, Arizona, who lived within the 100-year floodplain (the area of land which is predicted to flood in a storm that has a 1 percent chance of occurring in any given year), and asked questions about their thoughts on the advantages and disadvantages of the area, awareness of a flood, and personal responses to it. The residents were found to be generally satisfied with their neighborhoods, without noting real disadvantages; although they appreciated the risk for the floodplain as a whole, they did not recognize the danger in their neighborhoods. McPherson and Saarinen discussed that these observations induce researchers and resource managers to realize "the importance of understanding the attitudes, perceptions, opinions and motives of the flood plain users to the hazard if effective management strategies are to be implemented at the local level" (p. 26).

Structure of affective meanings

We learn the environment by actively experiencing it, and through the experience we have certain feelings about the environment. Or, conversely, our feelings about the environment direct (or limit) our explorations of the environment and structure (or restructure) the knowledge about the environment. Thus, environmental perception entails the development of feelings about the environment, as well as the acquisition of knowledge about the environment. What is known about our feelings about the environment, then? What is the structure of affects and emotions pertaining to the environment that we live in? Early studies of affects and emotions in the literature of psychology were conducted through a method called the *semantic differential.* The method was developed by Osgood, Suci, and Tannenbaum (1957) in their analysis of "meaning" in general. Osgood et al. were interested in the process in which a certain stimulus (a sign, not necessarily an object that it signifies) produces certain behavior in an organism. They called it a representational mediation process, and discussed that the representation that mediates a stimulus and a behavior forms the basis of meaning in the psychological sense of the term. In the semantic differential, participants are presented with a concept and a set of scales defined by pairs of bipolar adjectives: participants indicate on each scale their thoughts about the concept in question. For example, for concepts such as *mother*, *snow*, *time*, and *symphony*, participants are given scales consisting of bipolar adjective pairs such as good–bad, strong–weak, or active–passive, and indicate the direction and intensity of their judgment on a 7-point scale, by choosing one number from seven alternatives (e.g., 1 = *extremely good*; 2 = *quite good*; 3 = *slightly good*; 4 = *neither good nor bad*; 5 = *slightly bad*; 6 = *quite bad*; and 7 = *extremely bad*). Participants' responses are examined through factor analysis to find the dimensions (or orthogonal axes) that define the semantic space for the concept.

From the analyses of a variety of concepts across different domains (e.g., persons, physical objects, events, sonar signals, paintings, or political concepts), Osgood et al. (1957) identified three major factors: evaluation, potency, and activity. The first factor, *evaluation*, is evaluative in nature, either morally, aesthetically, socially, or emotionally, showing judgments about good or bad, pleasant or unpleasant, and positive or negative. It is generally extracted as the most dominant factor and explains over 50 percent of the total variance. The second factor, *potency*, concerns power, size, weight, or toughness (e.g., large–small, strong–weak, or heavy–light), and the third factor, *activity*, concerns quickness, excitement, or warmth (e.g., fast–slow, active–passive, or hot–cold). One can thus quantitatively analyze the semantic space, or the space of human affects, emotions, and meanings; it is multidimensional and explained by the three dimensions of evaluation, potency, and activity.

Russell and colleagues extended the semantic differential analysis of affective meanings to an examination of affective qualities of spatial environments (Russell and Pratt, 1980;

Russell, Ward, and Pratt, 1981). Russell et al. (1981) picked out 323 environments of various kinds, ranging from natural to human-made and from indoor to outdoor environments (e.g., a beach, a garden, a street, a restaurant, a grocery store, a laboratory, an airport, a museum, a city bus, a hospital, or an elevator), and developed a list of 105 adjectives that were commonly used to describe landscapes (e.g., pleasant, lovely, or enjoyable). They visited the 323 environments individually and asked one person in each environment to assess on an 8-point scale the degree to which the 105 adjectives accurately described the environment (1 = *extremely inaccurate*; 8 = *extremely accurate*). A factor analysis of the 323 responses to the 105 scales revealed three factors, in correspondence to the three major semantic differential factors of evaluation, activity, and potency, which respectively explained 29 percent, 11 percent, and 8 percent of the total variance. Adjectives that loaded on the first factor (evaluation) included *pleasant*, *inviting*, *happy*, *joyful*, *enjoyable*, and *good*; adjectives on the second factor (activity) included *hurried*, *active*, *hectic*, *lively*, *rushed*, and *alive*; and adjectives on the third factor (potency) included *overwhelming*, *overpowering*, *frightening*, and *powerful*. Through a cluster analysis of the intercorrelations among the 105 adjectives, they were classified into 21 groups. Figure 9.3 shows representative adjectives corresponding to the 21 groups in the first two dimensions of evaluation and activity.

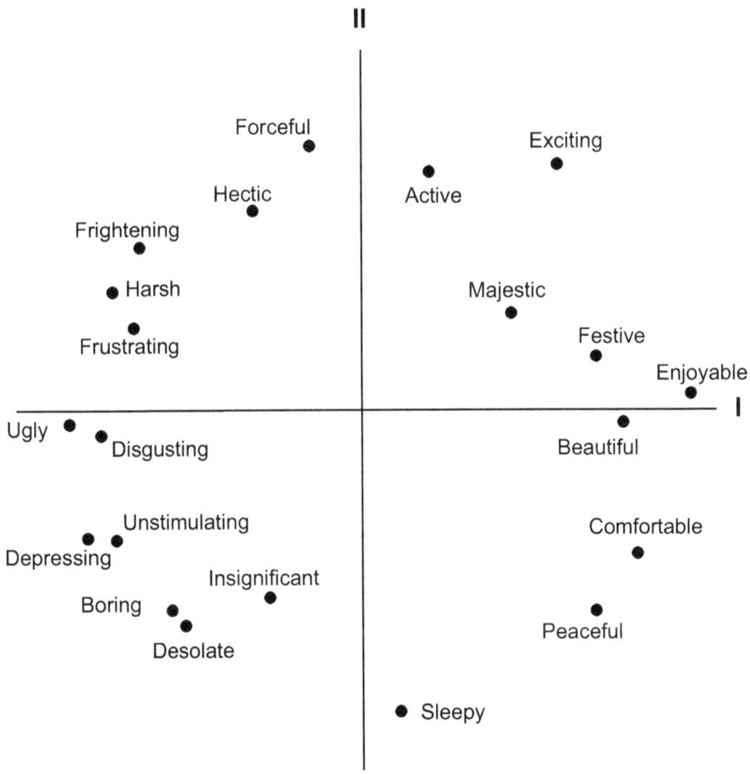

Figure 9.3 **Three factors were revealed in a factor analysis of responses to scales of affective qualities of spatial environments: evaluation, activity, and potency. This figure shows major adjectives that correspond to 21 groups identified in the first two dimensions, evaluation and activity. Reprinted from Russell and Pratt, 1980, p. 312.**

Russell et al. (1981) summarized these results as a two-dimensional representation of the affective qualities of places, and discussed the components of affective meanings of environments (Figure 9.4). The two dimensions, labeled *pleasure* and *arousal*, are bipolar ("pleasing–displeasing" and "arousing–sleepy") and independent of each other. In this psychological space of affective meanings, affective qualities attributed to environments are conceived of as coordinates of values for the degrees of pleasure and arousal, with any affect term represented as a two-dimensional vector from the origin. Figure 9.4 shows eight major affective descriptors arranged in a circle in the affective space, from *pleasant, exciting, arousing, distressing, unpleasant, gloomy, sleepy*, to *relaxing*. Russell and Pratt (1980) developed an assessment scale based on these eight affective qualities, and reported that it was reliable for the quantification of people's affective responses to various environments, with an inter-rater reliability of 0.97. Responses to the scale validly replicated the circular placement of the eight descriptors in the two-dimensional affective space.

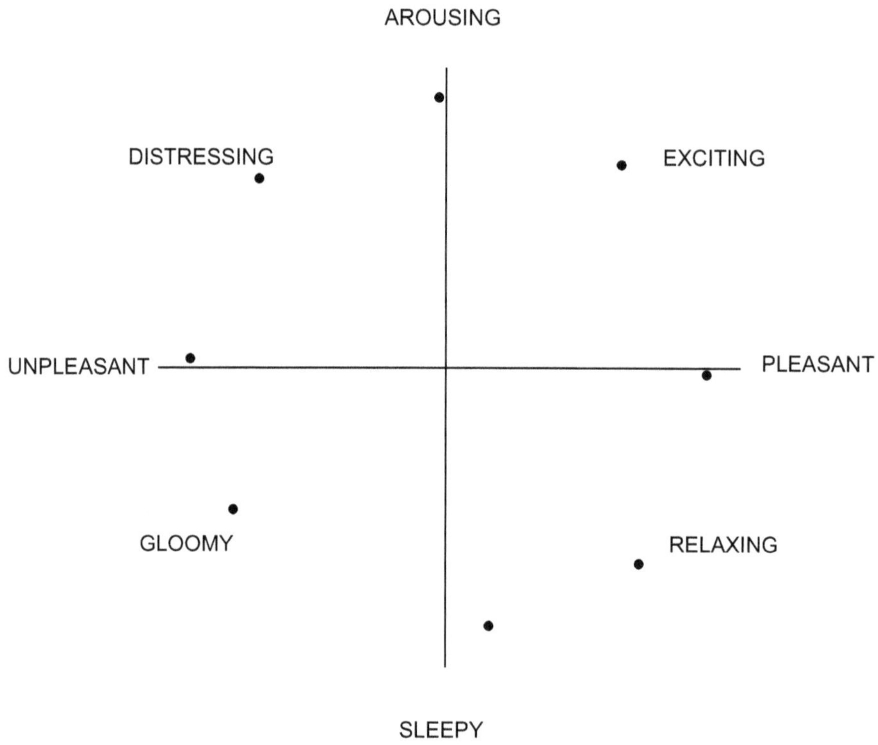

Figure 9.4 **In the two bipolar dimensions of pleasure (pleasing–displeasing) and arousal (arousing–sleepy), eight major affective descriptors are arranged in a circle, ordering from pleasant, exciting, arousing, distressing, unpleasant, gloomy, sleepy, to relaxing. Reprinted from Russell and Pratt, 1980, p. 313.**

Collative properties of the environment

Discussions in the previous section about affective components of environmental perception show that there is an identifiable structure in the emotions about environments. Now the question is: Why do people feel differently in different environments? In other words, what are the characteristics of environments that evoke different emotions in people? This question has been pursued in the literature of experimental aesthetics from the perspective of the observer's exploratory behavior. Research in this line examines the relationship between the properties of environmental scenes that are visually presented to the observer and the observer's aesthetic responses to the scenes. Importantly, Berlyne (1966) discussed aesthetic responses in terms of information search by animals in the environment. Higher animals interact with their environments and the environments provide the animals with a certain level of stimulation, information, and challenge. Animals, Berlyne argues, have a natural tendency to seek the "right" amount of stimulation to the nervous system, or information that is neither too easy nor too difficult to assimilate (p. 26). In one case, animals have the feeling of discomfort because they find the information available is inadequate. Then they are motivated to search for additional information that reduces uncertainty and conflict in the information sources (p. 26). In another case, the exploration is aimed at finding stimulation that offers "an optimum amount of novelty, surprisingness, complexity, change, or variety" (p. 26) and is not necessarily preceded by the lack or inadequacy of information in a stimulus (p. 27).

Berlyne (1960) explained the selection of a stimulus by an animal (or its exploratory behavior) by the concepts of novelty, uncertainty, conflict, and complexity. Something is perceived to be novel or not depending on a person's total experience or recent experience with it; namely, *novelty* of a stimulus is judged in comparison with a particular set of previous experiences. The stimulus attracts a person's attention most strongly when it is "rather like something well known but just distinct enough from it to be 'interesting'" (p. 21). Something may be regarded as certain or not depending on the range of alternative possibilities or the deviation of its probability of occurrence from 0 (never) and 1 (always); the greater the range or deviation, the higher the degree of uncertainty. *Uncertainty* has significance in a person's decision making when he or she is confronted with a set of alternatives that lead to different consequences, incompatible with each other—uncertainty is of particular psychological importance to a person's exploratory behavior when information available is in *conflict* (p. 29). Finally, stimuli differ in *complexity* depending on the number of distinguishing elements, the degree of dissimilarity between elements, and the extent to which the elements are treated as a unit.

Berlyne (1958) examined how these properties of a stimulus relate to the observer's search behavior, and discussed aesthetic appraisal of visual stimuli from the perspective of the observer's exploratory behavior. Berlyne conducted an experiment in which participants were presented with pairs of animal pictures. In one condition, one picture was the same on every trial and the other picture changed from one trial to the next. Berlyne found that participants looked at the recurrent picture for a shorter time and looked at the changing pictures for a longer time (Figure 9.5). In another condition, participants were presented with pairs of visual patterns of different degrees of complexity side by

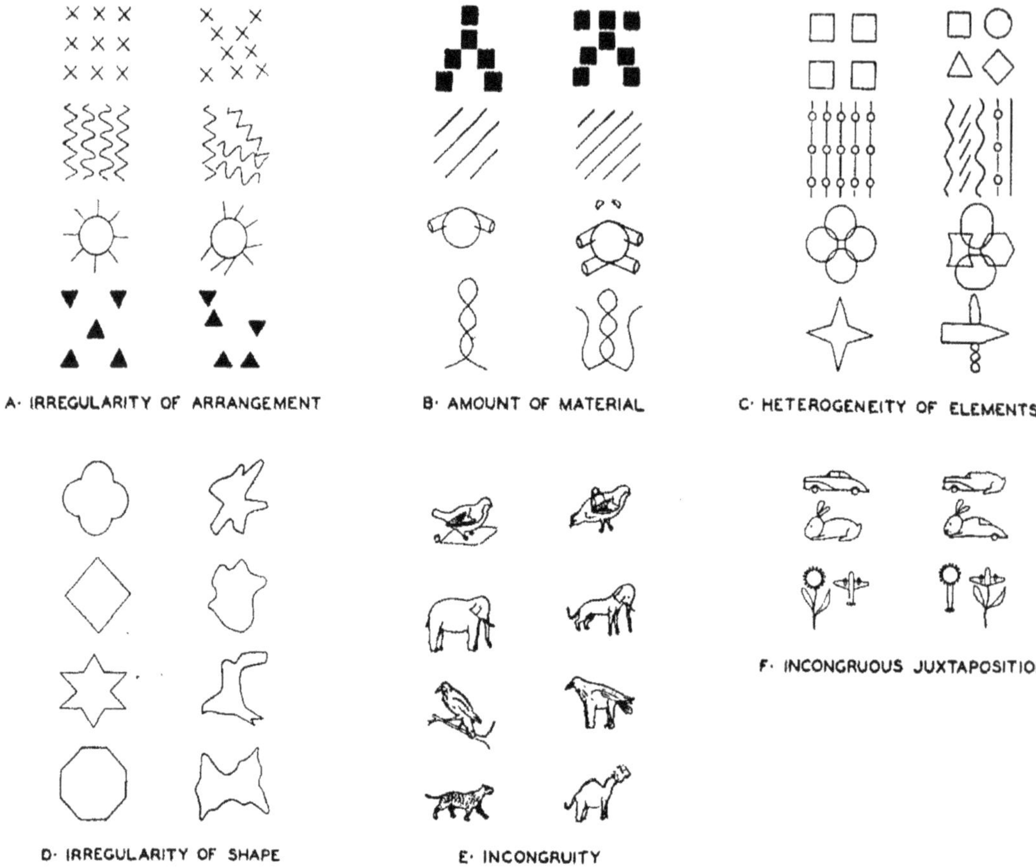

Figure 9.5 **When presented with pairs of animal pictures, one being the same on every trial and the other changing from one trial to the next, participants looked at the changing pictures for a longer time When presented with pairs of visual patterns of different degrees of complexity, participants looked at the more complex pattern for a longer time. Participants also looked at pictures of animals that were made up of parts from different species for a longer time. Reprinted from Berlyne, 1958, p. 291.**

side. Then they looked at the more complex or irregular pattern in the pair for a longer time. Also, participants looked at pictures of animals that were made up of parts from different species (e.g., a bird with legs of a dog) for a longer time than pictures of animals that people usually see. These results pointed to the influence of the degrees of novelty and complexity of visual stimuli and conflict with normal expectations (and hence uncertainty) on the observer's exploration of the stimuli. Berlyne called them *collative properties* as they pertain to comparison or collation of stimulus elements, and explained exploratory behavior in terms of novelty, surprisingness, incongruity, complexity, variability, and puzzlingness (p. 30). A major characteristic shared by these collative properties concerns a conflict engendered by the stimulus: the engendered conflict increases the observer's motivation and arousal, and in cases of excessive novelty or surprisingness, provokes a feeling of terror or fear.

Berlyne's discussions of aesthetic appraisal from the perspective of stimulus exploration stimulated researchers to examine the relationship between arousal level and environmental preference. For example, Wohlwill (1968) discussed landscape appraisal in relation to stimulus complexity. Wohlwill presented slides of various environmental scenes (e.g., a grassy field, the sea, a farmstead, a city square, and an urban skyline) with different levels of complexity to participants, and asked them to assess on a 7-point scale how much they liked each slide. The levels of complexity of the environmental scenes were judged by another group of participants with respect to the variations in color, shape, texture, directions of dominant lines, and natural–artificial composition. Wohlwill found that the number of gazes at the slides increased monotonically as a function of the degree of complexity. Preference ratings were related to complexity in a curvilinear, inverted U-shaped pattern, with a maximum at an intermediate level of complexity. Thus people tended to like environments that were neither too simple nor too complex. Research has also shown that preference ratings are affected by the types or contents of environments (e.g., water and foliage), with natural scenes generally preferred to urban scenes (Kaplan, Kaplan, and Wendt, 1972). Researchers discussed this pattern of preference in relation to the degree of complexity: natural environments produce a desirable level of complexity with irregular and curvilinear lines and continuous gradations of shape and color (Hartig and Evans, 1993; Wohlwill, 1983).

Habitat theory of environmental aesthetics

In Chapter 6, we looked at an evolutionary description of spatial ability in the hunter-gatherer model, which states that the labor division across the evolutionary time (men hunt and women forage) relates to the male superiority on mental rotation and the female superiority on object and location memory. Environmental preference has also been discussed from an evolutionary perspective. It goes like this: For animals, the act of living in the environment entails escaping the dangers of hazards (e.g., being caught by predators) and adapting to varied conditions of nature (the processes of assimilation and accommodation); animals therefore should show preference for environments that are amenable to their survival or "environments in which their species prospers" (Kaplan, 1987, p. 15). Hartig and Evans (1993) discussed that evolutionary approaches consider environmental preference in terms of beneficial value. Individuals who can quickly and unconsciously judge the sustenance values of environmental conditions are more likely to survive. Hartig and Evans stated, "This operation of natural selection remains manifest to some degree in contemporary environmental preferences, [which] retain benefit value in that they signal positive states (e.g., aesthetic satisfaction, lack of anxiety)" (p. 437).

On the basis of such a habitat-based argument of natural selection and environmental preference, Appleton (1975) proposed a framework of environmental aesthetics, known as *prospect-refuge theory*. The theory identifies environmental perception as the key to adaptive behavior, because adaptation is only made possible by the information that animals acquire from the environment. Animals prefer environments that support the acquisition of information necessary and the retrieval of the information when needed for survival (Appleton, 1988, p. 31). Appleton discussed two properties of environments that animals find favorable for adaptation, as they are immediately relevant to improving the chances of survival. One concerns the "opportunity to keep open the channels by which we receive

Figure 9.6 **Two properties of environments that animals find favorable for adaptation, the opportunity to receive environmental information (prospects, *left*) and the opportunity to achieve concealment (refuges, *right*). Prospects allow an animal to see what is out there and refuges provide an animal with the opportunities to protect itself. Photographs from https://pixabay.com/**

environmental information" (p. 31), and the other concerns the "opportunity to achieve concealment" (p. 33). Appleton referred to the former as *prospects* and the latter *refuges* (Figure 9.6). Prospects are views outward from a vantage point, which allow an animal to see what is out there and figure out possible dangers, leading to increased chances of adjusting to environmental conditions. Refuges are hiding places or shelters from hazards such as predators, bad weather, or obstacles, providing an animal with the opportunities to protect itself. In essence, these two environmental properties offer the observer the ability to see without being seen, which is imperative for an animal's survival (think about an animal that has an open view around and also can hide from potential predators). Environments that support this ability, therefore, produce the feeling of aesthetic satisfaction in the animal, inasmuch as they satisfy its biological needs of survival. The theory has been widely discussed and applied as a guiding framework to landscape planning and architectural design. At the same time, some meta-analytic studies pointed out the descriptive and qualitative nature of the theory and inconsistency in the predictability of environmental preference by the concepts of prospects and refuges (Dosen and Ostwald, 2016; Stamps, 2008).

Informational model of environmental preference

Research in the tradition of experimental aesthetics views environmental scenes as visual stimuli and discusses landscape preference in relation to collative stimulus properties such as complexity, but environmental preference is affected by the contents of environments; for example, natural scenes tended to be preferred to urban scenes (Kaplan et al., 1972). Kaplan et al. found that preference ratings of environmental scenes correlated 0.37 with the level of complexity; but when the scenes were grouped into natural and urban scenes, the correlation increased to 0.69 and 0.78, respectively. Kaplan (1987) noted the relatively low predictability of environmental preference by complexity and a large variability in the preference ratings for natural scenes; he thus suggested other possible predictors of environmental preference. In particular, Kaplan found that environmental scenes that were preferred fell into two categories, namely scenes that "either contained a trail that disappeared around a bend or . . . depicted a brightly lit clearing partially obscured from view by intervening foliage" (p. 8).

Given these patterns of preferences, Kaplan (1987) proposed a framework in which environmental preference is described in relation to the characteristics of environmental information and human dispositions. The framework says that humans are inclined to search for new information to enrich and update their knowledge (or cognitive maps) for survival, and at the same time to avoid a situation that is too unfamiliar to understand with the existing knowledge. Kaplan described these two human dispositions as the desire to understand what is going on and the desire to explore the setting further (Kaplan, Kaplan, and Brown, 1989). That is, the framework considers environmental preference in the context of human–environment interaction and relates aesthetic appraisal to the process of extracting information from the environment: "The primary distinction here is between two affectively important informational outcomes. These outcomes are 'Understanding' (comprehending or making sense of a scene) and 'Exploration' (being held by the setting, being attracted by or pulled toward sources of additional information)" (Kaplan, 1987, p. 10).

Kaplan (1987, pp. 10–12) discussed four properties of environmental scenes that people prefer, composed of two dimensions. In Kaplan's terms, one dimension contrasts the desire to understand with the desire to explore, or the ease of achieving understanding (making sense) with the attraction to exploration (involvement). The other dimension concerns the availability of information in the scene, whether it is immediately available on the surface or it may be predicted or inferred in three dimensions. You may feel ill at ease in an environment that lacks a coherent structure, with meandering streets, limited views outwards, or buildings of different colors, facades, and heights. On the other hand, you experience a sense of boredom in an environment that is too uniform in appearance, with regularly spaced streets forming a grid and houses having the same look. Also, as seen in the discussion of imageability and legibility by Lynch (1960), understanding the surrounding space and becoming oriented in the space provide us with a sense of security. And, finally, if the environment contains unknown or hidden parts, you may feel like exploring it to see what is there.

Kaplan and Kaplan (1989) explained these feelings with the concepts of coherence, complexity, legibility, and mystery (Figure 9.7). *Coherence* contributes to the ease of comprehension, with information being immediately present in the scene and supporting mental organization in the perceiver's mind. *Complexity* signifies that information is immediately available in the scene and invites the perceiver to further process the information in the picture plane. *Legibility* concerns the type of information that is inferred from the scene and enhances understanding; it predicts the perceiver's ability to remain oriented in three dimensions when wandering further into the scene. *Mystery*, as in the case of complexity, concerns further processing of information, but the information is inferred from what is in the scene, rather than immediately present on the surface; the scene allures the perceiver to a further exploration and the exploration promises a gain of more information. A partial view of a duomo through a wandering path takes you further into it.

The concepts of coherence, complexity, legibility, and mystery have been applied to the prediction of people's preference judgments for various natural and urban environments (e.g., Herzog, 1985, 1992). Stamps (2004) conducted a meta-analysis of past studies for a total of 6,288 participants and 1,820 scenes, and reported a wide variability in the

Figure 9.7 **Four properties of environmental scenes that are conceived by two human dispositions: the desire to understand what is going on and the desire to explore the setting further: (a) coherence, (b) complexity, (c) legibility, and (d) mystery. Photographs from https://pixabay.com/**

correlations of coherence, complexity, legibility, and mystery with preference judgments, ranging from –0.33 to 0.84, –0.11 to 0.97, –0.70 to 0.58, and –0.45 to 0.95, respectively. Finally, environmental preference is affected by a person's attributes, such as residential experience, familiarity, culture, and socioeconomic status. With such a caveat, the Kaplans' framework helps planners to design environments by taking people's perceptions into account. Kaplan, Kaplan and Ryan (1998) discussed possible ways to apply these concepts to the practice of landscape design; for example, creating coherent regions or providing distinctive landmarks to increase understanding and providing partial views through gateways or along curved trails to foster exploration.

Evolutionary perspective and habitat theory

The aforementioned frameworks are based on the idea that through the history of evolution, humans have formed a tendency to prefer environments that support their survival and prosperity. Preference for such environments is thus an inherited trait and should reveal itself in aesthetic appraisal quickly and unconsciously. A dominant view on human evolution states that human ancestors (modern humans) appeared in Africa around 200,000 to 300,000 years ago. Their habitats back in those days were a savanna and woodland environment, composed of open areas of grasslands and scattered dense areas of trees with large but open canopies and small trunks (Figure 9.8). Such a savanna-like environment offers (or affords, in the Gibsonian sense of the term) humans within it

Figure 9.8 **The habitats for human ancestors (modern humans) were a savanna and woodland environment, composed of open areas of grasslands and scattered dense areas of trees. Such a savanna-like environment affords humans the ability to see without being seen. Photographs from https://pixabay.com/**

the ability to see without being seen, which is the key to their survival. Habitat theory argues that humans feel aesthetically satisfied when they perceive environments that possess favorable conditions for their survival, because of evolutionary benefits. This line of argument forms the basis of prospect-refuge theory, which predicts preference for environments that provide open views and shelter from hazards; savanna-like environments with grasslands and trees fit this description quite well. A question of interest is whether people indeed exhibit preference for savanna-like environments in the evaluation of landscape aesthetics.

Balling and Falk (1982) empirically examined this question by presenting to a large sample of American people (105 third graders, 77 sixth graders, 96 ninth graders, 130 college students, 126 adults, and 14 retired senior citizens) slides of five different natural environments (tropical rain forests, deserts, savannas, temperate deciduous forests, and temperature coniferous forests), and asked them to rate how much they liked each slide on a 6-point scale (1 = *extremely undesirable*; 6 = *extremely desirable*). Balling and Falk hypothesized that if the preference for savanna-like environments is existent as an evolutionary inherited trait, responses would be made spontaneously and unconsciously, and the tendency would be observed more strongly among children and become weaker with age. The amount of experience with various different environments increases with age, and accordingly the effect of variables other than an inherited trait would be larger.

Balling and Falk (1982) found that the participants showed the highest preference for savannas, deciduous forests, and coniferous forests (with no significant difference in preference among the three); the next highest preference for rain forests; and the lowest preference for deserts. A further analysis of the three most preferred environments revealed age differences in preference judgments: the third and sixth graders (8- and 11-year-olds) showed a higher preference for savannas than for deciduous and coniferous forests, whereas the other age groups rated the three environments equivalently. These results suggest, Balling and Falk discussed, that the preference for savanna-like environments is innate, expressing itself most strongly in childhood and being modified through experiences with age. Falk and Balling (2010) conducted a replication study with a sample of three populations of West Africans in Rivers State, Nigeria. They asked 63 secondary school children and 32 college students to rate the same set of slides of five natural environments as above, and found that the participants showed the highest preference for savannas. Thus the tendency to prefer savanna-like environments was observed across cultures.

The feeling of place attachment

Another aspect of people's emotions about environments concerns the feeling of attachment toward a particular place. *Place attachment* is a broad concept and is given a variety of definitions, and sometimes referred to by different labels, but basically it signifies the emotional bond between people and their meaningful environments (Scannell and Gifford, 2010). The meaningfulness to a person emerges as a result of his or her interactions with the environment; experiences in the environment structure and organize mental representations of the environment cognitively and affectively, and imbue the environment with meanings and values. In other words, with experiences and accordingly memories in the environment, an abstract *space* is turned into a *place*. It is filled with specific meanings and values to the person, thus creating in the person's mind a feeling of attachment to the place (Tuan, 1977).

Scannell and Gifford (2010) discussed the concept of place attachment in a framework of *persons*, *processes*, and *places*. First, place attachment is considered from the perspective of persons at the individual and group levels. Place attachment at the individual level focuses on personal connections to a particular place—places that are filled with personal memories evoke the feeling of attachment to a higher degree, for example, places where you grew up, attended school, or met your partner. Such memorable and important places often provide the person with a stable sense of self, or self-definitions; in such a case, the concept of place attachment is equated with *place identity*. It defines who you are and of what value you are in relation to places of importance to you (Proshansky, Fabian, and Kaminoff, 1983, p. 74). Personal experiences with a place thus play an integral part in the formation of the feeling of attachment. Place attachment may also be considered at the group level, emphasizing symbolic meanings shared among the members of a culture, gender, or religion, for example forests with special meanings for specific cultural groups or revered places with the status of the sacred for religious groups.

The Scannell and Gifford (2010) framework discusses the psychological processes of place attachment, by looking at the relationship between people and places from affective, cognitive, and behavioral points of view. An affective component of place attachment concerns an emotional connection to a particular place, which yields in a person's mind a sense of belonging, well-being, and pride (Brown, Perkins, and Brown, 2003). The significance of such an emotional connection is often revealed in the event of forced displacement owing to natural disasters or neighborhood redevelopment. Fried (1963) reported symptoms of grief displayed by a group of residents who were forced to move out of a familiar neighborhood because of a redevelopment project. The residents felt that the relocation fragmented the sense of spatial identity (this is our "home" area) and the sense of group identity (social networks and continuity) that they had fostered in the neighborhood. The loss of an important place, for example a familiar landscape or a social setting, thus often causes a feeling of sadness among the people who cherished it.

A cognitive component of place attachment concerns the knowledge, memories, beliefs, and meanings that people associate with their important places (Scannell and Gifford, 2010, p. 3). Cognition thus connects a person to a particular place through memorable events and creates the sense of self-identity, belonging, and community. Examples include a house where you are raised, a park where you met your partner for the first time, or a place where you saw a friend off (Manzo, 2005). The affective and cognitive components of place attachment lead the person to exhibit specific behaviors or actions in and toward the place. For example, positive feelings about a place induce the person to maintain closeness to that place (Hidalgo and Hernández, 2001), which is revealed by a longer length of residence (Hay, 1998), a desire to return to one's hometown after relocation for work (Riemer, 2000), and reconstructing a neighborhood as before or relocating to a similar place after natural disasters (Kates et al., 2006).

A third perspective of the Scannell and Gifford (2010) framework concerns places. The characteristics of places to which people feel attached have been studied with respect to social connections, or the sense of community that people foster through interactions with other people in the place (Fried, 1963; Perkins and Long, 2002). Scannell and Gifford discussed that attachment to a place means attachment to people who live there and to the social interactions afforded by the place (p. 5). Neighborhood bonds thus created through social and interpersonal connections often lead to communities that are homogeneous in terms of status, interest, and lifestyles, and contribute to the establishment of a sense of place identity.

Scannell and Gifford (2010) discussed that the function of place attachment is to offer a sense of security and comfort to a person. Such a sense allows the person to engage in self-reflection and maintain self-continuity, and provides restorative effects such as stress relief. Environmental or neighborhood qualities of this sort are also discussed in the literature from the perspective of *social capital*, a resource that is actually or potentially available to the residents. The concept of social capital has been defined in various ways, but Putnam (2000) stated, "Whereas physical capital refers to physical objects and human capital refers to properties of individuals, social capital refers to connections among individuals—social networks and the norms of reciprocity and trustworthiness that arise from them" (p. 19). The amount of social capital may be measured by survey questionnaires, asking about the degrees of connections to the community, for example, participation in the local community, a sense of personal and collective efficacy, feelings of trust and safety, and connections with neighborhood, family, and friends (Duke et al., 2009; Lewicka, 2005; Onyx and Bullen, 2000). An empirical analysis of people's responses to survey questionnaires has shown that social capital is positively correlated with civic engagement (Duke et al., 2009), perceived safety (Dallago et al., 2009; Vieno et al., 2010), and physical and mental health (Wood and Giles-Corti, 2008; Ziersch et al., 2005). These findings point to the practical significance of the concept of place attachment.

Human values

We have discussed in this chapter various affective components of environmental perception, and lastly we now see a basic human trait that concerns evaluation in various domains and influences our emotions, beliefs, perceptions, behaviors, and interpretations in our life; namely *values*. Schwartz (2016) defined values as desirable goals that transcend specific situations and guide the selection and evaluation of competing alternative decisions and actions. When a person is confronted by a set of alternative actions, values are activated in the person and specific feelings are evoked that motivate the person to pursue the goals defined by the values. On the basis of judgments made in reference to the priorities ordered by relative importance of the values, the person takes a specific action (pp. 63–64).

Schwartz (1994) developed a survey questionnaire with a view to assessing and revealing the structure of values, on the assumption that human values are composed of different types, some of which are compatible and some of which are in conflict with each other (e.g., pursuit of self-interest and consideration of others). The scale consists of 56 items, each of which shows a motivational goal in a personal or social context; the respondent indicates how important it is as a guiding principle in his or her life on a 9-point scale, ranging from *opposed to my values* to *of supreme importance* (Struch, Schwartz, and van der Kloot, 2002, p. 27):

1. Equality (equal opportunity for all)
2. Inner harmony (at peace with myself)
3. Social power (control over others, dominance)
4. Pleasure (gratification of desires)
5. Freedom (freedom of action and thought)
6. A spiritual life (emphasis on spiritual not material matters)
7. Sense of belonging (feeling that others care about me)
8. Social order (stability of society)
9. An exciting life (stimulating experiences)
10. Meaning in life (a purpose in life)
11. Politeness (courtesy, good manners)
12. Wealth (material possessions, money)
13. National security (protection of my nation from enemies)
14. Self-respect (belief in one's own worth)
15. Reciprocation of favors (avoidance of indebtedness)
16. Creativity (uniqueness, imagination)
17. A world at peace (free of war and conflict)
18. Respect for tradition (preservation of time-honored customs)
19. Mature love (deep emotional and spiritual intimacy)
20. Self-discipline (self-restraint, resistance to temptation)
21. Detachment (from worldly concerns)
22. Family security (safety for loved ones)
23. Social recognition (respect, approval by others)
24. Unity with nature (fitting into nature)
25. A varied life (filled with challenge, novelty, and change)
26. Wisdom (a mature understanding of life)

27. Authority (the right to lead or command)
28. True friendship (close, supportive friends)
29. A world of beauty (beauty of nature and the arts)
30. Social justice (correcting injustice, care for the weak)
31. Independent (self-reliant, self-sufficient)
32. Moderate (avoiding extremes of feeling and action)
33. Loyal (faithful to my friends, group)
34. Ambitious (hard-working, aspiring)
35. Broadminded (tolerant of different ideas and beliefs)
36. Humble (modest, self-effacing)
37. Daring (seeking adventure, risk)
38. Protecting the environment (preserving nature)
39. Influential (having an impact on people and events)
40. Honoring of parents and elders (showing respect)
41. Choosing own goals (selecting own purposes)
42. Healthy (not being sick physically or mentally)
43. Capable (competent, effective, efficient)
44. Accepting my portion in life (submitting to life's circumstances)
45. Honest (genuine, sincere)
46. Preserving my public image (protecting my "face")
47. Obedient (dutiful, meeting obligations)
48. Intelligent (logical, thinking)
49. Helpful (working for the welfare of others)
50. Enjoying life (enjoying food, sex, leisure, etc.)
51. Devout (holding to religious faith and belief)
52. Responsible (dependable, reliable)
53. Curious (interested in everything, exploring)
54. Forgiving (willing to pardon others)
55. Successful (achieving goals)
56. Clean (neat, tidy).

An analysis of responses to the questionnaire by diverse groups of participants identified ten basic values, and Schwartz (1994) represented them as a circular continuum of related motivational goals: security, conformity, tradition, benevolence, universalism, self-direction, stimulation, hedonism, achievement, and power (Figure 9.9). Schwartz explained how compatible value types are related and competing value types are opposed to each other in reference to their locations along two bipolar dimensions. In Schwartz's terms, one dimension is "openness to change" versus "conservation," which contrasts values that emphasize independence of thought and readiness for change (self-direction and stimulation) with values that emphasize self-restriction, preservation of the past, and resistance to change (security, conformity, and tradition). The other dimension is "self-transcendence" versus "self-enhancement," which contrasts values that emphasize concern for the welfare and interest of others (benevolence and universalism) with values that emphasize one's own interest and dominance over others (achievement and power). In this two-dimensional space, hedonism is related to both openness to change and self-enhancement, but typically closer to the former.

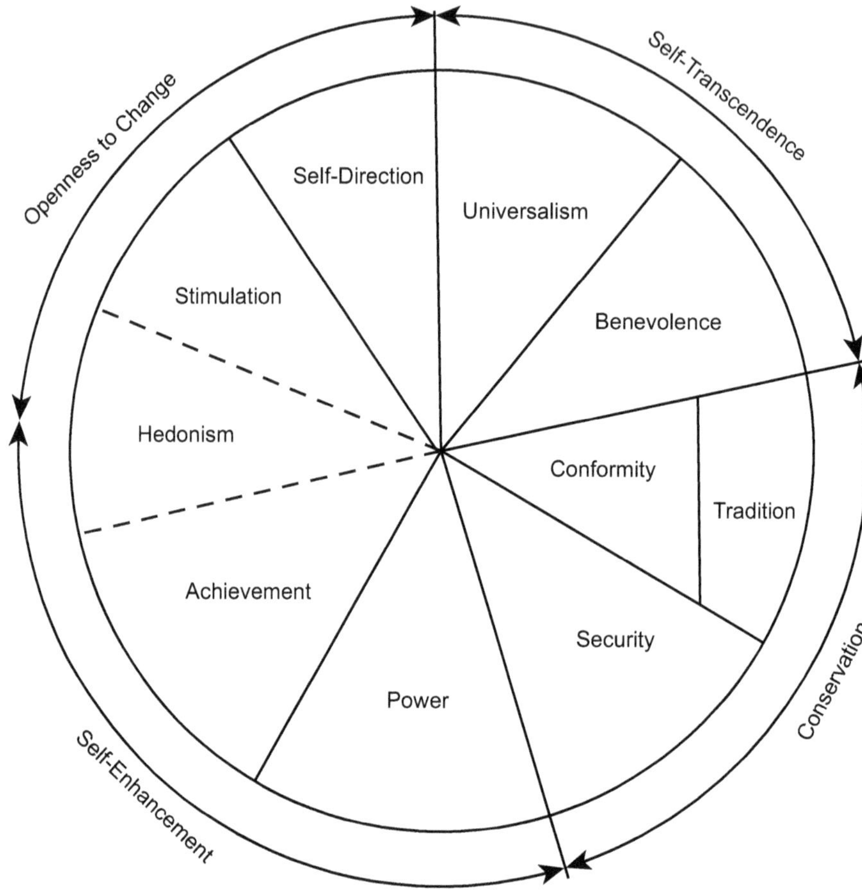

Figure 9.9 **An analysis of people's responses to questionnaires of human values identified ten basic values, represented here as a circular continuum of related motivational goals: security, conformity, tradition, benevolence, universalism, self-direction, stimulation, hedonism, achievement, and power. Reprinted with permission of John Wiley & Sons, Inc., from Schwartz, 1994, p. 24.**

On the basis of empirical studies conducted in various different countries, Schwartz (1994) suggested that the basic structure of different types of values, particularly the arrangement of the values along the two dimensions, is universal and stable. Furthermore, Schwartz (2016) reported that the values assessed by the questionnaire have significant relationships with various kinds of personal and social behaviors. For example, environmentally friendly behavior, prosocial behavior, and work for voluntary organization were correlated positively with benevolence and universalism, and negatively with power. Attendance to religious services was correlated positively with conformity and tradition, and negatively with self-direction, stimulation, and hedonism. Adoption of technological innovations and creative behavior were correlated positively with self-direction and stimulation, and negatively with security, conformity, and tradition. The patterns of these correlations conform to the interpretations of the ten values based on their circular positioning in Figure 9.9.

Review box for Chapter 9

1. Discuss the locational information and the attributive information in our cognitive maps.
2. Explain the concept of mental maps as discussed by Gould and White (1986) with respect to people's locational preferences.
3. Discuss the characteristics of hazard perception by laypersons.
4. Discuss the three factors found in the analysis of meaning with the method of semantic differential by Osgood (1957).
5. Describe what collative properties are and how they are related to aesthetic appraisal of visual stimuli.
6. Explain habitat theory of environmental aesthetics. What is an evolutionary perspective on environmental preference?
7. Discuss the four properties of environmental scenes that are posited by the Kaplans' informational model of environmental preference.
8. Show a set of photographs of various environmental scenes to people and ask them to evaluate the attractiveness of each photograph on a 5-point scale. Can you explain their evaluation scores by any of the frameworks discussed in this chapter?
9. How do you define the construct of place attachment? How is it related to our feelings and behavior in our living environments?
10. How do you define human values? How are they related to personal and social behaviors?

Chapter 10

Space, information, and society

Spatially enabled society

A common theme for the varieties of topics discussed in this book is the recognition that space matters—our cognition and experience is a result of our interactions with the environment and most of our daily thinking and behavior entails spatial components (it *is* spatial). We live in space and space surrounds us. We locate ourselves and other things in space. You remember where you left the keys in the house or where you parked your car in a large parking lot. You locate your favorite pair of red socks in the third drawer from the bottom of your dresser. For a change, you sometimes take a new route to work that is different from your usual one. By the same token, a good deal of information around us concerns place and time (it is spatiotemporal). You will meet a friend of yours, you saw a car accident happen, and you plan to have a surprise party for your mother—when and where? You will meet your friend at the local bar tomorrow night, you saw the car accident at Fairview Street and Parkway Avenue yesterday, and you plan to have the party for your mother at Jeff's on Wednesday evening. Moreover, information about space needs to be specified at an appropriate level of precision. The bar is on Green Street—where on Green Street? My office is in Ellison Hall—where in Ellison Hall? You call an ambulance—where are you now?

The last set of examples addresses the issue of place identification. You need to tell your partner *where* you are (Figure 10.1). It seems to be an easy task on the surface. You may mention a street address (101 Green Street), the name of a building (Ellison Hall), a room number (Room 406), or a landmark you see (across from the drug store). You may also give coordinate values to a place, for example, by reading off latitude and longitude values on a GPS-enabled smartphone (34° 41′ N and 119° 85′ W). You can thus use many different labels to describe your current location: they are all "correct" descriptions of where you are. Stated differently, to say "I am here" involves some sort of ambiguity, which is inherent in the identification of locations—specified at what resolution, in which frame of reference, and by what label? In some cases, you need to tell which street you are on. In others, you need to tell more specifically what building or which room you are in. Some countries have

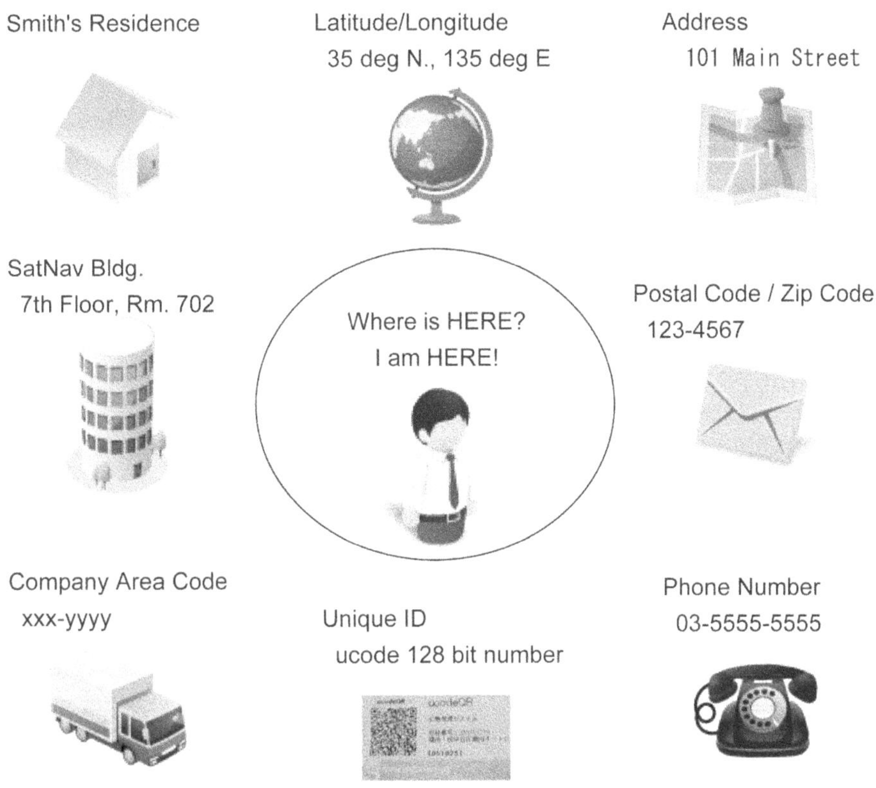

Figure 10.1 **Description of a location or place identification. There are multiple ways to identify a location (or the same place can be described in different ways), some of which are easy to understand and commonly used in our daily life and some of which are not. It is not as straightforward as you may think to describe *where you are*. Reprinted with permission of Springer Nature from Ishikawa, 2016a, p. 133.**

a street-based address system, and some countries have an area-based address system. Some places are called by names that are shared by the majority of people, but some have names that are only recognized by local residents. And sometimes the same place is called by different names, for example, by a formal name and a traditional name. The existence of multiple names for a single place suggests the existence of different meanings of the place to the residents; the consideration of which name is appropriate in what situation requires careful thought (Williams and Stewart, 1998). In our daily verbal communication, it is enough (or preferable) to say "turn right at the intersection," instead of "make a 62° turn to the right at the intersection," although the former is less precise.

These examples point to the ambiguity owing to the existence of multiple methods of identifying a place. Important for the present discussion, people find some of them easy to understand and some of them difficult. For instance, it would be unusual to say "see you at 35° 42′ north and 139° 45′ east tomorrow." Moreover, the same location can take different values of latitude, longitude, and elevation depending on how the shape of the earth is conceived in the computation of coordinate values, because the earth is not completely spherical or smooth (for example, you may have heard the term WGS 84 [World Geodetic System] in the use of a GPS receiver). You would not see latitude and longitude lines running on the surface of the earth if you were to look at it from outer space in a spaceship; they are conceptually or mathematically superimposed on the earth by humans (Figure 10.2). It is thus not always easy to choose an appropriate identifier, nor is it easy to translate between different identifiers and relate one identifier to another (e.g., what is the street address for 35° 42′ N and 139° 45′ E, or what building or room number corresponds to the latitude and longitude). You are definitely *here*, but it is not as straightforward as you may think to describe *where here is*.

Figure 10.2 **As this satellite image shows, the earth has no latitude or longitude lines running on it physically. Conceptualizing the shape of the earth as a sphere (or an ellipsoid) and laying out a network of coordinates on it, for the sake of knowing where we are, has a long history (see Sobel, 1995). Image from NASA, https://earthobservatory.nasa.gov/images/2181/the-blue-marble/**

Our society now offers possible solutions to these problems with technological and computational capabilities. In particular, we have a variety of navigation applications and location-based services available today. We can view a digital map on a tablet while walking and use the functions of route navigation and street viewing. With an input of information about where you are and where you want to go, a smartphone will guide you to the destination along the shortest, fastest, or least expensive route. We can also have access to real-time information about road traffic, public transport, and parking spaces. Since the 1990s, national and global efforts to develop a coordinated framework for the production, distribution, management, and use of geospatial data have been made; for example, the National Spatial Data Infrastructure (NSDI) in the United States, INSPIRE in the EU, and NSDI Act in Japan. And back in 2004 the U.S. Department of Labor identified geospatial technologies as one of the most important emerging and evolving fields (Gewin, 2004). With these technological advances and societal interest, we now live in a society where advanced geospatial technologies and locational data infrastructures support universal access to information whenever and wherever we want it (the concept of "anytime, anywhere, anyone"). This may be called a spatially enabled society.

What makes these locational service applications possible is the capacity for location identification and positioning, brought about by the advances in information technologies and the development of geospatial location databases. For outdoor positioning, satellite navigation systems are available (Hofmann-Wellenhof, Lichtenegger, and Wasle, 2008), including the Global Positioning System (GPS), the Global Navigation Satellite System (GLONASS), Galileo, or the Quasi-Zenith Satellite System (QZSS). But precise indoor positioning requires additional identification devices. For example, radio frequency identification (RFID) tags and infrared beacons are installed in actual places to form part of large sensor networks. Thus, in a spatially enabled society, computational capabilities are not confined to cyberspace but extended to real space, and systems of location identification allow a seamless transition between indoor and outdoor spaces, merging the digital and real worlds. It is a vision that has been pursued in the framework of pervasive or ubiquitous computing. The idea of ubiquitous computing is characterized by its aim to create a society in which people have universal access to information through computers installed everywhere and in everything. In essence, computational capabilities are ubiquitously available and support all sorts of human activities by automatic identification of real-world objects, places, and situations (real-time context awareness). Currently, in a similar vein, the concept of the Internet of Things (IoT) envisions a world in which objects with embedded computational capabilities communicate with each other through the network, enabling intelligent interconnections between humans, objects, and information (Figure 10.3). Open data movements can also be considered in relation to the development of geospatial databases, particularly in the context of digital mapping. Many governments and agencies now offer various kinds of geographic data for public use, such as base maps, addresses, road networks, land uses, topographies, and political boundaries (e.g., Ordinance Survey's OS OpenData). Also, geographic data collected and shared by individual users contribute to the development of geospatial databases, which some researchers call volunteered geographic information (Goodchild, 2007) or crowdsourcing geospatial data (Heipke, 2010). Mapping services by OpenStreetMap are an example of user-contributed open geographic data.

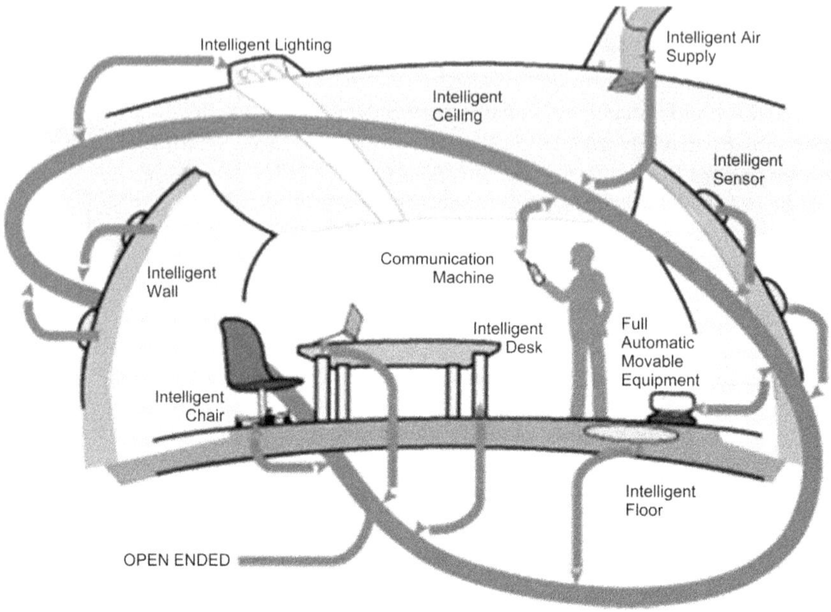

Figure 10.3 **Concepts of ubiquitous computing and the Internet of Things (IoT). These concepts aim to create a society in which people have universal access to information through computers installed everywhere and in everything. Illustration by courtesy of YRP Ubiquitous Networking Laboratory, https://www.ubin.jp/**

Real-world applications of these concepts are seen in the domains of retailing, agriculture, and healthcare, and also in city environments in the attempts to develop smart cities and intelligent buildings. Navigation applications and location-based services are major examples of applied geospatial technologies. Behind the applications of digital maps, mobile navigation assistance, in-car navigation systems, and even autonomous cars nowadays, is the technology of geographic positioning by satellite navigation, based on signals from a system of satellites. As seen above, for pedestrian navigation, however, there are situations where signals from satellites are not available, for example indoors or underground. This is why a network of place identification not entirely based on satellite navigation is required. Installment of positioning capabilities in the environment would enable seamless navigation and translation between indoor or outdoor spaces, and such an integrated system of place identification contributes to linking different place identifiers associated with the same location computationally (see Figure 10.1).

The basic requirement for place identification is that a location is uniquely identified. It is realized by geocoding systems superimposed on the surface of the earth, for example, postal codes, street addresses, or latitude and longitude. But for more precise and fine-grained positioning, either indoors or outdoors, a location is preferably identified by an object located at the position. That is, a place is uniquely identified by a "code" (or number) assigned to the place. Consider the case of an autonomous delivery of a product, to a specific desk in some room of a particular building. All desks in the room and all rooms in the building have the same postal code and street address; but you need to identify the specific desk uniquely. With a view to accomplishing such very precise location identification, an informational infrastructure that consists of 128-bit identification numbers and a network-based identification architecture, called *ucodes* (for ubiquitous codes), has been developed for use in real-world applications (Figure 10.4). In ucodes any information may be stored about objects, places, concepts, and the relations among them (e.g., "A belongs to B," "C is located next to D," or "E is manufactured by F"). To serve as place identifiers, ucodes may be installed on any kind of tag or marker (e.g., an RFID tag, a quick response [QR] code, or an infrared beacon). The tag or marker is then installed in a particular place (or an object at the place, such as a desk or a street lamp) to offer information about the place (e.g., locational information such as latitude and longitude, a street address, and a floor number, or store information such as business hours and sales information), which the user can access through the Internet with mobile devices such as smartphones or wristwatches. The ucode identification architecture and service descriptions are standardized by the International Telecommunication Union, ITU-T F.771, H.621, and H.642. The method of place identification and interrelating different place identifiers is also standardized by the International Organization for Standardization, ISO 19155. The methods of identifying "where is here" are thus now developed in a standardized way, which is regarded as an attempt to make a "placename dictionary" that specifies and interrelates different place identifiers.

Locational information and place identification are the key to developing geospatial data infrastructures. For example in Japan, the Ministry of Land, Infrastructure and Transport and the Geospatial Information Authority of Japan have established more than 20,000 intelligent control

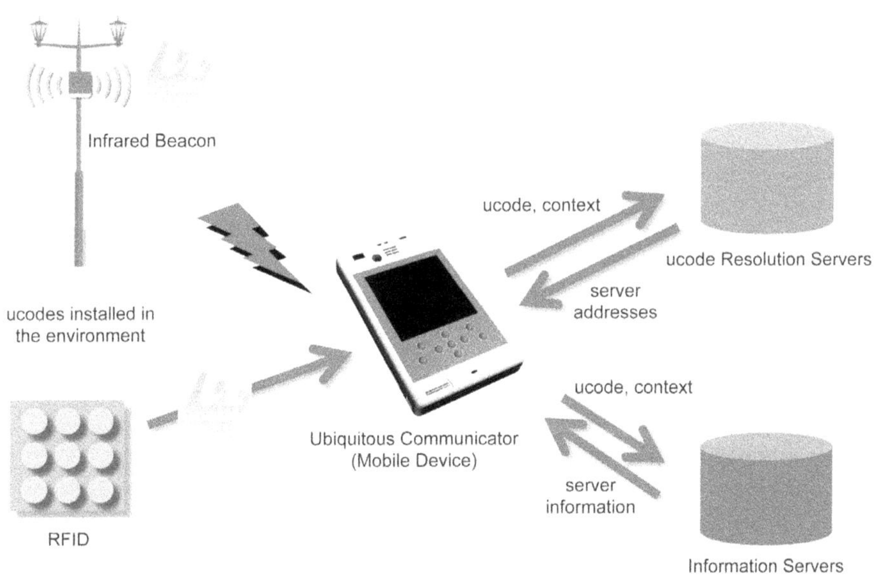

Figure 10.4 **Network-based identification architecture and ucodes. These are developed as an informational infrastructure for a unique identification of objects and locations and used in various real-world applications, especially for precise location identification in city environments. Illustration by courtesy of YRP Ubiquitous Networking Laboratory. Reprinted with permission of Springer Nature from Ishikawa, 2016a, p. 130.**

Figure 10.5 **Examples of real-world applications of ucodes, providing through a mobile device information about stores or sightseeing spots and supporting pedestrian navigation in multiple formats and languages. Images by courtesy of YRP Ubiquitous Networking Laboratory. Reprinted with permission of Springer Nature from Ishikawa, 2016a, p. 131.**

points over city areas. These control points have ucodes installed inside and store information about their locations in a digital format of latitude, longitude, and elevation values. Ucodes are also installed in many cities across Japan for application purposes by local governments. They enable real-world location-based services that offer information about stores, sightseeing spots, historic places, artworks, or train operations, and support pedestrian navigation in multiple formats and languages, tailored to specific contexts and situations of use (Figure 10.5).

Spatial thinking as a broad concept

As we discussed in Chapter 6, spatial ability constitutes an important component of human abilities and underlies students' learning in various science and engineering disciplines. For example, think about a chemist envisioning the structures of molecules in 3-D, a mathematician imagining the intersection of two solids, or an engineer drawing a plan of a new product. Skills related to such disciplinary thinking, and also a range of everyday thinking, are thus *spatial* in nature (remember the term spatial behavior we discussed in Chapter 1). But, at the same time, they are broader in scope than the basic spatial ability pertaining to the perception, retention, and manipulation of visual images (Eliot and Smith, 1983; Lohman, 1979). Among the disciplines that tap into advanced spatial thinking, physics and mathematics are identified (Wai, Lubinski, and Benbow, 2009), but one should also note geography and geoscience as other major spatial disciplines, especially because they deal with the spatial environments that we live in, hence geographic, not merely spatial (Golledge, 2002; Goodchild, 2001).

Golledge (2002) discussed that the nature of geographic knowledge changed in the latter part of the twentieth century, from an inventory of knowledge about the distributions and patterns of phenomena in space to an understanding of the processes behind observed spatial distributions and patterns. Studies of spatial thinking in geography have particularly focused on the concepts of geography, or geospatial concepts. Golledge, Marsh, and Battersby (2008) offered a conceptual framework for a discussion of spatial thinking in geography that consists of five levels of geospatial concepts: primitive, simple, difficult, complicated, and complex. The most basic and primitive concepts include identity, location, magnitude, and space-time, which serve as building blocks for the understanding of higher-order concepts, such as distance, direction, order, proximity, shape, arrangement, and distribution (simple concepts); change, growth, spread, clusters, centers, grids, and links (difficult concepts); scale, buffers, surfaces, gradients, and profiles (complicated concepts); and projections, distortion, interpolation, activity space, subjective space, and central places (complex concepts). Similar conceptual structures have been proposed by other researchers, as shown in Table 10.1 (Gersmehl and Gersmehl, 2007; Janelle and Goodchild, 2011; Kuhn, 2012).

Table 10.1 Geospatial concepts proposed by researchers

Gersmehl & Gersmehl (2007)	*Janelle & Goodchild (2011)*	*Kuhn (2012)*	*Golledge et al. (2008)*
Location	Location	Location	Location
	Network	Network	Link, Network
Aura			Buffer
Region	Neighborhood and region	Neighborhood	Area, Cluster, Social region
Hierarchy			Central place, Hierarchical order
Transition			Change, Growth
Pattern			Arrangement, Distribution
Association	Spatial dependence/ heterogeneity		Interpolation, Similarity, Correlation
	Objects and fields	Object, Field	Identity, Space-time
	Distance		Distance, Proximity
	Scale	Granularity	Scale
		Event	Identity, Occurrences
		Accuracy	Data and representation
		Meaning	Time-space context
		Value	Values
Comparison			Comparisons
Analogy			Similarity
	Overlays		Map comparisons

Source: Adapted from Ishikawa 2016b, p. 89. Reprinted with permission of Taylor & Francis.

Ishikawa (2013) discussed the characteristics of spatial thinking in geography, by empirically examining the relationship between college students' performances on major geospatial thinking tasks and their spatial abilities. Students' geospatial thinking was assessed by questions concerning spatial distributions, geographic coordinate systems, map projection, map scale, and the earth's movement. Their answers were evaluated with respect to the accuracy with which they understood basic geographic concepts and the degree to which they accurately reasoned about the processes behind observed patterns. For example, in the question about spatial distributions, students viewed three maps illustrating the locations of traffic accidents that happened in three different regions (Figure 10.6); students discussed the characteristics of spatial distributions on the maps and possible reasons for the three different patterns of spatial distributions (Table 10.2). Students' spatial abilities were assessed by the Card Rotations Test, Cube Comparisons Test, Surface Development Test, and Paper Folding Test (see Figures 6.1–6.5). Ishikawa found that students' scores on spatial ability tests correlated strongly with each other, while their scores on geospatial thinking tasks did not. The two sets of abilities, spatial ability and geospatial thinking, were correlated somehow, but not significantly, and were basically separate except for the geospatial thinking task of map projection (Figure 10.7).

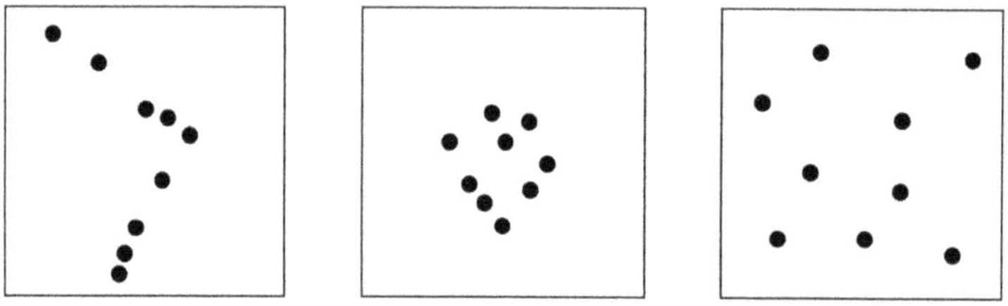

Figure 10.6 **Students viewed map representations of the locations of traffic accidents that happened in three different regions. They described the characteristics of spatial distributions and discussed possible reasons for the three different patterns of spatial distributions. Reprinted with permission of Taylor & Francis from Ishikawa, 2013, p. 640.**

Table 10.2 **Questions about spatial distributions in the geospatial thinking test**

Suppose that the three maps above (Figure 10.6) show the locations of traffic accidents that happened in three different regions (within an area of 1km by 1km) for the past one month. One dot represents one accident.

(1) Explain the characteristics of the three maps in as much detail as possible, by comparing them with each other.
(2) What do you think are the possible reasons underlying these three maps of traffic accidents?

Source: Adapted from Ishikawa 2013, p. 639. Reprinted with permission of Taylor & Francis.

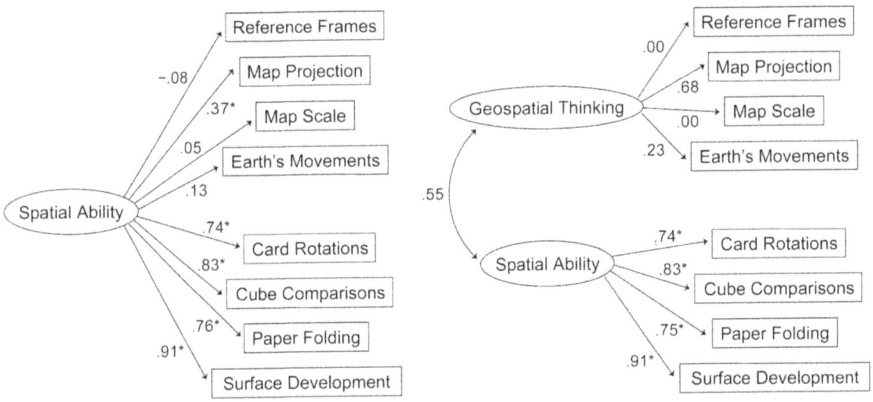

Figure 10.7 **Scores on spatial ability tests correlated with each other strongly, whereas scores on geospatial thinking tasks did not. Geospatial thinking consists of different components; some of them, but not others, share properties with abilities assessed by psychometric spatial tests. Thus, having high spatial ability cannot be considered in itself a sufficient condition for the mastery of geospatial thinking skills. Reprinted with permission of Taylor & Francis from Ishikawa, 2013, p. 643.**

Ishikawa (2016b) further examined the characteristics of spatial thinking in geography, focusing on students' geospatial reasoning about four sets of thematic maps, which depicted (a) coffee buyers markets in Manhattan in the daytime and nighttime; (b) diffusion patterns of cholera in the United States in 1832, 1849, and 1866 (Figure 10.8); (c) the crude birth rates, crude death rates, natural increase rates, and infant mortality rates for countries of the world; and (d) global distributions of the types of soils and agriculture. Participants read these maps and explained what they thought about them in as much detail as possible; their responses were scored in terms of whether they correctly understood or identified (a) the themes of the maps (what the maps represented), (b) the map symbols and colors (correspondence between symbols or colors and represented variables), (c) places that have characteristic values of the represented variables (correspondence between symbols and maps), (d) the spatial or temporal distributions of the depicted phenomena (identifications of spatial-temporal patterns), (e) interrelations among maps within each set (comparison or cross-referencing of multiple maps of the same theme), and (f) possible reasons for the spatial patterns observed on the maps (geospatial reasoning about underlying processes).

Ishikawa (2016b) found that all students accurately understood the themes of the maps and the map symbols and colors, and correctly identified places that have characteristic values of the represented variables. However, their performance varied on the more advanced tasks of identifying spatial distributions or patterns, comparing multiple maps, and reasoning about processes underlying observed patterns. Good answers identified spatial and temporal patterns by comparing maps within the set, and considered reasons for the patterns; whereas poor answers focused on visual aspects of the maps and did not go beyond what is seen on the maps. Moreover, the more advanced tasks differed in the relationship with spatial ability: a significant correlation with spatial visualization

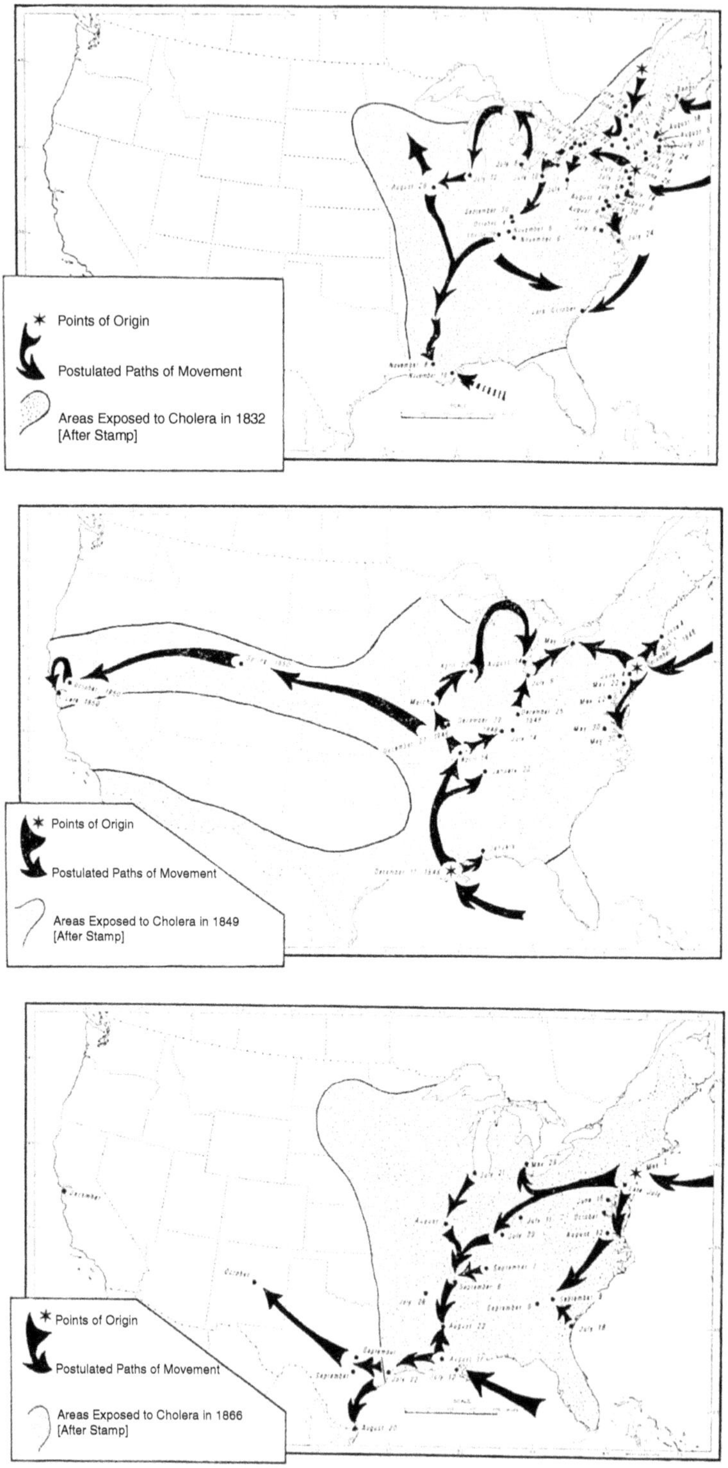

Figure 10.8 **Maps of the diffusion of cholera in the United States in 1832, 1849, and 1866. Students reasoned about spatial-temporal patterns shown on these maps. Reprinted with permission of John Wiley & Sons, Inc., from Pyle, 1969, pp. 63, 68, and 72.**

ability was observed for the identification of spatial-temporal patterns and comparison of multiple maps, but not for the reasoning about underlying processes. That is, students with higher spatial visualization ability tackled the task of thematic map reading by identifying more spatial distributions and comparing multiple maps more frequently. However, they did not necessarily make more statements about possible reasons for the observed patterns.

These results show that geospatial thinking is neither identical with, nor completely separate from, spatial ability. Geospatial thinking consists of different components, and some of them have properties that are shared by abilities assessed by psychometric spatial tests (concepts and reasoning about map projection, identifying spatial distributions on a thematic map, and comparing multiple maps of the same theme). That is, having high spatial ability cannot be considered in itself a sufficient condition for the mastery of geospatial thinking skills. Geospatial thinking and expertise in geospatial science span a wide range of concepts and reasoning, some of which are related to spatial ability and some are not.

Kastens and Ishikawa (2006) discussed spatial thinking in geoscience by itemizing major spatial tasks that geoscience experts engage in and juxtaposing them with the findings in the cognitive science literature. The first task they discussed was describing and interpreting the shapes of objects. For example, structural geologists describe the shape of a folded sedimentary layer, and mineralogists describe the angular relationships between the faces of a crystal. The descriptions of the shapes are then applied to the task of classifying objects by shape. Paleontologists classify fossils, and geomorphologists classify landforms, according to their morphology; similarly, mineralogists identify minerals by shape, color, and texture. Another task that geoscientists do is ascribe meanings to the shapes of natural objects, or reason about why the objects have the particular shapes. For example, on the basis of observations of distortions in the shapes of crystals and fossils, structural geologists infer the strain that a body of rock has undergone. Sedimentologists note in a sedimentary stratum the presence of ripples, and infer from the orientation and asymmetry of the ripples the formation processes of the stratum.

The data that geoscientists typically collect and examine are displayed as images, and so their task is often to identify significant patterns in data displays, for example, seismic reflection profiles, side-looking sonar data, or remotely sensed images. Data collection by geoscientists is also often conducted in the field. Geologists, for example, describe the orientation of a planar surface relative to the earth by measuring, with a geological compass, strike (the compass azimuth of the line defining the intersection between the surface to be described and the horizontal plane) and dip (the angle between the horizontal plane and the surface to be described, measured within a vertical plane perpendicular to the strike line). Furthermore, geoscientists' observations are basically collected in one or two dimensions, and then interpreted in three dimensions for objects, structures, and processes. Field geologists examine rock layers exposed above the earth's surface in outcrops, such as road cuts or stream cuts. From these multiple planar views at individual outcrop locations, they mentally envision an internal three-dimensional structure of the large rock body. Finally, geologists figure out the sequence of events of folding and faulting that has created the observed structure in a three-dimensional space and over a long period of time (Figure 10.9). Field observation requires a geoscientist to read and make maps and stay oriented in space and time that often span a large extent (hundreds of kilometers and millions of years).

Figure 10.9 Spatial tasks that geologists do. They examine rock layers exposed above the earth's surface in outcrops, and measure the orientation of a planar surface relative to the earth by measuring strike and dip. From these multiple planar views at individual outcrop locations, they mentally envision an internal three-dimensional structure of the large rock body. Photographs reprinted with permission of Taylor & Francis from Ishikawa and Kastens, 2005, p. 185.

Kastens and Ishikawa (2006) discussed how these geoscience tasks are related to spatial concepts, abilities, and thinking discussed in the literature of cognitive science, including conceptions of space at the topological, projective, or Euclidean level, theories of categorization and schemata, expert problem solving, spatial frames of reference, psychometric spatial tests of mental rotation, perspective taking, paper folding, embedded figures, or water level, and spatial orientation and wayfinding in the environment. The spatial sciences of geography and geoscience offer various tasks that involve advanced spatial thinking, and thus provide insights into the study of spatial thinking and benefit from basic research findings on spatial thinking as they are applied to the classroom.

Spatial literacy in a spatially enabled society

It has been clarified in this book that accurate spatial orientation and successful wayfinding require an understanding of metric spatial relations between objects and places in a fixed, global frame of reference, and that this poses difficulty especially for people with a poor sense of direction. Importantly, individuals differ greatly in the ability to learn the layout of the environment metrically and configurationally and to stay orientated in the environment while they move around in it—some people accomplish this almost effortlessly and some people find it very difficult. This is one of the reasons why digital maps and navigation applications are used so widely and popularly now. They certainly benefit people who easily get lost in an unfamiliar environment. Furthermore, navigation assistance that takes information from multiple senses into account undoubtedly provides opportunities of independent traveling for people with, for example, visual impairments.

As we saw in the discussion of environment perception in Chapter 3, information about the environment is multimodal (Ittelson, 1973). Visual information may be dominant in many cases, but other sources of information are equally important, notably auditory information and tactile information. People can localize auditory cues in space in terms of the direction and distance from a sound source, and this human ability is used in navigation systems that convey spatial information by sound (Bujacz et al., 2011). Also, from a questionnaire survey and interviews with people with visual impairments, Koutsoklenis and Papadopoulos (2014) identified major haptic cues that these people used for wayfinding and traveling in the environment, such as changes in the texture of a walking surface, sidewalks, slopes, and curb ramps. Tactile information has been found to present and communicate information about space effectively, for example in the form of tactile maps (Lobben, 2015; Miele et al., 2006) and vibration signals (Gkonos, Giannopoulos, and Raubal, 2017; Mahmud et al., 2014; Pielot and Boll, 2010; Rümelin, Rukzio, and Hardy, 2011; Van Erp, 2004). Rümelin et al. developed a system that conveyed information to the pedestrian user about the distance to the next intersection and which way to turn at the intersection by different patterns of duration, intensity, roughness, and rhythm of vibration. These examples show the significance of geospatial navigation technologies and attest to the efficacy of providing navigation assistance through advanced systems.

At the same time, we have also seen that satellite navigation tools are not necessarily a good navigator or a facilitator of spatial learning (see Chapter 7)—the use of mobile navigation tools interferes with spatial learning and memory. The user moves through space to the destination physically, but is disconnected from the space cognitively (Farman, 2015; Kalin and Frith, 2016). Kalin and Frith argued that the use of a smartphone, which mediates our physical mobility, is not readily separable from the act of walking in, thinking about, or interacting with the environment; navigational directions automatically generated by the smartphone induced users to "feel as if they knew the city without remembering the physical space" (p. 225).

We are certainly aware of the benefits and convenience provided by geospatial technologies in general and navigation tools in particular. Do the benefits and convenience, however, change the manner in which we interact with the environment at all? In other words, do our spatial thinking, behavior, and awareness change in the age of satellite navigation? This is a question that has attracted attention from many people, whether researchers, practitioners, or laypersons, and stimulated various discussions about possible effects, potentially negative, of the use of satellite navigation on spatial thinking. Frean (2006) expressed a concern about negative effects of the use of satellite navigation systems on people's geospatial literacy and awareness. Axon, Speake, and Crawford (2012) pointed out that satellite navigation systems are used popularly now and perceived differently from traditional maps, and argued that they could potentially change people's wayfinding behavior and the practice of navigation. Axon et al. observed that people were aware of potential negative consequences of being accustomed to navigation tools, for example a possible deterioration of spatial orientation or map skills, but they preferred, and were happy, to use them (see also Speake, 2015; Speake and Axon, 2012). Furthermore, McKinlay (2016) warned against excessive reliance on the tools of automatic navigation and a potential "erosion" of human navigational abilities.

These arguments contrast with the statement that is often made these days, namely, "it is time to throw away paper maps." Reflecting the public interest in the question of whether to throw away our maps, featured articles have appeared in science media with suggestive titles such as "Don't Throw Away Your Paper Maps Just Yet" (McKinney, 2010), "Uncharted Territory: Where Digital Maps Are Leading Us" (Austen, 2013), "Do Our Brains Pay a Price for GPS?" (Neyfakh, 2013), "Smartphones and the Uncertain Future of 'Spatial Thinking'" (Grabar, 2014), "Ignore the GPS: That Ocean Is Not a Road" (Milner, 2016a), or "Use or Lose Our Navigation Skills" (McKinlay, 2016). A number of tragic, and somewhat unbelievable, GPS-related incidents have also been reported; for example, people driving their car into a lake, getting their car stuck on a cliff or in a narrow alley, or traveling in the wrong direction hundreds of kilometers across the border, as a result of passively following GPS directions (Milner, 2016b).

The foregoing discussions point to the possibility that the ability to learn and explore the environment, or the essential human aptitude of sense of direction, might degenerate if we are accustomed to passively following the directions given by the tool and leave navigational decision making and route planning to the technology. Although these concerns about negative effects of accustomed use of satellite navigation are legitimate, they await empirical examination. In particular, although existing research examined momentary or short-term effects of using satellite navigation tools, long-term impacts of using such tools have not been systematically investigated. In the empirical studies mentioned in Chapter 7, the performance on wayfinding and spatial orientation was assessed concurrently with or soon after the use of navigation systems. Likewise, with respect to cognitive processing and attention span in general, the use of mobile tools has been shown to have a detrimental effect (Barr et al., 2015; Sparrow, Liu, and Wegner, 2011), but empirical evidence about lasting impacts is scarce (Wilmer, Sherman, and Chein, 2017). The question is whether and how regular or accustomed use of navigation systems alters the user's spatial cognition and behavior in the environment.

With that issue in mind, Ishikawa (2019) conducted an empirical study to examine possible negative impacts, in the sense of long-term effects, that the accustomed use of satellite navigation systems could have on human wayfinding and spatial orientation. Specifically, it examined the relationship between the experience of using different navigation tools and the performance on wayfinding and spatial learning. Since the study focused on long-term effects, the experience of tool use was considered in terms of the length of time and frequency of using navigation tools, particularly the amount of accumulated experience given by the product of the length of time and frequency. The tools examined were the two major types of tools in popular use—pedestrian navigation systems and in-car navigation systems—and the major tool of navigation of traditional use—paper maps.

In the study, 249 adults responded to survey questionnaires about the use of different navigation tools (the length of time, frequency, and purposes of using a pedestrian navigation system, an in-car navigation system, and a map) and spatial aptitudes (sense of direction, route and survey navigation strategies, and mental rotation ability). Of these respondents, 74 people further participated in the second phase of the study, real-world navigation, in which they traveled two environmental routes using a mobile tool and a paper map (Figure 10.10). The participants were recruited from a wide range of ages (21–69), so that their spatial aptitudes and experiences of using different navigation tools varied widely.

The mobile tool, run on a 9.7-inch (24.6-cm) tablet, directed participants from the start to the end point along the shortest path through two landmark locations. On the tablet screen, a map of the surrounding area was shown in a "heads up" orientation, with the participant's current position and a route to take dynamically updated as their movement in space (Figure 10.10a). Due to the limited size of the tablet screen, the route was not visible in its entirety, but shown in a piecemeal fashion. The paper map (with a scale of 1:2,200, printed on A4-sized paper) showed the locations of the start, end point, and two landmarks, without any specific routes indicated (Figure 10.10b). Participants were allowed to take a route of their own choice, as long as they passed the two landmarks in the predetermined order. Participants' spatial behavior in the real-world navigation was examined in terms of two types of variables. One concerned the performance on wayfinding, assessed by (a) travel time, (b) the number of stops (the times that they got lost or disoriented), (c) the number of deviations (wandering off a direct route or in the opposite direction to the end point), and (d) the distance traveled. Another variable concerned the performance on spatial orientation, assessed by the accuracy of configurational understanding of routes: absolute errors of direction estimates from the end point to the start and two landmarks.

An analysis of participants' responses to the survey questionnaires showed that frequent users of pedestrian navigation systems tended to be low on sense of direction and mental rotation (r = –0.29 and –0.20, respectively). In contrast, long-time users of maps tended to be high on sense of direction and favor survey navigation strategies (r = 0.13 and 0.16,

Figure 10.10 **Participants traveled two routes using a mobile tool and a paper map. (a) On the tablet screen, the participant's position and a route to take were shown in a "heads up" orientation dynamically updated as their movement in space. (b) The paper map showed the locations of the start, end point, and two landmarks, without any specific routes indicated. Reprinted with permission of Taylor & Francis from Ishikawa, 2019, p. 201.**

respectively). In relation to real-world navigation performance, people who had more accumulated experience of using in-car navigation systems made more wayfinding errors and learned the configurations of traveled routes less accurately with the paper map (correlations around 0.40–0.60).

To further examine the long-term effects of satellite navigation use on wayfinding and spatial orientation, Ishikawa (2019) considered three models of the relationship between spatial aptitudes, experience of using navigation tools, and performance on real-world navigation. A first model (Figure 10.11a) posits that the experience of using different navigation tools affects spatial aptitudes (sense of direction and mental rotation ability), which in turn affect the performance on wayfinding and spatial orientation with the mobile tool and the paper map. This model intended to examine whether the use of different navigation tools directly influences spatial aptitudes, or the plausibility of the statement that the reliance on satellite navigation jeopardizes our sense of direction, as suggested by Axon et al. (2012) and McKinlay (2016). A second model (Figure 10.11b) posits that spatial aptitudes determine the use of different navigation tools, and accumulated use of different navigation tools affects the performance on wayfinding and spatial orientation. This model hypothesized that people's spatial aptitudes would affect which navigation tools they use and the use of different navigation tools would influence, as a

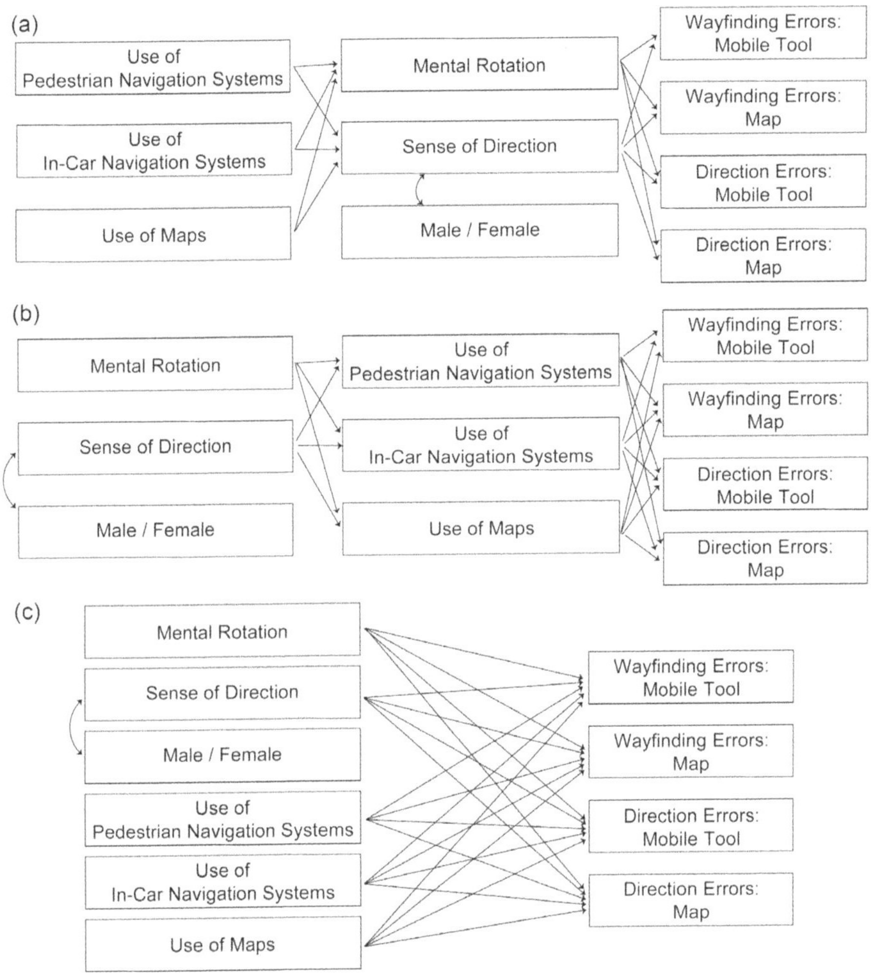

***Figure 10.11* Three models of the relationship between spatial aptitudes, experience of using navigation tools, and performance on real-world navigation. (a) A model that posits that the experience of using different navigation tools affects spatial aptitudes, which in turn affect the performance on wayfinding and spatial orientation. (b) A model that posits that spatial aptitudes determine the use of different navigation tools, and accumulated use of different navigation tools affects the performance on wayfinding and spatial orientation. (c) A model that posits that the experience of using different navigation tools and spatial aptitudes independently affect the performance on wayfinding and spatial orientation. Reprinted with permission of Taylor & Francis from Ishikawa, 2019, p. 205.**

mediator, wayfinding and spatial orientation. A third model (Figure 10.11c) posits that the experience of using different navigation tools and spatial aptitudes independently affect the performance on wayfinding and spatial orientation. This model intended to examine the possibility that the effects of using different navigation tools on wayfinding and spatial orientation are separate from (or over and above) the effects of sense of direction and mental rotation.

The third model was one that received empirical support in structural equation modeling analysis (Figure 10.12). The accumulated experience of using in-car navigation systems negatively affects our wayfinding and spatial orientation performances separately from spatial aptitudes. Importantly, the size of the negative impact of navigation use is larger than the degree of facilitation by a better sense of direction and higher mental rotation ability (path coefficients of 0.32–0.42 vs. 0.19–0.29 in size). The statement that the use of satellite navigation tools jeopardizes our sense of direction is not viable, at least not in the sense of direct influence, as the first model posits. But there *is* a long-term negative impact of accumulated use of navigation tools. Thus, the oft-reported concern with satellite navigation (Axon et al., 2012; McKinlay, 2016) is now provided with empirical data in the long-term sense, and specified in the form that allows a comparison with the effects of spatial ability.

Being accustomed to passively following turn-by-turn instructions and lacking in conscious decision making and route planning possibly has a negative effect on our spatial cognition in the long run, independent of our sense of direction. The lack of necessity to comprehend the environment when one is guided by technology might also be related. This does not, however, simply argue against satellite navigation—it has a good potential to benefit people with low levels of spatial aptitude and assist people with visual impairments in independent traveling. These results nonetheless support and stimulate the continued discussion of satellite navigation from the perspective of the user's cognition and behavior. An attempt to provide navigational directions that prompt the user to engage in the environment is a promising approach (Waters and Winter, 2011). Spatial thinking, after all, is an important part of human cognition and behavior (see the case for Inuit hunters in Igloolik described by Aporta and Higgs, 2005), which could desirably be supported, but not entirely replaced, by technology.

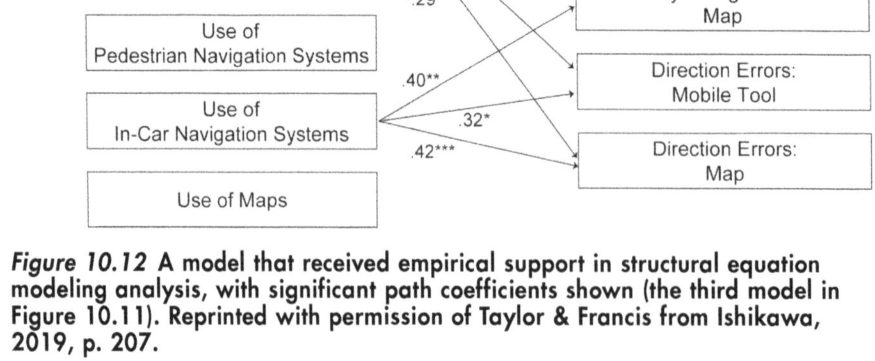

***Figure 10.12* A model that received empirical support in structural equation modeling analysis, with significant path coefficients shown (the third model in Figure 10.11). Reprinted with permission of Taylor & Francis from Ishikawa, 2019, p. 207.**

Conclusion and commencement

We have now explored various major regions of internally represented environments, or the space in our minds. What will you report to your boss about your mission? First, people live in space and time, and their thinking and behavior inherently involve spatial components. People perceive surrounding environments, and environments are represented in people's minds. Environments that people perceive or interact with span a range of scales, from figural, vista, environmental to geographical space (Montello, 1993). Thus, people need to coordinate perceptions of spaces at different scales; notably people's living environments (e.g., a neighborhood, campus, town, city, or county) are larger than the human body and surround it (Ittelson, 1973). In such large-scale spaces, people are explorers, not merely observers, and need to integrate observations or views at individual places over space and time. This is why spatial orientation and wayfinding in the environment pose difficulty for many, especially people who have a poor sense of direction (Hegarty et al., 2006). Stated differently, people's thinking and behavior in the environment (or spatial behavior) relates to, and is affected by, internally represented environments, as well as physical environments (Golledge and Timmermans, 1990). What people think and how they feel about the environment thus have theoretical and practical significance, and should be considered in the context of mutual dependence and interaction of people and the environment (Gibson, 1979/1986).

Importantly, the structures of internally represented environments, or cognitive maps, show systematic patterns of distortion across people, and also individual variations between people and within a person. Systematic biases reflect how people reason about space. A major tendency is to organize spatial information hierarchically (Stevens and Coupe, 1978). This may be interpreted as a human response to the complexity of the environment with the limited capacity of information processing (reminiscent of the concept of satisficing; Simon, 1956) and is related to the basic human trait of categorization (Rosch, 1978). Distortions in cognitive maps are also observed in the knowledge of environmental distance, and together with the idea of distance decay (Tobler, 1970), have attracted researchers' attention in the context of the prediction of human travel behavior (Cadwallader, 1981; MacKay and Olshavsky, 1975; Thompson, 1963) and the discussion of inconsistencies in spatial knowledge (Hirtle, 1998; Montello, 1992; Tversky, 1993).

Individual variations between people in spatial cognition are particularly the focus of recent research in relation to the questions of microgenetic and ontogenetic development of spatial knowledge, taking individual differences in spatial abilities and aptitudes into account (Newcombe, 2010; Wolbers and Hegarty, 2010). People's spatial knowledge about a new environment does not proceed universally from lower to higher levels of accuracy and coordination, in either a stagelike or a continuous manner (Ishikawa and Montello, 2006). Some people attain extremely high accuracy relatively easily, while others stay at a rudimentary level in terms of metric and configurational accuracy. On the premise that maps in the head direct people's behavior, and knowledge about the environment guides (or limits) people's interpretations and explorations, the existence of large individual differences in cognitive mapping has significant consequences. Researchers have sought the reasons for, or the correlates of, individual differences in cognitive mapping in spatial abilities (e.g., mental rotation and perspective taking; Hegarty and Waller, 2004; Liben and Downs, 1993), spatial aptitudes (particularly sense of direction; Hegarty et al., 2002;

Kozlowski and Bryant, 1977), navigational preferences or tendencies (route-based vs. survey strategies; Lawton and Kallai, 2002), kinesthetic, proprioceptive, or vestibular senses (Cullen and Taube, 2017; Loomis et al., 1993; Sholl, 1989), neural activities (Maguire et al., 2006; Schinazi et al., 2013), personality traits (Bryant, 1982; Condon et al., 2015), and language use (Levinson, 2003). Thus, the question of significance to researchers now is not so much whether cognitive maps are maplike, but how and why people differ in the accuracy of spatial knowledge acquired about the environment and the flexibility in reasoning about spatial relations and orientation in the environment.

The structure of physical environments and the type of externally provided information such as navigation assistance also influence people's spatial cognition and behavior. For example, are the streets in your neighborhood laid out in a gridlike pattern or composed of meandering paths and cul-de-sacs? Are the streets, highways, or railroads that are referred to as northbound and southbound in your city really aligned with the north–south direction or tilted with respect to the orthogonal axes? Are there any global landmarks in the area, such as a tower, mountains, or the ocean? Do you prefer navigation instructions given in the form of maps or verbal directions? Do you usually hold a map in a north-up orientation or rotate it so that the upward direction on the map matches the direction of travel? Do you give verbal navigational directions in terms of front-back-left-right or north-south-east-west? Which do you prefer when you receive navigational directions?

Affects and emotions in the environment are another important component of spatial cognition and experience. Can subjective feelings (affects, emotions, or meanings) be objectively studied? If so, what has been found in the literature about the structure of affective judgments of concepts and environmental scenes? Can aesthetic responses to visual stimuli and environmental scenes be explained by the characteristics of information in the stimuli? What types of environments do people prefer and how is the preference related to the evolutionary history of the human species? How can we define the concepts of place attachment and human values as they relate to our experiences in the environment? These are all pertinent questions because we react to and behave in the environment as we feel about it.

The development of geospatial technologies and data infrastructures, and accordingly ubiquitous access to information and real-time sensing of situations, will merge information space and real space, and eventually create a city environment that is fully augmented by computational capabilities, just like a map with a scale of 1:1. The goal of a spatially enabled society—the realization of the concept of "anytime, anywhere, anyone" by place identification and context awareness—is truly meaningfully achieved by a careful consideration of who needs the information, when and where, for what purposes and in what situation (Figure 10.13). How can you design navigation tools that help the user not only find the way to a destination but also learn the navigated space and enjoy the navigation itself? How can advanced tools effectively adapt the format of navigational information to the attributes of the user and the context and situation of tool use? How can we use geospatial technologies to foster the natural and inherited human sense of place, beyond merely improving the efficiency and convenience of our society? How do we live in a spatially enabled society? What is the sort of spatial literacy that is required in such a society and what will our spatial awareness be like in the near future? The potential benefits of advanced geospatial technology and society will be fully realized by considering these questions from the perspective of people who use the technology and live in the society. To that end, research into spatial cognition and behavior from the perspective of human–space–information interaction will make a substantial contribution.

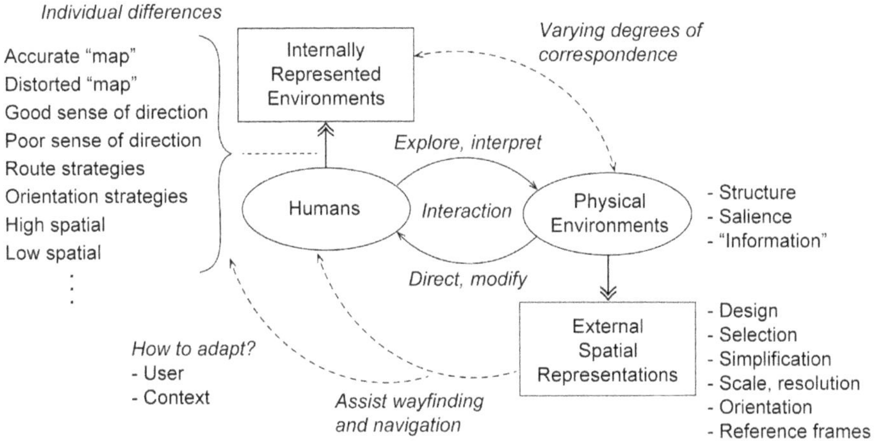

Figure 10.13 **Spatial behavior entails interaction between humans, the environment, and representations of the environment, both internal (knowledge and feelings about the environment) and external (e.g., maps, navigation tools, and verbal descriptions). The aim of developing a spatially enabled society (everyone has access to information whenever and wherever they need it) will be meaningfully achieved through the consideration of who needs the information, when and where, for what purposes and in what situation. Adapted with permission of Alexandrine Press from Ishikawa, 2018b, p. 201.**

Now you are ready to pursue the issues that we have discussed in this book and other related issues in more detail by yourself. I hope that you will continue the journey in search of cognized space. Have fun.

Review box for Chapter 10

1. List the topics that you want to examine further about human spatial cognition and experience.
2. Dig into the literature to find relevant studies. You can start with articles in the reference list in this book. Then go into studies that are cited in those articles and studies that cite those articles.
3. Refine your research ideas in reference to the studies you find in the literature.
4. Think about research design. How do you examine the questions that you want to pursue? For example, do you employ within-subject design or between-subject design?
5. Think about hypotheses and possible results.
6. How do you analyze data from your study? Do you take a quantitative approach or qualitative approach? Do you tackle your questions scientifically or in other nonscientific ways? Why?
7. You start it, with possible trials and errors, redesign, reanalysis, and additional data collection. It will be your own study, a welcome addition to the literature. Enjoy!

References

Acredolo, L. P. (1976). Frames of reference used by children for orientation in unfamiliar spaces. In G. T. Moore & R. G. Golledge (Eds.), *Environmental knowing: Theories, research, and methods* (pp. 173–184). Stroudsburg, PA: Dowden, Hutchinson & Ross.

Acredolo, L. P. (1977). Developmental changes in the ability to coordinate perspectives of a large-scale space. *Developmental Psychology, 13,* 1–8.

Acredolo, L. P., & Evans, D. (1980). Developmental changes in the effects of landmarks on infant spatial behavior. *Developmental Psychology, 16,* 312–318.

Aguirre, G. K., & D'Esposito, M. (1997). Environmental knowledge is subserved by separable dorsal/ventral neural areas. *Journal of Neuroscience, 17,* 2512–2518.

Allen, G. L., Kirasic, K. C., Dobson, S. H., Long, R. G., & Beck, S. (1996). Predicting environmental learning from spatial abilities: An indirect route. *Intelligence, 22,* 327–355.

Alonso, W. (1960). A theory of the urban land market. *Papers and Proceedings of the Regional Science Association, 6,* 149–157.

Amin, A., Townsend, S., Ossenbruggen, J. van, & Hardman, L. (2009). Fancy a drink in Canary Wharf? A user study on location-based mobile search. In T. Gross, J. Gulliksen, P. Kotzé, L. Oestreicher, P. Palanque, R. O. Prates, & M. Winckler (Eds.), *Human–computer interaction: INTERACT 2009* (Pt. 1, pp. 736–749). Heidelberg, Germany: Springer.

Aporta, C., & Higgs, E. (2005). Satellite culture: Global positioning systems, Inuit wayfinding, and the need for a new account of technology. *Current Anthropology, 46,* 729–753.

Appleton, J. (1975). *The experience of landscape.* London: Wiley.

Appleton, J. (1988). Prospects and refuges revisited. In J. L Nasar (Ed.), *Environmental aesthetics: Theory, research, and applications* (pp. 27–44). Cambridge: Cambridge University Press.

Appleyard, D. (1970). Styles and methods of structuring a city. *Environment and Behavior, 2,* 100–117.

Aristotle (1930). *Physica* (R. P. Hardie & R. K. Gaye, Trans.). In W. D. Ross (Ed.), *The works of Aristotle* (Vol. 2). Oxford: Clarendon Press.

Augustine (1955). *Confessions* (A. C. Outler, Trans., Ed.). Philadelphia, PA: Westminster Press.

Austen, K. (2013, January 16). Uncharted territory: Where digital maps are leading us. *New Scientist.* http://www.newscientist.com/article/mg21729002-300-uncharted-territory-where-digital-mapsare-leading-us/

Axon, S., Speake, J., & Crawford, K. (2012). "At the next junction, turn left": Attitudes towards Sat Nav use. *Area, 44*(2), 170–177.

Baddeley, A. (1986). *Working memory*. Oxford: Oxford University Press.

Baddeley, A. (2003). Working memory: Looking back and looking forward. *Nature Reviews Neuroscience, 4*, 829–839.

Bakdash, J. Z., Linkenauger, S. A., & Proffitt, D. (2008). Comparing decision-making and control for learning a virtual environment: Backseat drivers learn where they are going. *Proceedings of the Human Factors and Ergonomics Society Annual Meeting, 52*, 2117–2121.

Balling, J. D., & Falk, J. H. (1982). Development of visual preference for natural environments. *Environment and Behavior, 14*, 5–28.

Barr, N., Pennycook, G., Stolz, J. A., & Fugelsang, J. A. (2015). The brain in your pocket: Evidence that smartphones are used to supplant thinking. *Computers in Human Behavior, 48*, 473–480.

Bartlett, F. C. (1932). *Remembering: A study in experimental and social psychology*. Cambridge: Cambridge University Press.

Beck, R., & Wood, D. (1976). Comparative developmental analysis of individual and aggregated cognitive maps of London. In G. T. Moore & R. G. Golledge (Eds.), *Environmental knowing: Theories, research, and methods* (pp. 173–184). Stroudsburg, PA: Dowden, Hutchinson & Ross.

Berlyne, D. E. (1958). The influence of complexity and novelty in visual figures on orienting responses. *Journal of Experimental Psychology, 55*, 289–296.

Berlyne, D. E. (1960). *Conflict, arousal, and curiosity*. New York, NY: McGraw-Hill.

Berlyne, D. E. (1966). Curiosity and exploration. *Science, 153*, 25–33.

Berney, S., & Bétrancourt, M. (2016). Does animation enhance learning? A meta-analysis. *Computers & Education, 101*, 150–167.

Binet, M. A. (1894). Reverse illusions of orientation. *Psychological Review, 1*, 337–350.

Birchall, D. (2015). Spatial ability in radiologists: A necessary prerequisite? *British Journal of Radiology, 88*, 20140511.

Bisch-Knaden, S., & Wehner, R. (2003). Landmark memories are more robust when acquired at the nest site than en route: Experiments in desert ants. *Naturwissenschaften, 90*, 127–130.

Blades, M., & Spencer, C. (1994). The development of children's ability to use spatial representations. In H. W. Reese (Ed.), *Advances in child development and behavior* (Vol. 25, pp. 157–199). New York, NY: Academic Press.

Blaut, J., McCleary, G. S., Jr., & Blaut, A. S. (1970). Environmental mapping in young children. *Environment and Behavior, 2*, 335–349.

Borneman, E. (2014, October 9). The Geographer by Johannes Vermeer: Geography in fine art. *Geography Realm*. https://www.geographyrealm.com/geographer-johannes-vermeer-geography-fine-art/

Bremner, J. G., & Bryant, P. E. (1977). Place versus response as the basis of spatial errors made by young infants. *Journal of Experimental Child Psychology, 23*, 162–171.

Brown, B., Perkins, D. D., & Brown, G. (2003). Place attachment in a revitalizing neighborhood: Individual and block levels of analysis. *Journal of Environmental Psychology, 23*, 259–271.

Brown, W. (1932). Spatial integrations in a human maze. *University of California Publications in Psychology, 5*, 123–134.

Brown, W. (1937). Reorientation in a multiple-path maze. *University of California Publications in Psychology, 5*, 135–160.

Brunyé, T. T., Mahoney, C. R., Gardony, A. L., & Taylor, H. A. (2010). North is up(hill): Route planning heuristics in real-world environments. *Memory & Cognition, 38*, 700–712.

Bryant, K. J. (1982). Personality correlates of sense of direction and geographic orientation. *Journal of Personality and Social Psychology, 43*, 1318–1324.

Bujacz, M., Pec, M., Skulimowski, P., Strumillo, P., & Materka, A. (2011). Sonification of 3D scenes in an electronic travel aid for the blind. In P. Strumillo (Ed.), *Advances in sound localization* (pp. 251–268). Rijeka, Croatia: InTech.

Bullens, J., Iglói, K., Berthoz, A., Postma, A., & Rondi-Reig, L. (2010). Developmental time course of the acquisition of sequential egocentric and allocentric navigation strategies. *Journal of Experimental Child Psychology, 107*, 337–350.

Bunting, T. E., & Guelke, L. (1979). Behavioral and perception geography: A critical appraisal. *Annals of the Association of American Geographers, 69*, 448–462.

Burnett, G. E., & Lee, K. (2005). The effect of vehicle navigation systems on the formation of cognitive maps. In G. Underwood (Ed.), *Traffic and transport psychology: Theory and application* (pp. 407–418). Amsterdam: Elsevier.

Burroughs, W. J., & Sadalla, E. K. (1979). Asymmetries in distance cognition. *Geographical Analysis, 11*, 414–421.

Burton, I., & Kates, R. W. (1963). The perception of natural hazards in resource management. *Natural Resources Journal, 3*, 412–441.

Byrne, R. W. (1979). Memory for urban geography. *Quarterly Journal of Experimental Psychology, 31*, 147–154.

Cadwallader, M. T. (1976). Cognitive distance in intraurban space. In G. T. Moore & R. G. Golledge (Eds.), *Environmental knowing: Theories, research, and methods* (pp. 316–324). Stroudsburg, PA: Dowden, Hutchinson & Ross.

Cadwallader, M. (1981). Towards a cognitive gravity model: The case of consumer spatial behaviour. *Regional Studies, 15*, 275–284.

Carroll, J. B. (1993). *Human cognitive abilities: A survey of factor-analytic studies*. Cambridge: Cambridge University Press.

Carroll, L. (1893). *Sylvie and Bruno concluded*. London: MacMillan.

Carter, C., LaRussa, M., & Bodner, G. (1987). A study of two measures of spatial ability as predictors of success in different levels of general chemistry. *Journal of Research in Science Teaching, 24*, 645–657.

Chadwick, M. J., Jolly, A. E., Amos, D. P., Hassabis, D., & Spiers, H. J. (2015). A goal direction signal in the human entorhinal/subicular region. *Current Biology, 25*, 87–92.

Chomsky, N. (1959). Review of the book *Verbal behavior*, by B. F. Skinner. *Language, 35*, 26–58.

Chrastil, E. R., & Warren, W. H. (2014). From cognitive maps to cognitive graphs. *PLoS ONE, 9*(11), e112544.

Christaller, W. (1933/1966). *Central places in southern Germany* (C. W. Baskin, Trans.). Englewood Cliffs, NJ: Prentice-Hall.

Churchill, R. D., Curtis, J. M., Coombs, C. H., & Harrell, T. W. (1942). Effect of engineer school training on the Surface Development Test. *Educational and Psychological Measurement, 2*, 279–280.

Clarke, K. C. (2001). *Getting started with geographic information systems* (3rd ed.). Upper Saddle River, NJ: Prentice-Hall.

Condon, D. M., Wilt, J., Cohen, C. A., Revelle, W., Hegarty, M., & Uttal, D. H. (2015). Sense of direction: General factor saturation and associations with the Big-Five traits. *Personality and Individual Differences, 86*, 38–43.

Cooper, L. A., & Shepard, R. N. (1973). Chronometric studies of the rotation of mental images. In W. G. Chase (Ed.), *Visual information processing* (pp. 75–176). New York, NY: Academic Press.

Cornell, E. H., Heth, C. D., & Rowat, W. L. (1992). Wayfinding by children and adults: Response to instructions to use look-back and retrace strategies. *Developmental Psychology, 28*, 328–336.

Couclelis, H. (1998). Aristotelian spatial dynamics in the age of geographic information systems. In M. J. Egenhofer & R. G. Golledge (Eds.), *Spatial and temporal reasoning in geographic information systems* (pp. 109–118). New York, NY: Oxford University Press.

Cullen, K. E., & Taube, J. S. (2017). Our sense of direction: Progress, controversies and challenges. *Nature Neuroscience, 20,* 1465–1473.

Cullinane, M. J. (2011). Metric axioms and distance. *The Mathematical Gazette, 95,* 414–419.

Dallago, L., Perkins, D. D., Santinello, M., Boyce, W., Molcho, M., & Morgan, A. (2009). Adolescent place attachment, social capital, and perceived safety: A comparison of 13 countries. *American Journal of Community Psychology, 44,* 148–160.

Darwin, C. (1873). Origin of certain instincts. *Nature, 7,* 417–418.

DeLoache, J. S. (1987). Rapid change in the symbolic functioning of very young children. *Science, 238,* 1556–1557.

DeSilva, H. R. (1931). A case of a boy possessing an automatic directional orientation. *Science, 73,* 393–394.

Devlin, A. S. (1976). The "small town" cognitive map: Adjusting to a new environment. In G. T. Moore & R. G. Golledge (Eds.), *Environmental knowing: Theories, research, and methods* (pp. 58–66). Stroudsburg, PA: Dowden, Hutchinson & Ross.

Diez Roux, A. V. (2015). Ecological variables, ecological studies, and multilevel studies in public health research. In R. Detels, M. Gulliford, Q. A. Karim, & C. C. Tan (Eds.), *Oxford textbook of global public health* (6th ed., Vol. 2, pp. 411–429). Oxford: Oxford University Press.

Dillemuth, J. (2009). Navigation tasks with small-display maps: The sum of the parts does not equal the whole. *Cartographica 44,* 187–200.

Dosen, A. S., & Ostwald, M. J. (2016). Evidence for prospect-refuge theory: A meta-analysis of the findings of environmental preference research. *City, Territory and Architecture, 3,* 4.

Downs, R. M. (1981). Maps and mappings as metaphors for spatial representation. In L. S. Liben, A. H. Patterson, & N. Newcombe (Eds.), *Spatial representation and behavior across the life span* (pp. 143–166). New York, NY: Academic Press.

Downs, R. M., & Liben, L. S. (1991). The development of expertise in geography: A cognitive-developmental approach to geographic education. *Annals of the Association of American Geographers, 81,* 304–327.

Downs, R. M., & Stea, D. (1973). Cognitive maps and spatial behavior: Process and products. In R. M. Downs & D. Stea (Eds.), *Image and environment: Cognitive mapping and spatial behavior* (pp. 8–26). Chicago, IL: Aldine.

Dudchenko, P. A. (2010). *Why people get lost: The psychology and neuroscience of spatial cognition.* Oxford: Oxford University Press.

Duffy, S., Huttenlocher, J., Hedges, L. V., & Crawford, L. E. (2010). Category effects on stimulus estimation: Shifting and skewed frequency distributions. *Psychonomic Bulletin & Review, 17,* 224–230.

Duke, N. N., Skay, C. L., Pettingell, S. L., & Borowsky, I. W. (2009). From adolescent connections to social capital: Predictors of civic engagement in young adulthood. *Journal of Adolescent Health, 44,* 161–168.

Dumont, J. R., & Taube, J. S. (2015). The neural correlates of navigation beyond the hippocampus. *Progress in Brain Research, 219,* 83–102.

Ebbinghaus, H. (1885). *Memory: A contribution to experimental psychology* (H. A. Ruger & C. E. Bussenius, Trans.). New York, NY: Teachers College, Columbia University.

Ehrenberg, R. E. (Ed.). (2006). *Mapping the world: An illustrated history of cartography.* Washington, DC: National Geographic Society.

Ekstrom, R. B., French, J. W., Harman, H. H., & Dermen, D. (1976). *Kit of factor-referenced cognitive tests.* Princeton, NJ: Educational Testing Service.

Eliot, J., & Smith, I. M. (1983). *An international directory of spatial tests.* Windsor, UK: NFER-Nelson.

El Koussy, A. A. H. (1935). The visual perception of space. *British Journal of Psychology: Monograph Supplements, 20,* 1–80.

Etienne, A. S., & Jeffery, K. J. (2004). Path integration in mammals. *Hippocampus, 14,* 180–192.

Evans, G. W., Marrero, D. G., & Butler, P. A. (1981). Environmental learning and cognitive mapping. *Environment and Behavior, 13,* 83–104.

Evans, G. W., & Pezdek, K. (1980). Cognitive mapping: Knowledge of real-world distance and location information. *Journal of Experimental Psychology: Human Learning and Memory, 6,* 13–24.

Falk, J. H., & Balling, J. D. (2010). Evolutionary influence on human landscape preference. *Environment and Behavior, 42,* 479–493.

Farman, J. (2015). Stories, spaces, and bodies: The production of embodied space through mobile media storytelling. *Communication Research and Practice, 1,* 101–116.

Farrell, M. J., Arnold, P., Pettifer, S., Adams, J., Graham, T., & MacManamon, M. (2003). Transfer of route learning from virtual to real environments. *Journal of Experimental Psychology: Applied, 9,* 219–227.

Fleischmann, P. N., Christian, M., Müller, V. L., Rössler, W., & Wehner, R. (2016). Ontogeny of learning walks and the acquisition of landmark information in desert ants, *Cataglyphis fortis. Journal of Experimental Biology, 219,* 3137–3145.

Foo, P., Warren, W. H., Duchon, A., & Tarr, M. J. (2005). Do humans integrate routes into a cognitive map? Map-versus landmark-based navigation of novel shortcuts. *Journal of Experimental Psychology: Learning, Memory, and Cognition, 31,* 195–215.

Frank, A. U. (1996). Qualitative spatial reasoning: Cardinal directions as an example. *International Journal of Geographical Information Systems, 10,* 269–290.

Frean, A. (2006, December 27). Use satellite navigation and you'll miss the chance of finding your inner self. *The Times.* https://www.thetimes.co.uk/article/use-satellite-navigation-and-youll-miss-the-chance-of-finding-your-inner-self-qh0bwnsqw0x/

Freeman, F. N. (1916). Geography: Extension of experience through imagination. In *The psychology of common branches* (Chapter 8, pp. 161–178). Boston, MA: Houghton Mifflin.

Frick, A., Möhring, W., & Newcombe, N. S. (2012). Picturing perspectives: Development of perspective-taking abilities in 4- to 8-year-olds. *Frontiers in Psychology, 5,* 386.

Fried, M. (1963). Grieving for a lost home: Psychological costs of relocation. In J. Q. Wilson (Ed.), *Urban renewal: The record and the controversy* (pp. 359–379). Cambridge, MA: MIT Press

Friedman, A., & Kohler, B. (2003). Bidimensional regression: Assessing the configural similarity and accuracy of cognitive maps and other two-dimensional data sets. *Psychological Methods, 8,* 468–491.

Friedman, A., Montello, D. R., & Burte, H. (2012). Location memory for dots in polygons versus cities in regions: Evaluating the category adjustment model. *Journal of Experimental Psychology: Learning, Memory, and Cognition, 38,* 1336–1351.

Gallistel, C. R. (1990). *The organization of learning.* Cambridge, MA: MIT Press.

Gardner, H. (2006). *Multiple intelligences: New horizons.* New York, NY: Basic Books.

Gardony, A. L., Brunyé, T. T., Mahoney, C. R., & Taylor, H. A. (2013). How navigational aids impair spatial memory: Evidence for divided attention. *Spatial Cognition and Computation, 13,* 319–350.

Gartner, G., Cartwright, W., & Peterson, M. P. (Eds.). (2007). *Location based services and telecartography.* Berlin, Germany: Springer.

Gartner, G., & Hiller, W. (2010). Impact of restricted display size on spatial knowledge acquisition in the context of pedestrian navigation. In G. Gartner & K. Rehrl (Eds.), *Location based services and telecartography II: From sensor fusion to context models* (pp. 155–166). Berlin, Germany: Springer.

Gaunet, F., Vidal, M., Kemeny, A., & Berthoz, A. (2001). Active, passive and snapshot exploration in a virtual environment: Influence on scene memory, reorientation and path memory. *Cognitive Brain Research, 11,* 409–420.

Gersmehl, P. (2008). *Teaching geography* (2nd ed.). New York, NY: Guilford Press.

Gersmehl, P. J., & Gersmehl, C. A. (2007). Spatial thinking by young children: Neurologic evidence for early development and "educability". *Journal of Geography, 106,* 181–191.
Gescheider, G. A. (1997). *Psychophysics: The fundamentals* (3rd ed.). Mahwah, NJ: Erlbaum.
Gewin, V. (2004). Mapping opportunities. *Nature, 427,* 376–377.
Gibson, J. J. (1979/1986). *The ecological approach to visual perception.* Hillsdale, NJ: Erlbaum.
Gkonos, C., Giannopoulos, I., & Raubal, M. (2017). Maps, vibration or gaze? Comparison of novel navigation assistance in indoor and outdoor environments. *Journal of Location Based Services, 11,* 29–49.
Gladwin, T. (1970). *East is a big bird: Navigation and logic on Puluwat atoll.* Cambridge, MA: Harvard University Press.
Golledge, R. G. (1978). Learning about urban environments. In T. Carlstein, D. Parkes, & N. Thrift (Eds.), *Timing space and spacing time* (Vol. 1, pp. 76–98). London: Arnold.
Golledge, R. G. (1981). Misconceptions, misinterpretations, and misrepresentations of behavioral approaches in human geography. *Environment and Planning A, 13,* 1325–1344.
Golledge, R. G. (1991). Cognition of physical and built environments. In T. Gärling & G. W. Evans (Eds.), *Environment, cognition, and action: An integrated approach* (pp. 35–62). New York, NY: Oxford University Press.
Golledge, R. G. (1996). Geographical theories. *International Social Science Journal, 48,* 461–476.
Golledge, R. G. (2002). The nature of geographic knowledge. *Annals of the Association of American Geographers, 92,* 1–14.
Golledge, R. G., Briggs, R., & Demko, D. (1969). The configuration of distances in intra-urban space. *Proceedings of the Association of American Geographers, 1,* 60–65.
Golledge, R. G., & Hubert, L. J. (1982). Some comments on non-Euclidean mental maps. *Environment and Planning A, 14,* 107–118.
Golledge, R. G., Marsh, M., & Battersby, S. (2008). A conceptual framework for facilitating geospatial thinking. *Annals of the Association of American Geographers, 98,* 285–308.
Golledge, R. G., Rivizzigno, V. L., & Spector, A. (1976). Learning about a city: Analysis by multidimensional scaling. In R. G. Golledge & G. Rushton (Eds.), *Spatial choice and spatial behavior* (pp. 95–116). Columbus, OH: Ohio State University Press.
Golledge, R. G., Ruggles, A. J., Pellegrino, J. W., & Gale, N. D. (1993). Integrating route knowledge in an unfamiliar neighborhood: Along and across route experiments. *Journal of Environmental Psychology, 13,* 293–307.
Golledge, R. G., & Spector, A. N. (1978). Comprehending the urban environment: Theory and practice. *Geographical Analysis, 10,* 403–426.
Golledge, R. G., & Stimson, R. J. (1997). *Spatial behavior: A geographic perspective.* New York, NY: Guilford Press.
Golledge, R. G., & Timmermans, H. (1990). Applications of behavioural research on spatial problems I: Cognition. *Progress in Human Geography, 14,* 57–99.
Goodchild, M. F. (2001). A geographer looks at spatial information theory. In D. R. Montello (Ed.), *Spatial information theory* (pp. 1–13). Berlin, Germany: Springer.
Goodchild, M. F. (2006). The fourth R? Rethinking GIS education. *Arc News, 28*(3), 1, 5–7.
Goodchild, M. F. (2007). Citizens as sensors: The world of volunteered geography. *GeoJournal, 69,* 211–221.
Gould, P., & White, R. (1986). *Mental maps* (2nd ed.). London: Routledge.
Grabar, H. (2014, September 9). Smartphones and the uncertain future of 'spatial thinking'. *CityLab.* https://www.citylab.com/life/2014/09/smartphones-and-the-uncertain-future-of-spatial-thinking/379796/
Graham, E. (1976). What is a mental map? *Area, 8,* 259–262.
Grieves, R. M., & Jeffery, K. J. (2017). The representation of space in the brain. *Behavioural Processes, 135,* 113–131.

Guilford, J. P., & Zimmerman, W. S. (1948). The Guilford-Zimmerman Aptitude Survey. *Journal of Applied Psychology, 32,* 24–34.
Gulliver, F. P. (1908). Orientation of maps. *Bulletin of the American Geographical Society, 40,* 538–542.
Haken, H., & Portugali, J. (2003). The face of the city is its information. *Journal of Environmental Psychology, 23,* 385–408.
Halás, M., Klapka, P., & Kladivo, P. (2014). Distance-decay functions for daily travel-to-work flows. *Journal of Transport Geography, 35,* 107–119.
Hall-Wallace, M. K., & McAuliffe, C. M. (2002). Design, implementation, and evaluation of GIS-based learning materials in an introductory geoscience course. *Journal of Geoscience Education, 50,* 5–14.
Hanley, G. L., & Levine, M. (1983). Spatial problem solving: The integration of independently learned cognitive maps. *Memory & Cognition, 11,* 415–422.
Hart, R. A., & Moore, G. T. (1973). The development of spatial cognition: A review. In R. M. Downs & D. Stea (Eds.), *Image and environment: Cognitive mapping and spatial behavior* (pp. 246–288). Chicago, IL: Aldine.
Hartig, T., & Evans, G. W. (1993). Psychological foundations of nature experience. In T. Gärling & R. G. Golledge (Eds.), *Behavior and environment: Psychological and geographical approaches* (pp. 427–457). Oxford: North-Holland.
Hartley, T., & Harlow, R. (2012). An association between human hippocampal volume and topographical memory in healthy young adults. *Frontiers in Human Neuroscience, 6,* 338.
Hartley, T., Maguire, E. A., Spiers, H. J., & Burgess, N. (2003). The well-worn route and the path less traveled: Distinct neural bases of route following and wayfinding in humans. *Neuron, 37,* 877–888.
Hay, R. (1998). Sense of place in developmental context. *Journal of Environmental Psychology, 18,* 5–29.
Hazen, N. L., Lockman, J. J., & Pick, H. L., Jr. (1978). The development of children's representations of large-scale environments. *Child Development, 49,* 623–636.
Hegarty, M. (2010). Components of spatial intelligence. In B. H. Ross (Ed.), *Psychology of learning and motivation* (Vol. 52, pp. 265–297). San Diego, CA: Elsevier.
Hegarty, M., Keehner, M., Khooshabeh, P., & Montello, D. R. (2009). How spatial abilities enhance, and are enhanced by, dental education. *Learning and Individual Differences, 19,* 61–70.
Hegarty, M., Montello, D. R., Richardson, A. E., Ishikawa, T., & Lovelace, K. (2006). Spatial abilities at different scales: Individual differences in aptitude-test performance and spatial-layout learning. *Intelligence, 34,* 151–176.
Hegarty, M., Richardson, A. E., Montello, D. R., Lovelace, K., & Subbiah, I. (2002). Development of a self-report measure of environmental spatial ability. *Intelligence, 30,* 425–447.
Hegarty, M., & Waller, D. (2004). A dissociation between mental rotation and perspective-taking spatial abilities. *Intelligence, 32,* 175–191.
Hegarty, M., & Waller, D. (2005). Individual differences in spatial abilities. In P. Shah & A. Miyake (Eds.), *The Cambridge handbook of visuospatial thinking* (pp. 121–169). New York, NY: Cambridge University Press.
Heipke, C. (2010). Crowdsourcing geospatial data. *ISPRS Journal of Photogrammetry and Remote Sensing, 65,* 550–557.
Herman, J. F., Kail, R. V., & Siegel, A. W. (1979). Cognitive maps of a college campus: A new look at freshman orientation. *Bulletin of the Psychonomic Society, 13,* 183–186.
Hermer, L., & Spelke, E. S. (1994). A geometric process for spatial reorientation in young children. *Nature, 370,* 57–59.

Hermer-Vazquez, L., Moffet, A., & Munkholm, P. (2001). Language, space, and the development of cognitive flexibility in humans: The case of two spatial memory tasks. *Cognition, 79,* 263–299.

Herzog, T. R. (1985). A cognitive analysis of preference for waterscapes. *Journal of Environmental Psychology, 5,* 225–241.

Herzog, T. R. (1992). A cognitive analysis of preference for urban spaces. *Journal of Environmental Psychology, 12,* 237–248.

Hidalgo, M. C., & Hernández, B. (2001). Place attachment: Conceptual and empirical questions. *Journal of Environmental Psychology, 21,* 273–281.

Hillier, B. (2012). Studying cities to learn about minds: Some possible implications of space syntax for spatial cognition. *Environment and Planning B, 39,* 12–32.

Hirtle, S. C. (1998). The cognitive atlas: Using GIS as a metaphor for memory. In M. J. Egenhofer & R. G. Golledge (Eds.), *Spatial and temporal reasoning in geographic information systems* (pp. 263–271). New York, NY: Oxford University Press.

Hirtle, S. C., & Jonides, J. (1985). Evidence of hierarchies in cognitive maps. *Memory & Cognition, 13,* 208–217.

Höffler, T. N., & Leutner, D. (2007). Instructional animation versus static pictures: A meta-analysis. *Learning and Instruction, 17,* 722–738.

Hofmann-Wellenhof, B., Lichtenegger, H., & Wasle, E. (2008). *GNSS: Global Navigation Satellite Systems.* Vienna, Austria: Springer.

Hoinville, T., & Wehner, R. (2018). Optimal multiguidance integration in insect navigation. *Proceedings of the National Academy of Sciences of the United States of America, 115,* 2824–2829.

Holden, M. P., Newcombe, N. S., & Shipley, T. F. (2013). Location memory in the real world: Category adjustment effects in 3-dimensional space. *Cognition, 128,* 45–55.

Holden, M. P., Newcombe, N. S, Resnick, I., & Shipley, T. F. (2016). Seeing like a geologist: Bayesian use of expert categories in location memory. *Cognitive Science 40,* 440–454.

Holding, C. S., & Holding, D. H. (1988). Acquisition of route network knowledge by males and females. *Journal of General Psychology, 116,* 29–41.

Hudson, W. H. (1922). On the sense of direction. *The Century Magazine, 104,* 693–701.

Huff, D. L. (1963). A probabilistic analysis of shopping center trade areas. *Land Economics, 39,* 81–90.

Huttenlocher, J., Hedges, L. V., & Duncan, S. (1991). Categories and particulars: Prototype effects in estimating spatial location. *Psychological Review, 98,* 352–376.

Huttenlocher, J., Newcombe, N., & Sandberg, E. (1994). The coding of spatial location in young children. *Cognitive Psychology, 27,* 115–147.

Huttenlocher, J., Newcombe, N., & Vasilyeva, M. (1999). Spatial scaling in young children. *Psychological Science, 10,* 393–398.

Iaria, G., & Barton, J. J. S. (2010). Developmental topographical disorientation: A newly discovered cognitive disorder. *Experimental Brain Research, 206,* 189–196.

Ishikawa, T. (2013). Geospatial thinking and spatial ability: An empirical examination of knowledge and reasoning in geographical science. *The Professional Geographer, 65,* 636–646.

Ishikawa, T. (2016a). Maps in the head and tools in the hand: Wayfinding and navigation in a spatially enabled society. In R. H. Hunter, L. A. Anderson, & B. L. Belza (Eds.), *Community wayfinding: Pathways to understanding* (pp. 115–136). Cham, Switzerland: Springer International.

Ishikawa, T. (2016b). Spatial thinking in geographic information science: Students' geospatial conceptions, map-based reasoning, and spatial visualization ability. *Annals of the American Association of Geographers, 106,* 76–95.

Ishikawa, T. (2018a). Learning the environment: The acquisition of cognitive maps. In D. R. Montello (Ed.), *Handbook of behavioral and cognitive geography* (pp. 116–140). Cheltenham, UK: Elgar.

Ishikawa, T. (2018b). People in the environment and environments perceived by people: Cognitive-behavioral approaches to spatial cognition. *Built Environment, 44,* 192–204.

Ishikawa, T. (2019). Satellite navigation and geospatial awareness: Long-term effects of using navigation tools on wayfinding and spatial orientation. *The Professional Geographer, 71,* 197–209.

Ishikawa, T., Barnston, A. G., Kastens, K. A., & Louchouarn, P. (2011). Understanding, evaluation, and use of climate forecast data by environmental policy students. In A. D. Feig & A. Stokes (Eds.), *Qualitative inquiry in geoscience education research* (pp. 153–170). Boulder, CO: Geological Society of America (Special Paper 474).

Ishikawa, T., Fujiwara, H., Imai, O., & Okabe, A. (2008). Wayfinding with a GPS-based mobile navigation system: A comparison with maps and direct experience. *Journal of Environmental Psychology, 28,* 74–82.

Ishikawa, T., & Kastens, K. A. (2005). Why some students have trouble with maps and other spatial representations. *Journal of Geoscience Education, 53,* 184–197.

Ishikawa, T., & Montello, D. R. (2006). Spatial knowledge acquisition from direct experience in the environment: Individual differences in the development of metric knowledge and the integration of separately learned places. *Cognitive Psychology, 52,* 93–129.

Ishikawa, T., & Takahashi, K. (2013). Relationships between methods for presenting information on navigation tools and users' wayfinding behavior. *Cartographic Perspectives, 75,* 17–28.

Ittelson, W. H. (1973). Environment perception and contemporary perceptual theory. In W. H. Ittelson (Ed.), *Environment and cognition* (pp. 1–19). New York, NY: Seminar Press.

Jacobs, L. F., Arter, J., Cook, A., & Sulloway, F. J. (2015). Olfactory orientation and navigation in humans. *PLoS ONE 10*(6), e0129387.

Jacobs, L. F., & Menzel, R. (2014). Navigation outside of the box: What the lab can learn from the field and what the field can learn from the lab. *Movement Ecology, 2,* 3.

Janelle, D. G., & Goodchild, M. F. (2011). Concepts, principles, tools, and challenges in spatially integrated social science. In T. Nyerges, H. Couclelis, & R. McMaster (Eds.), *The SAGE handbook of GIS and society* (pp. 27–45). London: Sage.

Johnson, M. (1987). *The body in the mind: The bodily basis of meaning, imagination, and reason.* Chicago, IL: University of Chicago Press.

Just, M. A., & Carpenter, P. (1985). Cognitive coordinate systems: Accounts of mental rotation and individual differences in spatial ability. *Psychological Review, 92,* 137–172.

Kail, R., & Park, Y.-S. (1990). Impact of practice on speed of mental rotation. *Journal of Experimental Child Psychology, 49,* 227–244.

Kalin, J., & Frith, J. (2016). Wearing the city: Memory p(a)laces, smartphones, and the rhetorical invention of embodied space. *Rhetoric Society Quarterly, 46,* 222–235.

Kant, I. (1768/1991). On the first ground of the distinction of regions in space (J. Handyside, Trans.). In J. Van Cleve & R. E. Frederick (Eds.), *The philosophy of right and left: Incongruent counterparts and the nature of space.* (pp. 27–33). Dordrecht, The Netherlands: Kluwer Academic.

Kant, I. (1781/1787/1998). *Critique of pure reason* (P. Guyer & A. W. Wood, Trans.). Cambridge: Cambridge University Press.

Kaplan, R., & Kaplan, S. (1989). *The experience of nature.* New York, NY: Cambridge University Press.

Kaplan, R., Kaplan, S., & Brown, T. (1989). Environmental preference: A comparison of four domains of predictors. *Environment and Behavior, 21,* 509–530.

Kaplan, R., Kaplan, S., & Ryan, R. L. (1998). *With people in mind: Design and management of everyday nature.* Washington, DC: Island Press.
Kaplan, S. (1987). Aesthetics, affect, and cognition: Environmental preference from an evolutionary perspective. *Environment and Behavior, 19,* 3–32.
Kaplan, S., Kaplan, R., & Wendt, J. S. (1972). Rated preference and complexity for natural and urban visual material. *Perception & Psychophysics, 12,* 354–356.
Kastens, K. A., & Ishikawa, T. (2006). Spatial thinking in the geosciences and cognitive sciences: A cross-disciplinary look at the intersection of the two fields. In C. A. Manduca & D. W. Mogk (Eds.), *Earth and mind: How geologists think and learn about the Earth* (pp. 53–76). Boulder, CO: Geological Society of America (Special Paper 413).
Kates, R. W., Colten, C. E., Laska, S. B., & Leatherman, S. P. (2006) Reconstruction of New Orleans after Hurricane Katrina: A research perspective. *Proceedings of the National Academy of Sciences of the United States of America, 103,* 14653–14660.
Keehner, M. M., Tendick, F., Meng, M. V., Anwar, H. P., Hegarty, M., Stoller, M. L., & Duh, Q. (2004). Spatial ability, experience, and skill in laparoscopic surgery. *The American Journal of Surgery, 188,* 71–75.
Kiefer, P., Huang, H., Van de Weghe, N., & Raubal, M. (Eds.). (2018). *Progress in location based services 2018.* Cham, Switzerland: Springer International.
Klatzky, R. L., Loomis, J. M., Beall, A. C., Chance, S. S., & Golledge, R. G. (1998). Spatial updating of self-position and orientation during real, imagined, and virtual locomotion. *Psychological Science, 9,* 293–298.
Klatzky, R. L., Loomis, J. M., Golledge, R. G., Cicinelli, J. G., Doherty, S., & Pellegrino, J. W. (1990). Acquisition of route and survey knowledge in the absence of vision. *Journal of Motor Behavior, 22,* 19–43.
Klippel, A., Freksa, C., & Winter, S. (2006). You-are-here maps in emergencies: The danger of getting lost. *Journal of Spatial Science, 51,* 117–131.
Koláčný, A. (1969). Cartographic information: A fundamental concept and term in modern cartography. *The Cartographic Journal, 6,* 47–49.
Kosslyn, S. M., Ball, T. M., & Reiser, B. J. (1978). Visual images preserve metric spatial information: Evidence from studies of image scanning. *Journal of Experimental Psychology: Human Perception and Performance, 4,* 47–60.
Koutsoklenis, A., & Papadopoulos, K. (2014). Haptic cues used for outdoor wayfinding by individuals with visual impairments. *Journal of Visual Impairment & Blindness, 108,* 43–53.
Kozhevnikov, M., & Hegarty, M. (2001). A dissociation between object manipulation spatial ability and spatial orientation ability. *Memory & Cognition, 29,* 745–756.
Kozhevnikov, M., Motes, M. A., & Hegarty, M. (2007). Spatial visualization in physics problem solving. *Cognitive Science, 31,* 549–579.
Kozlowski, L. T., & Bryant, K. J. (1977). Sense-of-direction, spatial orientation, and cognitive maps. *Journal of Experimental Psychology: Human Perception and Performance, 3,* 590–598.
Kuhn, W. (2012). Core concepts of spatial information for transdisciplinary research. International *Journal of Geographical Information Science, 26,* 2267–2276.
Küpper, A. (2005). *Location-based services: Fundamentals and operation.* Chichester, UK: Wiley.
Lakoff, G. (1987). *Women, fire, and dangerous things: What categories reveal about the mind.* Chicago, IL: University of Chicago Press.
Lakoff, G., & Johnson, M. (1980). *Metaphors we live by.* Chicago, IL: University of Chicago Press.
Lambrey, S., Doeller, C., Berthoz, A., & Burgess, N. (2012). Imagining being somewhere else: Neural basis of changing perspective in space. *Cerebral Cortex 22,* 166–174.
Landau, B., Spelke, E., & Gleitman, H. (1984). Spatial knowledge in a young blind child. *Cognition, 16,* 225–260.

Lawton, C. A., & Kallai, J. (2002). Gender differences in wayfinding strategies and anxiety about wayfinding: A cross-cultural comparison. *Sex Roles, 47,* 389–401.

Learmonth, A. E., Nadel, L., & Newcombe, N. S. (2002). Children's use of landmarks: Implications for modularity theory. *Psychological Science, 13,* 337–341.

Learmonth, A. E., Newcombe, N. S., & Huttenlocher, J. (2001). Toddlers' use of metric information and landmarks to reorient. *Journal of Experimental Child Psychology, 80,* 225–244.

Lee, T. (1970). Perceived distance as a function of direction in the city. *Environment and Behavior, 2,* 40–51.

Lee, J., & Bednarz, R. (2009). Effect of GIS learning on spatial thinking. *Journal of Geography in Higher Education, 33,* 183–198.

Leshed, G., Velden, T., Rieger, O., Kot, B., & Sengers, P. (2008). In-car GPS navigation: Engagement with and disengagement from the environment. In *Proceeding of the SIGCHI conference on human factors in computing systems* (CHI 2008, pp. 1675–1684). New York, NY: ACM.

Levine, M. (1982). You-are-here maps: Psychological considerations. *Environment and Behavior, 14,* 221–237.

Levine, M., Jankovic, I. N., & Palij, M. (1982). Principles of spatial problem solving. *Journal of Experimental Psychology: General, 111,* 157–175.

Levine, M., Marchon, I., & Hanley, G. (1984). The placement and misplacement of you-are-here maps. *Environment and Behavior, 16,* 139–157.

Levinson, S. C. (1996). Frames of reference and Molyneux's question: Cross-linguistic evidence. In P. Bloom, M. A. Peterson, L. Nadel, & M. F. Garrett (Eds.), *Language and space* (pp. 109–169). Cambridge, MA: MIT Press.

Levinson, S. C. (2003). *Space in language and cognition: Explorations in cognitive diversity.* Cambridge: Cambridge University Press.

Lew, A. R., Foster, K. A., Crowther, H. L., & Green, M. (2004). Indirect landmark use at 6 months of age in a spatial orientation task. *Infant Behavior and Development, 27,* 81–90.

Lewicka, M. (2005). Ways to make people active: The role of place attachment, cultural capital, and neighborhood ties. *Journal of Environmental Psychology, 25,* 381–395.

Lewis, D. (1976). Observations on route finding and spatial orientation among the Aboriginal peoples of the Western Desert region of Central Australia. *Oceania, 46,* 249–282.

Li, P., & Gleitman, L. (2002). Turning the tables: Language and spatial reasoning. *Cognition, 83,* 265–294.

Liben, L. S. (1981). Spatial representation and behavior: Multiple perspectives. In L. S. Liben, A. H. Patterson, & N. Newcombe (Eds.). (1981). *Spatial representation and behavior across the life span* (pp. 3–36). New York, NY: Academic Press.

Liben, L. S., & Downs, R. M. (1989). Understanding maps as symbols: The development of map concepts in children. In H. W. Reese (Ed.), *Advances in child development and behavior* (Vol. 22, pp. 145–201). New York, NY: Academic Press.

Liben, L. S., & Downs, R. M. (1993). Understanding person-space-map relations: Cartographic and developmental perspectives. *Developmental Psychology, 29,* 739–752.

Liben, L. S., Kastens, K. A., & Stevenson, L. M. (2002). Real-world knowledge through real-world maps: A developmental guide for navigating the educational terrain. *Developmental Review, 22,* 267–322.

Liberman, N., Trope, Y., & Stephan, E. (2007). Psychological distance. In A. W. Kruglanski & E. T. Higgins (Eds.), *Social psychology: Handbook of basic principles* (2nd ed., pp. 353–381). New York, NY: Guilford Press.

Linde, C., & Labov, W. (1975). Spatial networks as a site for the study of language and thought. *Language, 51,* 924–939.

Lingwood, J., Blades, M., Farran, E. K., Courbois, Y., & Matthews, D. (2015). The development of wayfinding abilities in children: Learning routes with and without landmarks. *Journal of Environmental Psychology, 41,* 74–80.

Linn, M. C., & Petersen, A. C. (1985). Emergence and characterization of sex differences in spatial ability: A meta-analysis. *Child Development, 56,* 1479–1498.

Liu, S. (2018). *Navigation devices and usage.* New York, NY: Statista. https://www.statista.com/topics/2221/navigation-devices-and-usage/

Lobben, A. (2015). Tactile maps and mapping. *Journal of Blindness Innovation and Research, 5*(1). https://nfb.org/images/nfb/publications/jbir/jbir15/jbir050102.html

Lohman, D. F. (1979). *Spatial ability: A review and re-analysis of the correlational literature* (Tech. Rep. No. 8.). Stanford, CA: School of Education, Stanford University.

Loomis, J. M., Golledge, R. G., & Klatzky, R. L. (1998). System for the blind: Auditory display modes and guidance. *Presence, 7,* 193–203.

Loomis, J. M., Klatzky, R. L., Golledge, R. G., Cicinelli, J. G., Pellegrino, J. W., & Fry, P. A. (1993). Nonvisual navigation of blind and sighted: Assessment of path integration ability. *Journal of Experimental Psychology: General, 122,* 73–91.

Lorenz, C. A., & Neisser, U. (1986). *Ecological and psychometric dimensions of spatial ability* (Emory Cognition Project Rep. No. 10). Atlanta, GA: Department of Psychology, Emory University.

Lynch, K. (1960). *The image of the city.* Cambridge, MA: MIT Press.

MacEachren, A. M. (1980). Travel time as the basis of cognitive distance. *The Professional Geographer, 32,* 30–36.

MacEachren, A. M. (1995). *How maps work: Representation, visualization, and design.* New York, NY: Guilford Press.

MacKay, D. B., & Olshavsky, R. W. (1975). Cognitive maps of retail locations: An investigation of some basic issues. *Journal of Consumer Research, 2,* 197–205.

Maguire, E. A., Woollett, K., & Spiers, H. J. (2006). London taxi drivers and bus drivers: A structural MRI and neuropsychological analysis. *Hippocampus, 16,* 1091–1101.

Mahmud, N., Saha, R. K., Zafar, R. B., Bhuian, M. B. H., & Sarwar, S. S. (2014). Vibration and voice operated navigation system for visually impaired person. In *Proceedings of the 3rd international conference on informatics, electronics & vision* (ICIEV 2014).

Maki, R. H. (1981). Categorization and distance effects with spatial linear orders. *Journal of Experimental Psychology: Human Learning and Memory, 7,* 15–32.

Mandler, J. M. (1983). Representation. In P. H. Mussen (Ed.), *Handbook of child psychology* (4th ed.), Vol. 3: *Cognitive development* (J. H. Flavell & E. M. Markman, Vol. Eds., pp. 420–494). New York, NY: Wiley.

Mannings, R. (2008). *Ubiquitous positioning.* Norwood, MA: Artech House.

Manzo, L. C. (2005). For better or worse: Exploring multiple dimensions of place meaning. *Journal of Environmental Psychology, 25,* 67–86.

Marchette, S. A., Bakker, A., & Shelton, A. L. (2011). Cognitive mappers to creatures of habit: Differential engagement of place and response learning mechanisms predicts human navigational behavior. *Journal of Neuroscience, 31,* 15264–15268.

Marr, D. (1982). *Vision: A computational investigation into the human representation and processing of visual information.* New York, NY: Freeman.

Mathewson, J. H. (1999). Visual-spatial thinking: An aspect of science overlooked by educators. *Science Education, 83,* 33–54.

May, J. A. (1970). *Kant's concept of geography: And its relation to recent geographical thought.* Toronto, Canada: University of Toronto Press.

McBurney, D. H., Gaulin, S. J. C., Devineni, T., & Adams, C. (1997). Superior spatial memory of women: Stronger evidence for the gathering hypothesis. *Evolution and Human Behavior, 18,* 165–174.

McDermott, L. C., Rosenquist, M. L., & van Zee, E. H. (1987). Student difficulties in connecting graphs and physics: Examples from kinematics. *American Journal of Physics, 55,* 503–513.

McFadden, D. (2001). Economic choices. *American Economic Review, 91*(3), 351–378.

McGee, M. G. (1979). Human spatial abilities: Psychometric studies and environmental, genetic, hormonal, and neurological influences. *Psychological Bulletin, 86,* 889–918.

McKinlay, R. (2016). Use or lose our navigation skills. *Nature, 531,* 573–575.

McKinney, J. (2010, March 22). Don't throw away your paper maps just yet. *Pacific Standard.* https://psmag.com/social-justice/dont-throw-away-your-paper-maps-just-yet-11077/

McNamara T. P. (2003). How are the locations of objects in the environment represented in memory? In C. Freksa, W. Brauer, C. Habel, & K. F. Wender (Eds.), *Spatial cognition III* (pp. 174–191). Berlin, Germany: Springer.

McNamara, T. P., Hardy, J. K., & Hirtle, S. C. (1989). Subjective hierarchies in spatial memory. *Journal of Experimental Psychology: Learning, Memory, and Cognition, 15,* 211–227.

McNamara, T. P., Rump, B., & Werner, S. (2003). Egocentric and geocentric frames of reference in memory of large-scale space. *Psychonomic Bulletin & Review, 10,* 589–595.

McPherson, H. J., & Saarinen, T. F. (1977). Flood plain dwellers' perception of the flood hazard in Tucson, Arizona. *The Annals of Regional Science, 11*(2), 25–40.

Meek, T. J. (1935). *Excavations at Nuzi,* Vol.3: *Old Akkadian, Sumerian, and Cappadocian texts from Nuzi.* Cambridge, MA: Harvard University Press.

Meng, L., Zipf, A., & Winter, S. (Eds.). (2008). *Map-based mobile services: Design, interaction and usability.* Berlin, Germany: Springer.

Miele, J. A., Landau, S., & Gilden, D. (2006). Talking TMAP: Automated generation of audio-tactile maps using Smith-Kettlewell's TMAP software. *British Journal of Visual Impairment, 24,* 93–100.

Millard, A. R. (1987). Cartography in the ancient Near East. In J. B. Harley & David Woodward (Eds.), *The history of cartography* (Vol. 1, pp. 107–116). Chicago, IL: University of Chicago Press.

Milner, G. (2016a, February 11). Ignore the GPS: That ocean is not a road. *The New York Times.* https://www.nytimes.com/2016/02/14/opinion/sunday/ignore-the-gps-that-ocean-is-not-a-road.html

Milner, G. (2016b). *Pinpoint: How GPS is changing technology, culture, and our minds.* New York, NY: Norton.

Mittelstaedt, M. L., & Mittelstaedt, H. (1980). Homing by path integration in a mammal. *Naturwissenschaften, 67,* 566–567.

Moar, I., & Bower, G. H. (1983). Inconsistency in spatial knowledge. *Memory & Cognition, 11,* 107–113.

Moar, I., & Carleton, L. R. (1982). Memory for routes. *Quarterly Journal of Experimental Psychology A, 34,* 381–394.

Monmonier, M. (2018). *How to lie with maps* (3rd ed.). Chicago, IL: University of Chicago Press.

Montello, D. R. (1992). The geometry of environmental knowledge. In A. U. Frank, I. Campari, & U. Formentini (Eds.), *Theories and methods of spatio-temporal reasoning in geographic space* (pp. 136–152). Berlin, Germany: Springer.

Montello, D. R. (1993). Scale and multiple psychologies of space. In A. U. Frank & I. Campari (Eds.), *Spatial information theory* (pp. 312–321). Berlin, Germany: Springer.

Montello, D. R. (1997). The perception and cognition of environmental distance: Direct sources of information. In S. C. Hirtle & A. U. Frank (Eds.), *Spatial information theory* (pp. 297–311). Berlin, Germany: Springer.

Montello, D. R. (1998). A new framework for understanding the acquisition of spatial knowledge in large-scale environments. In M. J. Egenhofer & R. G. Golledge (Eds.), *Spatial*

and temporal reasoning in geographic information systems (pp. 143–154). New York, NY: Oxford University Press.
Montello, D. R., & Pick, H. L., Jr. (1993). Integrating knowledge of vertically aligned large-scale spaces. *Environment and Behavior, 25,* 457–484.
Montello, D. R., & Raubal, M. (2013). Functions and applications of spatial cognition. In D. Waller & L. Nadel (Eds.), *Handbook of spatial cognition* (pp. 249–264). Washington, DC: American Psychological Association.
Mook, D. G. (1983). In defense of external invalidity. *American Psychologist, 38,* 379–387.
Mou, W., Fan, Y., McNamara, T. P., & Owen, C. B. (2008). Intrinsic frames of reference and egocentric viewpoints in scene recognition. *Cognition, 106,* 750–769.
Moyer, R. S., & Bayer, R. H. (1976). Mental comparison and the symbolic distance effect. *Cognitive Psychology, 8,* 228–246.
Muhly, J. D. (1978). Ancient cartography: Man's earliest attempts to represent his world. *Expedition Magazine, 20*(2), 26–31.
Münzer, S., Zimmer, H. D., Schwalm, M., Baus, J., & Aslan, I. (2006). Computer-assisted navigation and the acquisition of route and survey knowledge. *Journal of Environmental Psychology, 26,* 300–308.
National Research Council. (2006). *Learning to think spatially.* Washington, DC: National Academies Press.
Neisser, U. (1976). *Cognition and reality: Principles and implications of cognitive psychology.* San Francisco, CA: Freeman.
Neisser, U. (1978). Memory: What are the important questions? In M. M. Grunenberg, P. E. Morris, & R. N. Sykes (Eds.), *Practical aspects of memory* (pp. 3–24). London: Academic Press.
Neisser, U. (1987). A sense of where you are: Functions of the spatial module. In P. Ellen & C. Thinus-Blanc (Eds.), *Cognitive processes and spatial orientation in animal and man* (Vol. 2, pp. 293–310). Dordrecht, The Netherlands: Nijhoff.
Newcombe, N. S. (2010). Picture this: Increasing math and science learning by improving spatial thinking. *American Educator, 34*(2), 29–43.
Newcombe, N. S. (2017). Cognitive development in comparative perspective: Exploring the role of language acquisition in spatial, quantitative, and memory development. In J. Call, G. M. Burghardt, I. M. Pepperberg, C. T. Snowdon, & T. Zentall (Eds.), *APA handbook of comparative psychology* (Vol. 1, pp. 403–425). Washington, DC: American Psychological Association.
Newcombe, N. S. (2018). Categorical influences on spatial bias. In T. L. Hubbard (Ed.), *Spatial biases in perception and cognition* (pp. 249–260). Cambridge: Cambridge University Press.
Newcombe, N. S., & Huttenlocher, J. (2000). *Making space: The development of spatial representation and reasoning.* Cambridge, MA: MIT Press.
Newcombe, N., Huttenlocher, J., Drummey, A. B., & Wiley, J. G. (1998). The development of spatial location coding: Place learning and dead reckoning in the second and third years. *Cognitive Development, 13,* 185–200.
Neyfakh, L. (2013, August 18). Do our brains pay a price for GPS? *Boston Globe.* https://www.bostonglobe.com/ideas/2013/08/17/our-brains-pay-price-for-gps/d2Tnvo4hiWjuybid5UhQVO/story.html
Okabe, A., Aoki, K., & Hamamoto, W. (1986). Distance and direction judgment in a large-scale natural environment: Effects of a slope and winding trail. *Environment and Behavior, 18,* 755–772.
O'Keefe, J., & Dostrovsky, J. (1971). The hippocampus as a spatial map: Preliminary evidence from unit activity in the freely-moving rat. *Brain Research, 34,* 171–175.
O'Keefe, J., & Nadel, L. (1978). *The hippocampus as a cognitive map.* Oxford: Oxford University Press.

Onyx, J., & Bullen, P. (2000). Measuring social capital in five communities. *Journal of Applied Behavioral Science, 36,* 23–42.

Osgood, C. E., Suci, G. J., & Tannenbaum, P. H. (1957). *The measurement of meaning.* Urbana, IL: University of Illinois Press.

Packard, M. G., & McGaugh, J. L. (1996). Inactivation of hippocampus or caudate nucleus with lidocaine differentially affects expression of place and response learning. *Neurobiology of Learning and Memory, 65,* 65–72.

Paivio, A. (1975). Perceptual comparisons through the mind's eye. *Memory & Cognition, 3,* 635–647.

Pallrand, G., & Seeber, F. (1984). Spatial ability and achievement in introductory physics. *Journal of Research in Science Teaching, 21,* 507–516.

Parush, A., Ahuvia, S., & Erev, I. (2007). Degradation in spatial knowledge acquisition when using automatic navigation systems. In S. Winter, M. Duckham, L. Kulik, & B. Kuipers (Eds.), *Spatial information theory* (pp. 238–254). Berlin, Germany: Springer.

Pellegrino, J. W., Alderton, D. L., & Shute, V. J. (1984). Understanding spatial ability. *Educational Psychologist, 19,* 239–253.

Pellegrino, J. W., & Kail, R., Jr. (1982). Process analysis of spatial aptitude. In R. J. Sternberg (Ed.), *Advances in the psychology of human intelligence* (Vol. 1, pp. 311–365). Hillsdale, NJ: Erlbaum.

Penn, A. (2003). Space syntax and spatial cognition: Or why the axial line? *Environment and Behavior, 35,* 30–65.

Perkins, D. D., & Long, D. A. (2002). Neighborhood sense of community and social capital: A multi-level analysis. In A. T. Fisher, C. C. Sonn, & B. J. Bishop (Eds.), *Psychological sense of community: Research, applications, and implications* (pp. 291–318). New York, NY: Kluwer Academic/Plenum.

Péruch, P., & Wilson, P. N. (2004). Active versus passive learning and testing in a complex outside built environment. *Cognitive Processing, 5,* 218–227.

Peterson, J. (1916). Illusions of direction orientation. *The Journal of Philosophy, Psychology and Scientific Methods, 13,* 225–236.

Piaget, J. (1983). Piaget's theory. In P. H. Mussen (Ed.), *Handbook of child psychology* (4th ed.), Vol. 1: *History, theory, and methods* (W. Kessen, Vol. Ed., pp. 103–128). New York, NY: Wiley.

Piaget, J., & Inhelder, B. (1948/1967). *The child's conception of space* (F. J. Langdon & J. L. Lunzer, Trans.). New York, NY: Norton.

Piburn, M. (1980). Spatial reasoning as a correlate of formal thought and science achievement for New Zealand students. *Journal of Research in Science Teaching, 17,* 443–448.

Pielot, M., & Boll, S. (2010). Tactile Wayfinder: Comparison of tactile waypoint navigation with commercial pedestrian navigation systems. In P. Floréen, A. Krüger, & M. Spasojevic (Eds.), *Pervasive computing* (pp. 76–93). Berlin, Germany: Springer.

Ploran, E. J., Rovira, E., Thompson, J. C., & Parasuraman, R. (2015). Underlying spatial skills to support navigation through large, unconstrained environments. *Applied Cognitive Psychology, 29,* 608–613.

Pocock, D. C. D. (1976). Some characteristics of mental maps: An empirical study. *Transactions of the Institute of British Geographers, 1,* 493–512.

Presson, C. C. (1982). Strategies in spatial reasoning. *Journal of Experimental Psychology: Learning, Memory, and Cognition, 8,* 243–251.

Presson, C. C., DeLange, N., & Hazelrigg, M. D. (1989). Orientation specificity in spatial memory: What makes a path different from a map of the path? *Journal of Experimental Psychology: Learning, Memory, and Cognition, 15,* 887–897.

Presson, C., & Hazelrigg, M. D. (1984). Building spatial representations through primary and secondary learning. *Journal of Experimental Psychology: Learning, Memory, and Cognition, 10,* 716–722.

Presson, C. C., & Montello, D. R. (1994). Updating after rotational and translational body movements: Coordinate structure of perspective space. *Perception, 23,* 1447–1455.

Proshansky, H. M., Fabian, A. K., & Kaminoff, R. (1983). Place-identity: Physical world socialization of the self. *Journal of Environmental Psychology, 3,* 57–83.

Proust, M. (1921–1927/2002). *In search of lost time* (C. Prendergast, Ed., L. Davis, J. Grieve, M. Treharne, J. Sturrock, C. Clark, P. Collier, & I. Patterson, Trans.). London: Penguin Books.

Putnam, R. D. (2000). Bowling alone: *The collapse and revival of American community.* New York, NY: Simon & Schuster Paperbacks.

Pyle, G. F. (1969). The diffusion of cholera in the United States in the nineteenth century. *Geographical Analysis, 1,* 59–75.

Ranck, J. B., Jr. (1984). Head direction cells in the deep cell layer of dorsal presubiculum in freely moving rats. *Society for Neuroscience Abstracts, 10,* 599.

Rand, G. (1969). Pre-Copernican views of the city. *Architectural Forum, 132*(9), 77–81.

Rappenglueck, M. A. (1999). Palaeolithic timekeepers looking at the Golden Gate of the ecliptic: The lunar cycle and the Pleiades in the cave of la-Tête-du-Lion (Ardéche, France)–21,000 BP. *Earth, Moon and Planets, 85/86,* 391–404.

Reitman, J. S., & Rueter, H. R. (1980). Organization revealed by recall orders and confirmed by pauses. *Cognitive Psychology, 12,* 554–581.

Retscher, G., & Kealy, A. (2006). Ubiquitous positioning technologies for modern intelligent navigation systems. *Journal of Navigation, 59,* 91–103.

Richardson, A. E., Montello, D. R., & Hegarty, M. (1999). Spatial knowledge acquisition from maps and from navigation in real and virtual environments. *Memory & Cognition, 27,* 741–750.

Riegelsberger, J. (2013, September). *User research at Google Maps.* Keynote speech at the 11th International Conference on Spatial Information Theory (COSIT 2013), Scarborough, UK.

Riemer, J. W. (2000). Job relocation, sources of stress, and sense of home. *Community, Work & Family, 3,* 205–217.

Rieser, J. J. (1989). Access to knowledge of spatial structure at novel points of observation. *Journal of Experimental Psychology: Learning, Memory, and Cognition, 15,* 1157–1165.

Rieser, J. J., Guth, D. A., & Hill, E. W. (1986). Sensitivity to perspective structure while walking without vision. *Perception, 15,* 173–188.

Rochford, K. (1985). Spatial learning disabilities and underachievement among university anatomy students. *Medical Education, 19,* 13–26.

Rosch, E. (1978). Principles of categorization. In E. Rosch & B. B. Lloyd (Eds.), *Cognition and categorization* (pp. 27–48). Hillsdale, NJ: Erlbaum.

Roskos-Ewoldsen, B., McNamara, T. P., & Shelton, A. L. (1998). Mental representations of large and small spatial layouts are orientation dependent. *Journal of Experimental Psychology: Learning, Memory, and Cognition, 24,* 215–226.

Rümelin, S., Rukzio, E., & Hardy, R. (2011). NaviRadar: A tactile information display for pedestrian navigation. In *Proceedings of the 24th annual ACM symposium on user interface software and technology* (UIST 2011, pp. 293–302). New York, NY: ACM.

Rushton, G. (1969). Analysis of spatial behavior by revealed space preference. *Annals of the Association of American Geographers, 59,* 391–400.

Russell, J. A., & Pratt, G. (1980). A description of the affective quality attributed to environments. *Journal of Personality and Social Psychology, 38,* 311–322.

Russell, J. A., Ward, L. M., & Pratt, G. (1981). Affective quality attributed to environments: A factor analytic study. *Environment and Behavior, 13,* 259–288.

Russell-Gebbett, J. (1984). Pupils' perceptions of three-dimensional structures in biology lessons. *Journal of Biological Education, 18,* 220–226.

Sadalla, E. K., & Magel, S. G. (1980). The perception of traversed distance. *Environment and Behavior, 12,* 65–79.

Sadalla, E. K., & Staplin, L. J. (1980). The perception of traversed distance: Intersections. *Environment and Behavior, 12,* 167–182.

Scannell, L., & Gifford, R. (2010). Defining place attachment: A tripartite organizing framework. *Journal of Environmental Psychology, 30,* 1–10.

Schinazi, V. R., Nardi, D., Newcombe, N. S., Shipley, T. F., & Epstein, R. A. (2013). Hippocampal size predicts rapid learning of a cognitive map in humans. *Hippocampus, 23,* 515–528.

Schwartz, S. H. (1994). Are there universal aspects in the structure and contents of human values? *Journal of Social Issues 50*(4):19–45.

Schwartz, S. H. (2016). Basic individual values: Sources and consequences. In T. Brosch & D. Sander (Eds.), *Handbook of value: Perspectives from economics, neuroscience, philosophy, psychology and sociology* (pp. 63–84). Oxford: Oxford University Press.

Shelton, A. L., & McNamara, T. P. (1997). Multiple views of spatial memory. *Psychonomic Bulletin & Review, 4,* 102–106.

Shelton, A. L., & McNamara, T. P. (2001). Systems of spatial reference in human memory. *Cognitive Psychology, 43,* 274–310.

Shepard, R. N., & Chipman, S. (1970). Second-order isomorphism of internal representations: Shapes of states. *Cognitive Psychology, 1,* 1–17.

Shepard, R. N., & Metzler, J. (1971). Mental rotation of three-dimensional objects. *Science, 171,* 701–703.

Sholl, M. J. (1989). The relation between horizontality and rod-and-frame and vestibular navigational performance. *Journal of Experimental Psychology: Learning, Memory, and Cognition, 15,* 110–125.

Sholl, M. J. (1996). From visual information to cognitive maps. In J. Portugali (Ed.), *The construction of cognitive maps* (pp. 157–186). Dordrecht, The Netherlands: Kluwer Academic.

Sholl, M. J., & Nolin, T. L. (1997). Orientation specificity in representations of place. *Journal of Experimental Psychology: Learning, Memory, and Cognition, 23,* 1494–1507.

Siegel, A. W., & White, S. H. (1975). The development of spatial representations of large-scale environments. In H. W. Reese (Ed.), *Advances in child development and behavior* (Vol. 10, pp. 9–55). New York, NY: Academic Press.

Siegler, R. S. (Ed.). (1978). *Children's thinking: What develops?* Hillsdale, NJ: Erlbaum.

Siegler, R. S. (1996). *Emerging minds: The process of change in children's thinking.* New York, NY: Oxford University Press.

Silverman, I., & Eals, M. (1992). Sex differences in spatial abilities: Evolutionary theory and data. In J. H. Barkow, L. Cosmides, & J. Tooby (Eds.), *The adapted mind: Evolutionary psychology and the generation of culture* (pp. 533–549). New York, NY: Oxford University Press.

Simon, H. A. (1956). Rational choice and the structure of the environment. *Psychological Review, 63,* 129–138.

Simon, H. A. (1990). Invariants of human behavior. *Annual Review of Psychology, 41,* 1–19.

Skinner, B. F. (1957). *Verbal behavior.* New York, NY: Appleton-Century-Crofts.

Skinner, B. F. (1985). Cognitive science and behaviourism. *British Journal of Psychology, 76,* 291–301.

Slocum, T. A., McMaster, R. B., Kessler, F. C., & Howard, H. H. (2009). *Thematic cartography and geovisualization* (3rd ed.). Upper Saddle River, NJ: Pearson.

Smith, I. M. (1964). *Spatial ability: Its educational and social significance.* San Diego, CA: Knapp.

Sobel, D. (1995). *Longitude: The true story of a lone genius who solved the greatest scientific problem of his time.* New York, NY: Walker.

Sparrow, B., Liu, J., & Wegner, D. M. (2011). Google effects on memory: Cognitive consequences of having information at our fingertips. *Science, 333,* 776–778.

Speake, J. (2015). "I've got my Sat Nav, it's alright": Users' attitudes towards, and engagements with, technologies of navigation. *The Cartographic Journal, 52,* 345–355.

Speake, J., & Axon, S. (2012). "I never use 'maps' anymore": Engaging with Sat Nav technologies and the implications for cartographic literacy and spatial awareness. *The Cartographic Journal, 49,* 326–336.

Spearman, C. (1904). "General intelligence," objectively determined and measured. *The American Journal of Psychology, 15,* 201–292.

Stamps, A. E., III. (2004). Mystery, complexity, legibility and coherence: A meta-analysis. *Journal of Environmental Psychology, 24,* 1–16.

Stamps, A. E., III. (2008). Some findings on prospect and refuge theory: II. *Perceptual and Motor Skills, 107,* 141–158.

Statista (2019). *Mobile audience reach of leading smartphone apps in the United States as of May 2019.* New York, NY: Statista. https://www.statista.com/statistics/281605/reach-of-leading-us-smartphone-apps/

Stern, E., & Leiser, D. (1988). Levels of spatial knowledge and urban travel modeling. *Geographical Analysis, 20,* 140–155.

Sternberg, R. J., & Grigorenko, E. L. (1997). Are cognitive styles still in style? *American Psychologist, 52,* 700–712.

Sternberg, R. J., Grigorenko, E. L., & Zhang, L.-F. (2008). Styles of learning and thinking matter in instruction and assessment. *Perspectives on Psychological Science, 3,* 486–506.

Stevens, A., & Coupe, P. (1978). Distortions in judged spatial relations. *Cognitive Psychology, 10,* 422–437.

Stevens, S. S. (1957). On the psychophysical law. *Psychological Review, 64,* 153–181.

Struch, N., Schwartz, S. H., & van der Kloot, W. A. (2002). Meanings of basic values for women and men: A cross-cultural analysis. *Personality and Social Psychology Bulletin, 28,* 16–28.

Subryan, H. (2010). *Tactile maps as navigational aids* (Design Resources DR-22). Buffalo, NY: Center for Inclusive Design and Environmental Access, University at Buffalo.

Takemiya, M., & Ishikawa, T. (2011). I can tell by the way you use your walk: Real-time classification of wayfinding performance. In M. Egenhofer, N. Giudice, R. Moratz, & M. Worboys (Eds.), *Spatial information theory* (pp. 90–109). Berlin, Germany: Springer.

Taube, J. S., Muller, R. U., & Ranck, J. B., Jr. (1990a). Head-direction cells recorded from the postsubiculum in freely moving rats. I. Description and quantitative analysis. *Journal of Neuroscience, 10,* 420–435.

Taube, J. S., Muller, R. U., & Ranck, J. B., Jr. (1990b). Head-direction cells recorded from the postsubiculum in freely moving rats. II. Effects of environmental manipulations. *Journal of Neuroscience, 10,* 436–447.

Taylor, H. A., & Tversky, B. (1992). Spatial mental models derived from survey and route descriptions. *Journal of Memory and language, 31,* 261–292.

Tenbrink, T. (2018). Natural language and geography: The meaning and use of spatial concepts in geographical contexts. In D. R. Montello (Ed.), *Handbook of behavioral and cognitive geography* (pp. 211–227). Cheltenham, UK: Elgar.

Thompson, D. L. (1963). New concept: "Subjective distance". *Journal of Retailing, 39,* 1–6.

Thorndyke, P. W., & Hayes-Roth, B. (1982). Differences in spatial knowledge acquired from maps and navigation. *Cognitive Psychology, 14,* 560–589.

Thurstone, L. L. (1934). The vectors of mind. *Psychological Review, 41,* 1–32.

Thurstone, L. L. (1938). *Primary mental abilities.* Chicago, IL: University of Chicago Press.

Tobler, W. R. (1965). Comparison of the correspondence of geographical patterns. *Papers of the Regional Science Association, 15,* 131–139.

Tobler, W. R. (1970). A computer movie simulating urban growth in the Detroit region. *Economic Geography, 46,* 234–240.

Tobler, W. R. (1994). Bidimensional regression. *Geographical Analysis, 26,* 186–212.

Tobler, W. (2004). On the first law of geography. *Annals of the Association of American Geographers, 94,* 304–310.

Tolman, E. C. (1948). Cognitive maps in rats and men. *Psychological Review, 55,* 189–208.

Tolman, E. C., & Honzik, C. H. (1930). Introduction and removal of reward, and maze performance in rats. *University of California Publications in Psychology, 4,* 257–275.

Tolman, E. C., Ritchie, B. F., & Kalish, D. (1946). Studies in spatial learning. I. Orientation and the short-cut. *Journal of Experimental Psychology, 36,* 13–24.

Trowbridge, C. C. (1913). On fundamental methods of orientation and "imaginary maps". *Science, 38,* 888–897.

Tuan, Y.-F. (1977). *Space and place: The perspective of experience.* Minneapolis, MN: University of Minnesota Press.

Tuckey, H., & Selvaratnam, M. (1993). Studies involving three-dimensional visualization skills in chemistry: A review. *Studies in Science Education, 21,* 99–121.

Tversky, B. (1981). Distortions in memory for maps. *Cognitive Psychology, 13,* 407–433.

Tversky, B. (1993). Cognitive maps, cognitive collages, and spatial mental models. In A. U. Frank & I. Campari (Eds.), *Spatial information theory* (pp. 14–24). Berlin, Germany: Springer.

Tversky, B., & Morrison, J. B. (2002). Animation: Can it facilitate? *International Journal of Human-Computer Studies, 57,* 247–262.

Uttal, D. H., Meadow, N. G., Tipton, E., Hand, L. L., Alden, A. R., Warren, C., & Newcombe, N. S. (2013). The malleability of spatial skills: A meta-analysis of training studies. *Psychological Bulletin, 139,* 352–402.

Vandenberg, S. G., & Kuse, A. R. (1978). Mental rotations, a group test of three-dimensional spatial visualization. *Perceptual and Motor Skills, 47,* 599–604.

Van Erp, J. B. F., & Van Veen, H. A. H. C. (2004). Vibrotactile in-vehicle navigation system. *Transportation Research Part F, 7,* 247–256.

Van Opstal, F., Gevers, W., De Moor, W., & Verguts, T. (2008). Dissecting the symbolic distance effect: Comparison and priming effects in numerical and nonnumerical orders. *Psychonomic Bulletin & Review, 15,* 419–425.

Vasilyeva, M., & Huttenlocher, J. (2004). Early development of scaling ability. *Developmental Psychology, 40,* 682–690.

Vieno, A., Nation, M., Perkins, D. D., Pastore, M., & Santinello, M. (2010). Social capital, safety concerns, parenting, and early adolescents' antisocial behavior. *Journal of Community Psychology, 38,* 314–328.

von Thünen, J. H. (1826/1966). *Von Thünen's isolated state* (C. M. Wartenberg, Trans., P. Hall, Ed.). Oxford: Pergamon Press.

Wai, J., Lubinski, D., & Benbow, C. P. (2009). Spatial ability for STEM domains: Aligning over 50 years of cumulative psychological knowledge solidifies its importance. *Journal of Educational Psychology, 101,* 817–835.

Wanzel, K. R., Hamstra, S. J., Anastakis, D. J., Matsumoto, E. D., & Cusimano, M. D. (2002). Effect of visual-spatial ability on learning of spatially-complex surgical skills. *Lancet, 359,* 230–231.

Warren, H. C. (1908). 'Magnetic sense' of direction. *Psychological Bulletin, 5,* 376–377.

Waters, W., & Winter, S. (2011). A wayfinding aid to increase navigator independence. *Journal of Spatial Information Science, 3,* 103–122.

Wehner, R., & Srinivasan, M. (1981). Searching behaviour of desert ants, genus *Cataglyphis* (Formicidae, Hymenoptera). *Journal of Comparative Physiology A, 142,* 315–338.

Weisberg, S. M., & Newcombe, N. S. (2016). Why do (some) people make a cognitive map? Routes, places, and working memory. *Journal of Experimental Psychology: Learning, Memory, and Cognition, 42,* 768–785.

Weisberg, S. M., Schinazi, V. R., Newcombe, N. S., Shipley, T. F., & Epstein, R. A. (2014). Variations in cognitive maps: Understanding individual differences in navigation. *Journal of Experimental Psychology: Learning, Memory, and Cognition, 40,* 669–682.

Wen, W., Ishikawa, T., & Sato, T. (2011). Working memory in spatial knowledge acquisition: Differences in encoding processes and sense of direction. *Applied Cognitive Psychology, 25,* 654–662.

Wen, W., Ishikawa, T., & Sato, T. (2013). Individual differences in the encoding processes of egocentric and allocentric survey knowledge. *Cognitive Science, 37,* 176–192.

Wen, W., Ishikawa, T., & Sato, T. (2014). Instruction of verbal and spatial strategies for the learning about large-scale spaces. *Learning and Individual Differences, 35,* 15–21.

Werner, H. (1957). The concept of development from a comparative and organismic point of view. In D. B. Harris (Ed.), *The concept of development: An issue in the study of human behavior* (pp. 125–148). Minneapolis, MN: University of Minnesota Press.

Werner, M. (2014). *Indoor location-based services: Prerequisites and foundations.* Cham, Switzerland: Springer International.

Werner, S., & Schmidt, K. (1999). Environmental reference systems for large-scale spaces. *Spatial Cognition and Computation, 1,* 447–473.

White, G. F. (1945). *Human adjustment to floods: A geographical approach to the flood problem in the United States* (Research Paper No. 29). Chicago, IL: Department of Geography, University of Chicago.

Wiener, S. I., & Taube, J. S. (Eds.). (2005). *Head direction cells and the neural mechanisms of spatial orientation.* Cambridge, MA: MIT Press.

Williams, D. R., & Stewart, S. I. (1998). Sense of place: An elusive concept that is finding a home in ecosystem management. *Journal of Forestry, 96*(5), 18–23.

Willis, K. S., Hölscher, C., Wilbertz, G., & Li, C. (2009). A comparison of spatial knowledge acquisition with maps and mobile maps. *Computers, Environment and Urban Systems, 33,* 100–110.

Wilmer, H. H., Sherman, L. E., & Chein, J. M. (2017). Smartphones and cognition: A review of research exploring the links between mobile technology habits and cognitive functioning. *Frontiers in Psychology, 8,* 605.

Wohlwill, J. F. (1968). Amount of stimulus exploration and preference as differential functions of stimulus complexity. *Perception & Psychophysics, 4,* 307–312.

Wohlwill, J. F. (1983). The concept of nature: A psychologist's view. In I. Altman & J. F. Wohlwill (Eds.), *Behavior and the natural environment* (pp. 5–37). New York, NY: Plenum Press.

Wolbers, T., & Hegarty, M. (2010). What determines our navigational abilities? *Trends in Cognitive Sciences, 14,* 138–146.

Wolbers, T., & Wiener, J. M. (2014). Challenges for identifying the neural mechanisms that support spatial navigation: The impact of spatial scale. *Frontiers in Human Neuroscience, 8,* 571.

Wood, L., & Giles-Corti, B. (2008). Is there a place for social capital in the psychology of health and place? *Journal of Environmental Psychology, 28,* 154–163.

Woodring, P. D. (1939). A technique for the investigation of direction orientation in human beings. *Papers of the Michigan Academy of Science, Arts and Letters, 24*(4), 147–152.

Wright, J. K. (1947). Terrae incognitae: The place of the imagination in geography. *Annals of the Association of American Geographers,* 37, 1–15.

Yeap, W. K., & Jefferies, M. E. (2000). On early cognitive mapping. *Spatial Cognition and Computation, 2,* 85–116.

Zacks, J. M., Mires, J., Tversky, B., & Hazeltine, E. (2000). Mental spatial transformations of objects and perspective. *Spatial Cognition and Computation, 2,* 315–332.

Zauberman, G., Kim, B. K., Malkoc, S. A., & Bettman, J. R. (2009). Discounting time and time discounting: Subjective time perception and intertemporal preferences. *Journal of Marketing Research, 46,* 543–556.

Zenrin (2018). *Survey on the use of maps 2018.* Tokyo: Zenrin Co. https://www.zenrin.co.jp/product/article/map-18/index.html

Ziersch, A. M., Baum, F. E., MacDougall, C., & Putland, C. (2005). Neighbourhood life and social capital: The implications for health. *Social Science & Medicine, 60,* 71–86.

Index

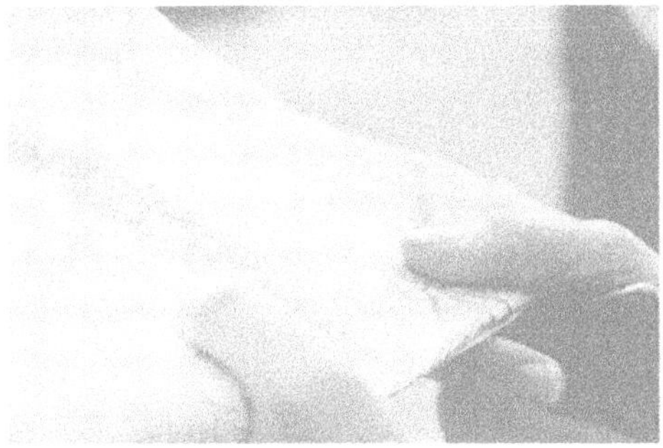

For Product Safety Concerns and Information please contact our EU
representative GPSR@taylorandfrancis.com
Taylor & Francis Verlag GmbH, Kaufingerstraße 24, 80331 München, Germany

www.ingramcontent.com/pod-product-compliance
Ingram Content Group UK Ltd.
Pitfield, Milton Keynes, MK11 3LW, UK
UKHW050052200926
475702UK00019B/2802
* 9 7 8 0 8 1 5 3 6 9 8 5 1 *